Mathematics for Industry

Volume 26

Aims & Scope

The meaning of "Mathematics for Industry" (sometimes abbreviated as MI or MfI) is different from that of "Mathematics in Industry" (or of "Industrial Mathematics"). The latter is restrictive: it tends to be identified with the actual mathematics that specifically arises in the daily management and operation of manufacturing. The former, however, denotes a new research field in mathematics that may serve as a foundation for creating future technologies. This concept was born from the integration and reorganization of pure and applied mathematics in the present day into a fluid and versatile form capable of stimulating awareness of the importance of mathematics in industry, as well as responding to the needs of industrial technologies. The history of this integration and reorganization indicates that this basic idea will someday find increasing utility. Mathematics can be a key technology in modern society.

The series aims to promote this trend by (1) providing comprehensive content on applications of mathematics, especially to industry technologies via various types of scientific research, (2) introducing basic, useful, necessary and crucial knowledge for several applications through concrete subjects, and (3) introducing new research results and developments for applications of mathematics in the real world. These points may provide the basis for opening a new mathematics oriented technological world and even new research fields of mathematics.

More information about this series at http://www.springer.com/series/13254

Hiromichi Itou · Masato Kimura
Vladimír Chalupecký · Kohji Ohtsuka
Daisuke Tagami · Akira Takada
Editors

Mathematical Analysis of Continuum Mechanics and Industrial Applications

Proceedings of the International Conference CoMFoS15

Springer

Editors
Hiromichi Itou
Tokyo University of Science
Tokyo, Tokyo
Japan

Masato Kimura
Institute of Science and Engineering
Kanazawa University
Kanazawa
Japan

Vladimír Chalupecký
Fujitsu Ltd.
Tokyo
Japan

Kohji Ohtsuka
Hiroshima Kokusai Gakuin University
Hiroshima
Japan

Daisuke Tagami
Kyushu University
Fukuoka
Japan

Akira Takada
Asahi Glass Co., Ltd.
Tokyo
Japan

ISSN 2198-350X ISSN 2198-3518 (electronic)
Mathematics for Industry
ISBN 978-981-10-2632-4 ISBN 978-981-10-2633-1 (eBook)
DOI 10.1007/978-981-10-2633-1

Library of Congress Control Number: 2016952521

Printed on acid-free paper

This Springer imprint is published by Springer Nature
The registered company is Springer Nature Singapore Pte Ltd.
The registered company address is: 152 Beach Road, #22-06/08 Gateway East, Singapore 189721, Singapore

Preface

The international conference CoMFoS15 was held in Fukuoka, Japan, at the Nishijin Plaza of Kyushu University, November 16–18, 2015. The name "CoMFoS" is derived from the research topics called "Continuum Mechanics Focusing on Singularities." The founding members held the first meeting, "Workshop on Fracture Criterion Viewed from Mathematics", in Ishikawa, Japan, January 27–29, 1995, to foster research cooperation among the mathematics, computer simulation, and continuum mechanics communities. The research group subsequently organized a series of CoMFoS conferences. In 2004 those activities blossomed into the foundation of a new division under the Japan Society for Industrial and Applied Mathematics (JSIAM). Finally, in 2010, the name of the division was renamed "Mathematical Sciences on Continuum Mechanics", aimed at broader exchanges among scientists and engineers, and subsequently the division has been steadily hosting a series of CoMFoS conferences.

The CoMFoS15 conference succeeded in bringing together a number of leading scientists in the field of mathematical and computational research on continuum mechanics as well as its peripheral domains such as physics, engineering, information, and experimentation. In consequence, academic scientists provided cutting-edge mathematical descriptions of phenomena in continuum mechanics for those working in industry and, in contrast, industrial researchers offered crucial aspects of core manufacturing technology to academics. In addition, the conference covered various technological aspects: fracture mechanics, shape optimization for product design, phenomena of earthquakes and tsunamis, viscoelasticity, materials science, interface mechanics, and industrial applications.

It is a particular pleasure to shed light on the important future prospects of the fields dealt with in the conference. First, several mathematical techniques developed in one field are applicable in another field. The theory and techniques either on shape optimization or on eigenvalue will be applied in a wider range of industrial problems. Second, cutting-edge mathematical theory and techniques will contribute to solving complex industrial problems in the near future. Several examples of complex industrial problems were discussed at the conference: (1) crack growth, propagation, or dislocation; (2) brittle, fragile, or viscoelastic behaviors of materials; and (3) phase transition or phase separation. Multiscale and multiphysics techniques, in particular, are becoming increasingly useful in industry. Third, it is important to nurture scientists and engineers who can translate from industrial problems to mathematical requirements and vice versa.

We would like to thank the Institute of Mathematics for Industry (IMI) of Kyushu University for operational support for organization, and for financial support of the IMI workshop of the Joint Research Projects.

Tokyo, Japan, May 2016 Akira Takada
On behalf of the Organizing Committee of CoMFoS15

Contents

Part I
Fracture Mechanics

Strong but Slippery Adhesion of Mushroom-Shaped Polysaccharide Gels

Yoshimi Tanaka and Teppei Nakamichi

Abstract We investigated the strength of capillary adhesion (i.e., adhesion caused by capillary effect of a liquid at an interface) of "mushroom-shaped" polysaccharide gels to clarify detailed mechanism under the enhancement of adhesive strength by the special shape. The used gels are not sticky. They can smoothly slide along the flat substrate, and weakly adheres to it due to the capillary adhesion by the water leached from the inside. However, if molded into the mushroom shape, the adhesive strength was remarkably increased. Experimental results show that there exists an optimal proportion of the mushroom shape to exhibit higher adhesive strength. Mechanical reason for the optimal proportion is discussed.

1 Introduction

Mechanical strength of connected interfaces depends not only on physicochemical nature of the interface [1] but the entire shape of jointed parts, especially the shape around the edge of the interface. This fact is well recognized in welding technology; various types of joint structure has been developed and utilized for assembling machines [2]. Recently, in soft matter science, mechanics of biological adhesion is an active topic, and development of bio-inspired attachment devices has been attempted [3–7]. Much attention [7] has been paid to the so-called mushroom-shape (Fig. 1a), an axially symmetrical geometry consisting of a long cylindrical shaft (with a radius R_1 and a length L) and a terminal disk part (with a radius R_2 and a thickness H). Previous investigations directly and indirectly indicate that if the disk part is adhered to a flat substrate, a larger pulling force should be applied on the shaft part to detach the interface, compared with other geometry such as the usual cylinder.

Y. Tanaka (✉) · T. Nakamichi
Graduate School of Environment and Information Sciences,
Yokohama National University, Yokohama, Japan
e-mail: tanaka-yoshimi-vm@ynu.ac.jp

T. Nakamichi
e-mail: teppei.1225@gmail.com

H. Itou et al. (eds.), *Mathematical Analysis of Continuum Mechanics and Industrial Applications*, Mathematics for Industry 26,
DOI 10.1007/978-981-10-2633-1_1

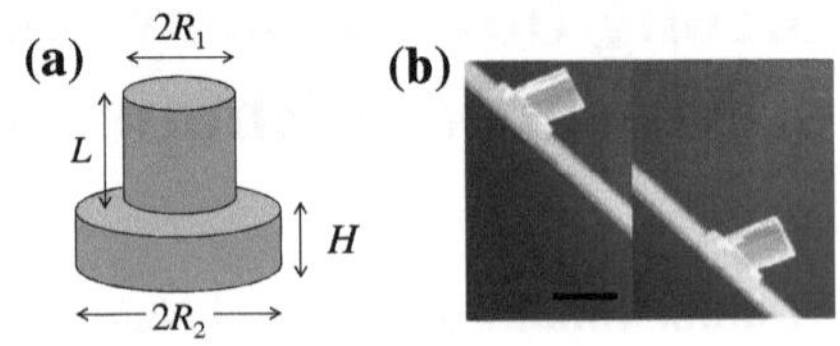

Fig. 1 **a** Definition of mushroom shape. **b** Slipping down of a mushroom-shaped polysaccharide gel. The *bar* represents 25 mm

So far, the mushroom effect has been investigated for adhesion of micro-patterned soft surfaces. Gorb et al. [8] prepared PVS (polyvinylsiloxane) surfaces where many small mushrooms (with a dimension of $R_1 \approx 20$ mm and $L \approx 100$ mm) exist, and showed that the micro-patterned surfaces realize a stronger adhesion to a flat glass substrate, compared with the flat PVS surface. Campo et al. [9] prepared PDMS (dimethylpolysiloxane) surfaces having many columnar projections with various terminal geometries (mushroom, flat circle, hemisphere etc.), and compared the adhesive strength (i.e., the critical force for detaching the interface) between the patterned surfaces and a rigid flat sapphire substrate. The highest adhesive strength was observed for the mushroom-shaped projections. These findings are followed by several detailed experiments [10] and theoretical and numerical investigations [6, 11, 12]. Among them, we emphasize that Spuskanyuk et al. found, based on FEM analysis, an interesting effect caused by a coupling of the mushroom geometry and a particular boundary condition of the interface: if the adhered interface is allowed to freely slip, the flange part of a mushroom is thrust to the substrate in spite of a pulling force exerted on the shaft of the mushroom. This effect reduces the possibility of the formation of detachment front at the periphery of the contact interface, resulting in the increasing of the adhesive strength, although it is not clear that the slippery boundary condition is adequate for the interface between PDMS and rigid substrates.

In this paper, we study the detailed mechanics under the mushroom effect (especially the case of the slippery interfaces), based on a different model system. The system proposed is isolated macroscopic mushrooms made of a polysaccharide gel with very slippery surface (Fig. 1b). In addition to the realization of frictionless boundary condition, our macroscopic approach has following two features: (1) The proportion of the mushroom shape can be systematically changed; (2) The mechanical state of the interface can be experimentally monitored by implanting special device and sensors to the substrate.

We measured the adhesive strength (critical pull-off force) for mushroom-shaped gels with different flange thickness H (see Fig. 1a) and performed visualization of the stress distribution over the contact interface. An important experimental finding is the existence of an optimal value of H that realizes a highest adhesive strength. The optimal value corresponds to a value of H for which FEM analysis predicts the maximum amplitude of contact stress on the periphery of the contact interface. We will discuss mechanical reasons for the thrust effect and for the existence of the optimal value of H.

2 Materials and Methods

2.1 Gel Preparation

In this study, we used a formulated gelling agent, GEL-UP G-F(F), hereafter GEL-UP, supplied from San-Ei-Gen F. F. I. Inc. (Osaka, Japan), which yields very slippery gels and is thus adequate in realizing nearly frictionless boundary condition for the interface. GEL-UP is a commercial product for food manufacturing. Its component is as follows: locust bean gum 45 wt%: carrageenan 33 wt%: sodium citrate 10 wt%: other food ingredient 12 wt%. Powder of GEL-UP was added in distilled water (2 wt%) and completely dissolved by heating (up to 90 °C) and by mechanical stirring for 10 min. After slowly cooling down the solution around 60 °C, the pre-gel solution was poured into special molds to make mushroom-shaped gel (see below) and left in a refrigerator (4 °C) for 24 h. The gels prepared by this procedure can adhere to flat surfaces of various hard solids (silicon glass, metals, and solid polymers) but the interface can slide along the surface: the gels slipped down when put on a sloped plate, as seen in Fig. 1b. This indicates that the adhesion is dominated by capillary adhesion due to the water leached from the inside to into the interface. (Actually, the gels were slightly turbid, indicating that the gel has inhomogeneity on the scale of visible light and thus high water permeability.)

2.2 Molding of Mushroom-Shaped Gels

Mushroom-shaped gels were prepared in a molds consisting of two acrylic pipes. The dimension of the gel specimens, R_1, R_2 and the total height $(L+H)$ were fixed at 9, 14, and 25 mm; while the thickness H was changed from 2.5 to 25 mm. For the convenience of the description blow, we introduce a dimensionless parameter $z = H/\Delta R$ ($\Delta R = R_2 - R_1$) to indicate the proportion of the mushroom structure.

2.3 Mechanical Tests

Before the mechanical tests, gels in the molds were kept at room temperature (20 ± 2 °C) for 1 h (and following all treatments and measurements were conducted at the temperature). Figure 2 shows a loading process of a gel specimen with $z = 1$ set in a testing machine, Instron 5564 (Instron Co.). In setting a gel specimen, the disk part of the gel was put on the lower substrate of acrylic resin plate (SHINKOLITE, Mitsubishi Rayon Co.). Then the top surface (with an area $\pi {R_1}^2$) was tightly glued to

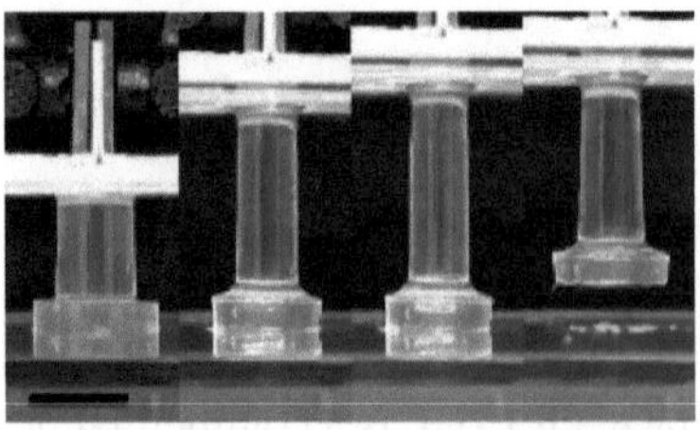

Fig. 2 Deformation of a mushroom-shaped gel (z = 0.5). The scale bar represents 20 mm

a rigid plate connected to the crosshead of the testing machine by alpha cyanoacrylate (Aron Alpha, Toagosei Co., Japan). Then the top surface was pulled up at a constant velocity of 0.1 mm/s and the force F of pulling up was detected. The substrate was wiped with ethanol every time setting a new specimen.

2.4 *Visualization of Stress Distribution at Interface*

As a separate experiment from the mechanical measurements, we visualized the distribution of the stress over the interface (hereafter "interface stress") by modifying the substrate: thin vertical holes were made on a substrate and a dyed viscous liquid, a mixture of glycerol and a black dye, was injected in the holes as the indicator for the stress distribution. (In the actual experiment, thin glass capillaries were used to realize sufficiently long liquid column; see Fig. 4a for the experiment set-up.) Due to the high viscosity of the injected liquid, the substrate behaves like the normal one just after the onset of the pulling up, but the liquid was gradually sucked up into the interface. The amount of the liquid left in the holes reflects the stress distribution over the interface.

2.5 *FEM Analysis*

To obtain detailed information on the stress distribution over the interface (and inside the gel), we carried out FEM analysis with commercial software, ADINA, ADINA R&D, Inc. In our FEM analysis, the gel was assumed to be an isotropic and incompressible linear elasticity material. Details of the simulation setup (boundary condition etc.) are presented in the next section.

3 Results

3.1 Mechanical Measurement

The photographs in Fig. 2 are side views of a specimen under loading. The area of the contact interface was shrunk in the earlier stage of the loading (Fig. 2a, b). In the later stage the area was almost constant, while the shaft part continued to be elongated (Fig. 2c). After the detachment of the interface, a residual strain remained in the gels (Fig. 2d).

After detachment, the substrate was often wetted with a small amount of water. It suggests that water was sucked out from the inside of the gel during the loading process.

Figure 3 shows representative loading curves for mushroom-shaped gels with different z. The horizontal axis represents the deformation of gels, i.e., the upward displacement of the top surface (from the un-deformed state) divided by the initial height of $L + H$, and the vertical axis Π shows the average "nominal stress" of the interface, i.e., the loading force F divided by the initial area of the contact interface (πR_2^2). The condition of $z = 5$ corresponds to the cylindrical specimen ($L + H$ was fixed to 25mm). From the initial slope of the curve with $z = 5$, the Young's modulus E of the gel was estimated as $E = 5.5 \times 10^3$ Pa. The peak of each curve corresponds to the detachment point. The adhesive strength Π_c (i.e., the peak value of Π) of a thin flange ($z = 1$) is almost 6 times larger than that for the cylindrical specimen.

Figure 4 shows a plot of the adhesive strength Π_c versus z. The plotted values are the average over 5 times repeated experiments. When the flange is thick ($z > 2$), the adhesive strength is low and weakly depends on z. With decreasing z, the adhesive strength steeply increases around $z = 2$, and accomplishes a peak around $z = 1$. Then it slowly decreases for further decrease in z. That is, there exists the optimum flange thickness to realize the largest adhesive strength.

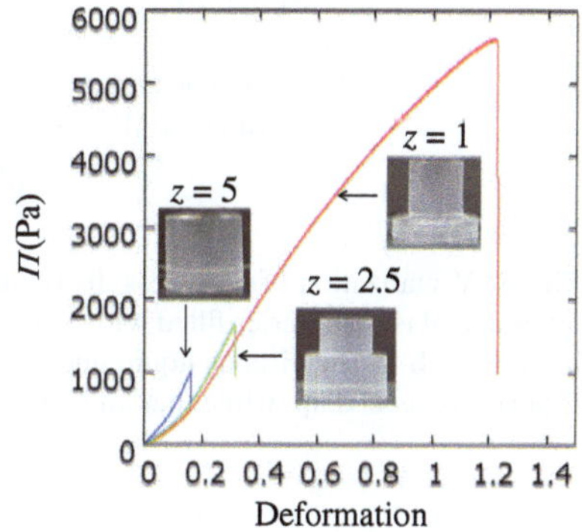

Fig. 3 A plot of the load Π versus the deformation; Π is defined by the pull-up force divided by the initial area of the contact interface and the deformation represents the upward distance of the top surface of the gels specimens divided by the initial height

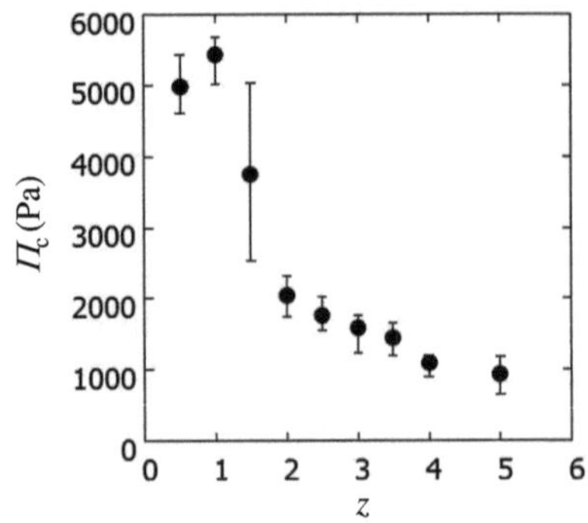

Fig. 4 A plot of the adhesive strength Π_c versus the reduced flange thickness z

3.2 Visualizing Distribution of Stress at Interface

The distribution of the stress over the interface is visualized in Fig. 5. The substrate was modified by implanting thin glass capillaries (0.9 mm in inner diameter) containing a mixture of glycerol and black dye (Fig. 5a). With pulling up the shaft part of the gel (Fig. 5b), the liquid under the shaft was sucked up, but that under the flange region was at rest, as shown in Fig. 5c. This indicates that the distribution of interface stress was uneven; only the central region carried strong interface stress.

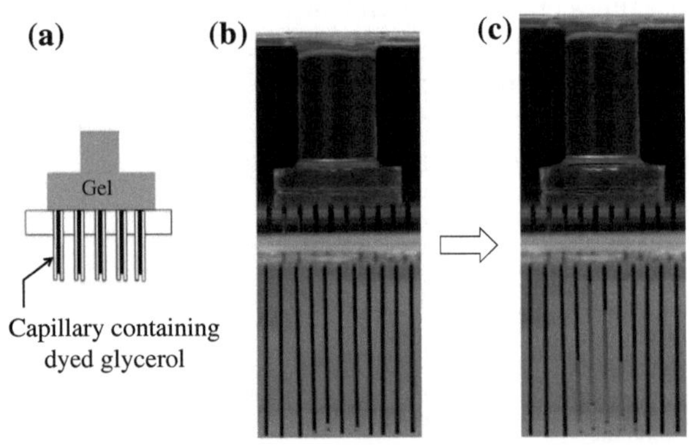

Fig. 5 Visualization of the stress distribution over the contact interface. **a** The substrate has thin holes and glass capillaries filled with a viscous liquid (glycerol containing black dye) were inserted in to them. **b–c** The viscous liquid under the central part of the interface (under the shaft of mushroom) was sucked up, while that under the flange part was at rest

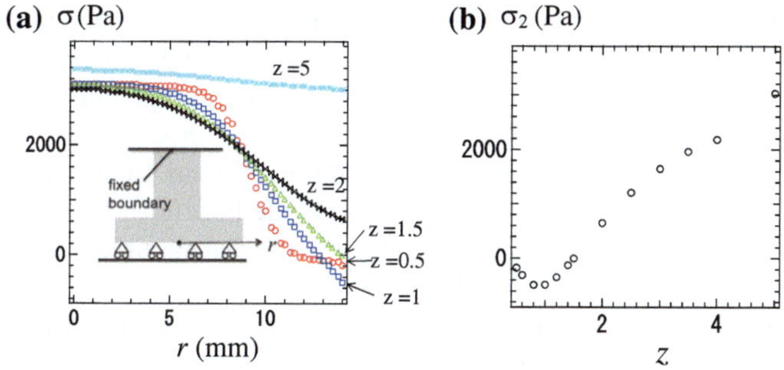

Fig. 6 **a** Calculated interface stress over the contacted interface as a function of the distance from the center of the interface r. The insert shows the boundary conditions in the FEM analysis. **b** A plot of the interface stress on the periphery of the interface versus the reduced flange thickness z

3.3 FEM Analysis

Figure 6a shows the calculated distribution of the interface stress at the interface, as a function of the radius r. The boundary conditions is illustrated in the insert: for the bottom surface (the contact interface), the vertical displacement is zero and the lateral displacement is freely allowed to vanish the shear component of the interface stress; the upper surface was fixed to a rigid plate and subjected to a constant lifting-up at 10 mm. The Young modulus E and Poisson ratio ν were chosen to be $E = 5.5 \times 10^3$ Pa and $\nu = 1/2$ (incompressibility condition), respectively. For z = 5 (corresponding to the cylindrical specimen), the interface stress $\sigma(r)$ is nearly constant (the slightly decrease in $\sigma(r)$ with r is due to the asymmetry of the boundary conditions of the top and bottom surfaces). For the mushroom-shaped specimens ($z \leq 2$), $\sigma(r)$ substantially decreases with r, and $\sigma(r)$ under the flange part (i.e., $r > R_1$) gets lower with decreasing z, the flange thickness. For $z = 1.5$, the interface stress on the periphery of the interface $\sigma_2 \equiv \sigma(R_2)$ (hereafter "periphery stress") is almost zero. Interestingly, σ_2 becomes negative (i.e., compressive) for $z = 1$, which represents the flange part is thrust to the substrate around the periphery of the interface.

Figure 6b shows a plot of σ_2 versus z. σ_2 is negative for $z < 1.5$. The amplitude of the negative periphery stress, $|\sigma_2|$, achieves the maximum at $z = 1$, and tends to zero with decreasing z further.

4 Discussion

In this paper, we have studied the effect of the mushroom shape on the adhesive strength of a polysaccharide gel, taking the thickness of flange of the mushroom shape as a control parameter. Since the contact interface can smoothly slide, the boundary condition is well approximated to be zero shear stress. Experimental results show that (i) the flange thickness should be small to realize high adhesive strength, but (ii) it should not be too small: there exists the optimal flange thickness to achieve the highest adhesive strength. The optimal thickness corresponds to the value of z for which the linear elastic FEM predicts the maximum amplitude of the thrust periphery stress, $|\sigma_2|$. Below we will discuss (1) the benefit of thin flanges, (2) the physical reasons for the thrusting contact stress, and (3) a possible interfacial process determining the adhesive strength Π_c.

(1) The principal benefit of thinner flange is screening of the stress transmitted from the upper surface of gels to the periphery of the contact interfaces, as illustrated in Fig. 7a (the curve arrows in the gel represent the lines of force for the stress vector with respective to the lateral the cross-sections of the gel). It is pretty certain that the detachment of an interface initiates at the periphery because the detachment from the inside of the contact interface inevitably involves mechanically costly processes such as cavitation. Thinner flanges reduce the possibility of nucleation of the detachment at the periphery. In other words, larger loads are required to initiate the detachment front at the periphery.

In addition to the above screening effect, an interesting behavior occurs for the frictionless boundary condition; as shown by linear FEM, the periphery stress σ_2 gets negative for smaller z, indicating that at the *earlier stage* of the deformation, the periphery of the interface is thrust to the substrate. This effect should further reduce the possibility of the detachment from the periphery, and explains the coincidence between the values of z giving the largest Π_c and the largest amplitude of the negative periphery stress, $|\sigma_2|$.

It should be noticed that the negative σ_2 for smaller z is a consequence of the linear FEM. On the other hand, actual gel specimens underwent fairly large deformations at the detachment (Fig. 2). The nonlinear effects brought by the large deformation, such as the pronounced shape change and the strain hardening, can violate the character-

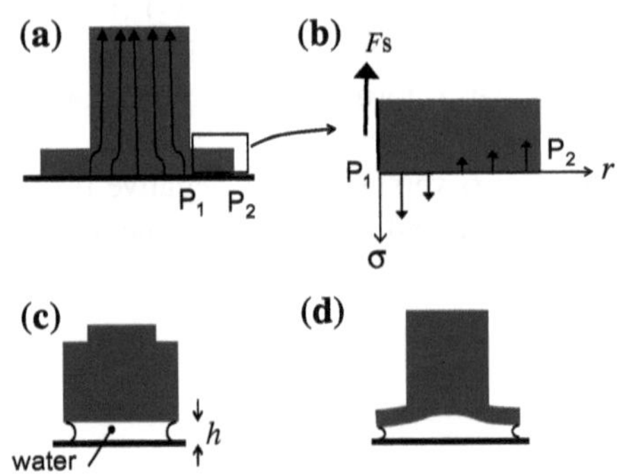

Fig. 7 **a** Schematic representation of the stress distributions by the lines of force in the case of thin flange. **b** Distributions of the interface stress on the bottom boundary. To satisfy the balance, the compressional interface stress should appear around the outside region. **c** Water layer under a thick flange. **d** Water layer under thin flange

istic of the stress distribution in the linear regime. In the actual loading processes of mushroom-shaped gels with thinner flanges the periphery stress σ_2 changes its sign from negative to positive and leads to the ultimate detachment.

(2) The reason for the negative σ_2, which never occurs if the non-slip boundary condition is imposed for the contact interface, is the balance of moment [11]. To clearly see the mechanical meaning of the moment balance, we virtually cut out a rectangular region from the flange part; see the dashed frame in Fig. 7a, b. (Below, we suppose that the cut-out region have a unit length in the direction perpendicular to the page.) The points of P_1 and P_2 in Fig. 7a, b correspond to $r = R_1$ and $r = R_2$, respectively. In Fig. 7b, the thin arrows on the bottom surface represent the interface stress $\sigma(r)$; the downward direction of the arrows says that $\sigma(r)$ is positive (tensional) because the direction of the interface stress is parallel to the outward normal of the bottom surface. On the left virtual cross-section (see the dashed line in Fig. 7b), a shear force $F_s = \tau H$ is exerted (H is the flange thickness and τ is the average shear stress), but the normal stress can be ignored because of the frictionless boundary condition of the bottom surface. The force balance for the rectangle is $F_s = \int_{R_1}^{R_2} \sigma(r)dr$. If $\sigma(r)$ is positive for all r in P_1P_2, the moment balance $\int_{R_1}^{R_2}(r - R_1)\sigma(r)dr = 0$ cannot be satisfied: the shear force has no contribution to the moment around the point P_1, while a positive $\sigma(r)$ produces a momentum in the clockwise direction around the point. That is, the thrusting (compressional) interface stress arises to fulfill the requirement of the moment balance (Fig. 7b).

To understand the existence of the optimal flange thickness, we carefully look at the $\sigma(r)$ curves in Fig. 6a. For thin flanges of $z \le 2$ in Fig. 6a, the $\sigma(r)$ curves intersect each other at almost the same point around $r = R_1$, but the slope at the intersection gets steeper with decreasing z. We can expect that in the limit of $z \to 0$ (the limit of vanishing the flange), the interface stress $\sigma(r)$ approaches to that of the limiting shape, a cylinder with the diameter R_1: $\sigma(r) = const.$ for $r < R_1$ and $\sigma(r) = 0$ for $r > R_1$. Thus, the negative $\sigma(r)$ for $r > R_1$ should collapse to zero for the thinner flange thickness. On the other hand, for sufficiently large flange thickness, the system is close to the cylindrical gel (z = 5) and the interface stress $\sigma(r)$ is positive and almost constant on the interface (the stress singularity at the contact edge should be absent due to the frictionless boundary condition). Thus, the amplitude of the negative periphery stress $|\sigma_2|$ achieves the maximum at an intermediate flange thickness.

(3) The smooth sliding of the contact interface indicates that there exists a water layer at the interface and that the adhesion is realized by capillary effect. Thus, the negative Laplace pressure in the water layer is the dominant factor for Π_c (Fig. 7c, d). For larger flange thickness, the interface stress is almost constant; the situation is similar to the normal capillary adhesion formed between two parallel rigid plate (Fig. 7c). The critical condition for the detachment is $\Pi_c \sim \gamma/h(t)$, where $\gamma = 0.073$ N/m is the surface energy of water and $h(t)$ is the thickness of water layer at the occurrence of the detachment. From the measured valued of $\Pi_c \sim 1 \times 10^3$ Pa, we have $h(t) \sim 0.1$ mm, indicating that a fairly large amount of water is sucked out. This estimation, which is consistent with the experimental observation that the substrate was wetted after the detachment, suggest that water is sucked our from the inside of the gel.

For small flange thickness, the situation is more complicated. The uneven interface stress $\sigma(r)$ means the existence of the gradient of hydrostatic pressure p in the water layer because $\sigma(r)$ balances with $-(p - P_{\mathrm{atm}})$, where P_{atm} is atmospheric pressure. The gradient of p causes a lateral water flow from the periphery side to the center of the interface (in addition to the water supply from the inside the gels). This lateral flow reduces the thickness of the water layer at the periphery (Fig. 7d) and enhances the adhesion strength Π_{c}. To qualitatively analyze the interfacial process and the detachment condition for small flange thickness, we need detailed experimental and theoretical investigations such as measurement of water permeability in the gel, in situ observations of the thickness of the interfacial water layer, and a treatment of the gel as poro-elastic material [13] and hydrodynamic analysis for the water layer. These detailed investigations remain as future challenges.

In summary, we investigated the strength of capillary adhesion of mushroom-shaped polysaccharide gels forming slippery contact interfaces with solid substrates. Experimental results show that for a particular flange thickness of the mushroom shapes, the adhesion strength takes the maximum. The optimal flange thickness is related to the occurrence of thrusting contact stress on the periphery of the contact interfaces. The physical reason for the thrust is the moment balance of the flange part.

Acknowledgments The authors thanks San-Ei-Gen F. F. I. Inc. for providing the gelling argent. They are grateful to K. Matsui for the assistance in FEM analysis and for helpful discussions. YT thanks T. Yamaguchi for helpful comments. This work was supported by JSPS KAKENHI Grant Number 24540432.

References

1. Israelachvili, J.N.: Intermolecular and surface forces: revised third edition. Academic press (2011)
2. Brandon, D.G., Wayne, D.: Kaplan. An Introduction, Wiley-VCH, Joining Processes (1997)
3. Arzt, E., Gorb, S., Spolenak, R.: From micro to nano contacts in biological attachment devices. PNAS **100**, 10603–10606 (2003)
4. Persson, B.N.J., Gorb, S.: The effect of surface roughness on the adhesion of elastic plates with application to biological systems. J. Chem. Phys. **119**, 11437–11444 (2003)
5. Glassmaker, N.J., Jagota, A., Hui, C.-Y., Noderer, W.L., Chaudhury, M.K.: Biologically inspired crack trapping for enhanced adhesion. PNAS **104**, 10786–10791 (2007)
6. Jagota, A., Hui, C.-Y.: Adhesion, friction, and compliance of bio-mimetic and bio-inspired structured interfaces. Mater. Sci. Eng., Rep. **72**, 253–292 (2011)
7. Heepe, L., et al.: First mushroom-shaped adhesive microstructure: a review. Theor. Appl. Mech. Lett. **2**, 014008–014008 (2012)
8. Gorb, S., et al.: Biomimetic mushroom-shaped fibrillar adhesive microstructure. J. R. Soc. Int. **4**, 271–275 (2007)
9. Campo, A.D., Greiner, C., Arzt, E.: Contact shape controls adhesion of bioinspired fibrillar surfaces. Langmuir **23**, 10235–10243 (2007)
10. Varenberg, M., Gorb, S.: Close-up of mushroom-shaped fibrillar adhesive microstructure: contact element behaviour. J. R. Soc. Int. **5**, 785–789 (2008)

11. Spuskanyuk, A.V., et al.: The effect of shape on the adhesion of fibrillar surfaces. Acta Biomater. **4**, 1669–1676 (2008)
12. Carbone, G., Pierro, E., Gorb, S.: Origin of the superior adhesive performance of mushroom-shaped microstructured surfaces. Soft Matter **7**, 5545–5552 (2011)
13. Sekimoto, K., Rabin, Y.: Role of mechanical pressure in permeation of a solvent through gels. Europhys. Lett. **27**, 445 (1994); Doi, M.: Gel dynamics. J. Phys. Soc. Jpn. **78**, 052001 (2009)

Bridging the Scales Between Discrete and Continuum Dislocation Models

Patrick van Meurs

Abstract We prove the many-particle limit passage of interacting particle systems described by gradient flows. The limiting equation is a gradient flow which describes the evolution of the particle density. Our proof methods rely on variational techniques such as Γ-convergence of the particle configuration energies and stability of gradient flows. The interacting particle systems under consideration model the motion of dislocations in metals. Since the collective motion of dislocations is the main driving force of plastic deformation of metals, we aim to contribute with our analysis to the current understanding of plasticity.

1 Disagreeing Plasticity Models

A large field of ongoing research studies plastic deformation of metals. Plastic deformation is an irreversible process, in which permanent changes within the material result in a macroscopic change of shape. These permanent changes are the collective behaviour of curve-like defects in the atomic lattice of the metal. Such defects are called dislocations. It is typical for metals to contain *many* dislocations (as much as 1000 km of dislocation curve in a cubic millimetre [19, p. 20]). Because of this large amount of dislocations, there is a general belief that plasticity can be described in terms of upscaled quantities such as the dislocation density.

This belief has led to several *different* dislocation density models in the engineering literature, see for instance [1, 16, 17]. The discrepancies between these models arise from using different phenomenological closure assumptions. These assumptions are needed to bridge the gap between dislocations interacting on the micro-scale and their collective behaviour on the macro-scale. As a consequence of these different continuum models, it remains unclear to what extent they approximate discrete dislocation dynamics. This brings us to the main question:

P. van Meurs (✉)
Kanazawa University, Kanazawa 920-1192, Japan
e-mail: pmeurs@staff.kanazawa-u.ac.jp

H. Itou et al. (eds.), *Mathematical Analysis of Continuum Mechanics and Industrial Applications*, Mathematics for Industry 26,
DOI 10.1007/978-981-10-2633-1_2

Which dislocation density models are the discrete-to-continuum limit of the dynamics of individual dislocations as the number of dislocations tends to infinity?

The main challenge in answering this question is to control the non-local and singular interactions between dislocations. Such interactions are not captured by similar discrete-to-continuum problems that are solved in, for example [6, 8, 18, 23, 27], because in those results the interactions are either local or bounded.

Instead, for non-local and singular interactions, different mathematical tools are needed to pass to the many-particle limit. Several such tools are developed, for instance, in [2, 11–15, 21]. We wish to expand this set of tools by connecting several more discrete dislocation models to their continuum counterparts, with the ultimate aim to validate the different models in the engineering literature. More precisely, we consider two scenarios for discrete dislocation dynamics: pile-ups of dislocation walls [25] in Sect. 2, and mixed positive and negative dislocations in two dimensions [16, 17] in Sect. 3.

2 Upscaling of Dislocation Walls

After introducing in Sect. 2.1 the model for the dynamics of dislocation walls, we describe in Sect. 2.2 the related upscaling results.

2.1 *Pile-Up of Dislocation Walls*

The idealised setting of dislocation walls (proposed in [20, 29]) allows to study a two-dimensional dislocation geometry by means of a one-dimensional problem. A dislocation wall is a vertically periodic arrangement of equidistant dislocations. We consider a pile-up of $n+1$ dislocation walls (based on the setting in [25]), which are trapped in a finite domain and subjected to a constant external force (see Fig. 1a).

The unknowns are the n horizontal positions of the dislocations walls given by

$$\Omega_n := \{x^n = (x_1^n, \ldots, x_n^n) \in \mathbb{R}^n : 0 =: x_0 < x_1^n < \cdots < x_n^n \leq L_n\}.$$

The energy related to a dislocation wall configuration is

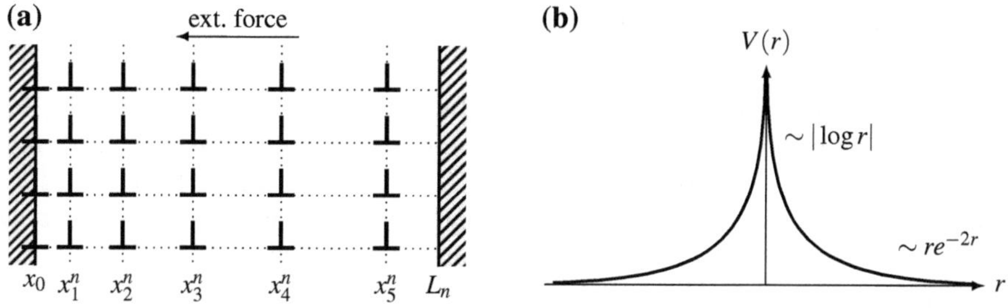

Fig. 1 **a** Pile-up of dislocations '⊥' arranged in wall structures. **b** Qualitative plot of the interaction potential V

$$E_n : \Omega_n \to [0, \infty], \qquad E_n = E_n^{\text{int}} + E_n^{\text{F}} + E_n^{\text{L}}, \tag{1}$$

$$E_n^{\text{int}}(x^n) = \frac{1}{n^2} \sum_{k=1}^{n} \sum_{j=0}^{n-k} \ell_n V\big(\ell_n (x_{j+k}^n - x_j^n)\big), \qquad E_n^{\text{F}}(x^n) = \frac{1}{n} \sum_{i=1}^{n} x_i^n,$$

$$E_n^{\text{L}}(x^n) = \chi_{\{x_n^n \le L_n/\ell_n\}} := \begin{cases} 0, & \text{if } x_n^n \le L_n/\ell_n, \\ \infty, & \text{otherwise.} \end{cases}$$

Here, V is the *interaction potential* between walls, which is illustrated in Fig. 1b and defined by

$$V : \mathbb{R} \to [0, \infty], \qquad V(r) := r \coth r - \log |2 \sinh r|. \tag{2}$$

The parameters L_n and ℓ_n correspond to the two characteristic length scales in the pile-up problem. Relative to the vertical distance between two neighbouring dislocations within a wall, L_n is the distance between the barriers, and ℓ_n is the length scale at which the dislocation walls spread when the interaction energy is balanced with the external forcing term.

Regarding the dynamics, we follow the well-known linear drag law for modelling dislocation movement, which results in the following gradient flow

$$\begin{cases} \dfrac{d}{dt} x^n(t) = -n \nabla E_n\big(x^n(t)\big), & t > 0, \\ x^n(0) = x_{\text{init}}^n, \end{cases} \tag{3}$$

for some initial condition $x_{\text{init}}^n \in \Omega_n$. Since E_n is strictly convex and has compact level sets, the minimisation problem of E_n over Ω_n and the gradient flow (3) have unique solutions.

2.2 Discrete-to-continuum Limit

We are interested in the limits of n-indexed sequences of gradient flows (3). For brevity, we assume for the asymptotic behaviour of the parameters L_n and ℓ_n that $1 \lesssim \ell_n \lesssim n$ and $L_n \gtrsim \ell_n$ as $n \to \infty$. In [31, 32] any scaling regime for L_n and ℓ_n is considered. The main difference for $\ell_n \ll 1$ and $\ell_n \gg n$ is that the interaction energy E_n^{int} is scaled differently, and for $L_n \ll \ell_n$ that x^n is replaced by $L_n x^n / \ell_n$.

To give a meaning to the convergence of gradient flows, we first define notions of convergence for a sequence of n-tuples (x^n) via the embedding

$$\pi_n : \Omega_n \to \mathscr{P}([0, \infty)), \qquad \pi_n(x^n) := \frac{1}{n} \sum_{i=1}^{n} \delta_{x_i^n},$$

where $\mathscr{P}([0, \infty))$ is the space of probability measures. We consider convergence in either the narrow (or weak) topology

$$\pi_n(x^n) \rightharpoonup \mu \quad :\Longleftrightarrow \quad \forall \varphi \in C_c^\infty([0, \infty)) : \int \varphi \, d(\pi_n(x^n)) \to \int \varphi \, d\mu,$$

or with respect to the stronger 2-Wasserstein distance denoted by W_2 (see e.g. [4] for an introduction to the Wasserstein distance).

Our first main result (which is an extension of [15, Theorem 1]) gives a precise meaning to the convergence of the sequence of energies E_n

Theorem 1 (Γ-convergence [15, 32]) *Let $1 \lesssim \ell_n \lesssim n$ and $L_n/\ell_n \to \gamma \in (0, \infty]$ as $n \to \infty$. Then for all sequences (x^n) such that $x^n \in \Omega_n$ and $E_n(x^n)$ bounded, it holds that $(\pi_n(x^n))$ has a narrowly converging subsequence. Moreover, E_n Γ-converges with respect to the narrow topology to*

$$E : \mathscr{P}([0, \infty)) \to [0, \infty], \qquad E(\mu) := E^{int}(\mu) + \int_0^\infty x \, d\mu(x) + \chi_{\{\operatorname{supp} \mu \subset [0, \gamma]\}}, \tag{4}$$

where E^{int} depends on the asymptotic behaviour of ℓ_n (see Table 1).

We remark that we impose the conditions $1 \lesssim \ell_n \lesssim n$ and $L_n \gtrsim \ell_n$ for brevity, and that the full result can be found in [15, 32].

We briefly mention the main arguments in the proof of Theorem 1. Prokhorov's Theorem gives a convenient characterisation of pre-compactness in the narrow topology, which is easily seen to be satisfied if $E_n(x^n)$ is bounded uniformly in n.

We prove Γ-convergence by establishing its defining inequalities

Table 1 Dependence of E^{int}, the interaction part of the limit energy (4), on the asymptotic behaviour of ℓ_n

Regime	$E^{\text{int}}(\mu)$
$\ell_n \to \ell$	$\frac{\ell}{2}\int_0^\infty \int_0^\infty V(\ell(x-y))\,d\mu(y)\,d\mu(x)$
$1 \ll \ell_n \ll n$	$\begin{cases} \left(\int_0^\infty V\right)\int_0^\infty \rho^2, & \text{if } d\mu(x) = \rho(x)dx, \\ \infty, & \text{otherwise} \end{cases}$
$\frac{\ell_n}{n} \to \alpha$	$\begin{cases} \alpha \int_0^\infty V_{\text{eff}}\left(\frac{\alpha}{\rho(x)}\right)\rho(x)\,dx, & \text{if } d\mu(x) = \rho(x)dx, \\ \infty, & \text{otherwise} \end{cases}$

We define $V_{\text{eff}}(r) := \sum_{k=1}^{\infty} V(kr)$

$$\forall\, \mu \in \mathscr{P}([0,\infty))\ \forall\, x^n \in \Omega_n \text{ such that } \pi_n(x^n) \rightharpoonup \mu : \qquad \liminf_{n\to\infty} E_n(x^n) \geq E(\mu), \tag{5a}$$

$$\forall\, \mu \in \mathscr{P}([0,\infty))\ \exists\, y^n \in \Omega_n \text{ such that } \pi_n(y^n) \rightharpoonup \mu : \qquad \limsup_{n\to\infty} E_n(y^n) \leq E(\mu). \tag{5b}$$

The main challenge in proving these inequalities is in controlling the interaction term E_n^{int} (1). The limsup-inequality (5b) is established by choosing the wall positions y_i^n to be 'locally equi-spaced', by which the high values of V around its singularity are avoided as much as possible.

For proving the liminf-inequality (5a), the scaling regime of ℓ_n is crucial. If $\ell_n \to \ell \in (0,\infty)$, then $\ell_n V(\ell_n \cdot) \to \ell V(\ell \cdot)$, and the basic properties of the narrow convergence of $\pi_n(x^n)$ are enough to show (5a). However, if $\ell_n \to \infty$, then $\ell_n V(\ell_n \cdot) \rightharpoonup (\int_{\mathbb{R}} V)\delta_0$ in the narrow topology, which is too weak for concluding (5a) with the basic properties of the narrow convergence of $\pi_n(x^n)$. Instead, the convexity of V on the positive half-line is exploited to prove (5a). When $\ell_n \to \infty$ fast enough as $n \to \infty$ (i.e. $\ell_n \sim n$), then part of the discreteness of E_n remains visible in the continuum limit through V_{eff} (see Table 1).

As a direct consequence of Theorem 1, the minimisers of E_n converge in the narrow topology to a minimiser of E. Moreover, Theorem 1 is crucial for the proof of the convergence of the gradient flows defined in (3). While Fig. 2 gives an intuitive meaning to the convergence concept for gradient flows, Theorem 2 states the precise notion of convergence.

Theorem 2 (Convergence of gradient flows [31]) *Let $\mu_{init} \in \mathscr{P}([0,\infty))$. Then for any $x_{init}^n \in \Omega_n$ such that $W_2(\pi_n(x_{init}^n), \mu_{init}) \to 0$ as $n \to \infty$, it holds for the solutions $x^n(t)$ and $\mu(t)$ to the gradient flows of E_n and E, respectively, that $W_2(\pi_n(x^n(t)), \mu(t)) \to 0$ pointwise as $n \to \infty$ for all $t > 0$.*

Instead of giving the definition of 'the gradient flow of E' in terms of an evolution variational inequality (cf. [4]), we list in (6) the formally derived evolution equations

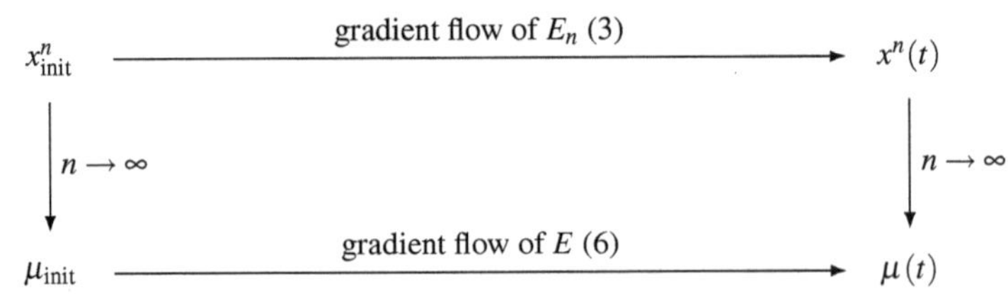

Fig. 2 The gradient flows of E_n converge to the gradient flow of E if this diagram commutes for all $x^n_{\text{init}} \in \Omega_n$ such that $\pi_n(x^n_{\text{init}}) \rightharpoonup \mu_{\text{init}}$

from these inequalities. These equations are equipped with an initial condition and zero-flux boundary conditions at both sides of the domain.

$$\frac{\partial}{\partial t}\rho_t = \big(\rho_t + \rho_t(\ell V(\ell\,\cdot) * \rho_t)'\big)' \qquad \text{if } \ell_n \to \ell, \tag{6a}$$

$$\frac{\partial}{\partial t}\rho_t = \Big(\rho_t + \Big[\int_{\mathbb{R}} V\Big]\rho_t\rho_t'\Big)' \qquad \text{if } 1 \ll \ell_n \ll n, \tag{6b}$$

$$\frac{\partial}{\partial t}\rho_t = \Big(\rho_t + \frac{\alpha^3}{\rho_t^2}V''_{\text{eff}}\Big(\frac{\alpha}{\rho_t}\Big)\rho_t'\Big)' \qquad \text{if } \frac{\ell_n}{n} \to \alpha. \tag{6c}$$

The proof of Theorem 2 relies, in addition to Theorem 1, on the convexity of E_n and E, which are the main conditions under which the general theory [5, Theorem 6.1] concerning the convergence of gradient flows applies. In addition to these conditions, it is also required that there exists a sequence (y^n) as in (5b) for which $\pi_n(y^n)$ converges in the 2-Wasserstein distance.

While Theorems 1 and 2 connect different continuum dislocation density models, they do not quantify how well any of the continuum models approximate the discrete model for a *fixed* number of dislocations. We plan to face this challenge in the near future.

3 Upscaling of Mixed Positive and Negative Dislocations

In Sect. 3.1 we describe the scenario of mixed positive and negative dislocations, which is closely related to [16, 17]. Section 3.2 treats the upscaling of the related energy functionals, and Sect. 3.3 concerns evolutionary convergence of the related gradient flows. The proofs of these results are documented in [30, Chaps. 5 and 8], which is intended to be submitted for publication with A. Garroni, M.A. Peletier, and L. Scardia.

3.1 Setting

In contrast to the dislocation walls in Fig. 1a, we now consider a finite number $n \in \mathbb{N}$ of dislocations without any imposed configuration. In this setting, a dislocation is characterised by its position $x_i^n \in \mathbb{R}^2$ (we do not consider barriers) and its *orientation* $b_i^n = \pm 1$. The result of dislocations having a different orientation is that dislocations with opposite orientation interact with *opposite* force with respect to two dislocations with the same orientation.

The naive approach for defining an interaction energy is to consider

$$\tilde{E}_n(\,\cdot\,; b^n) : (\mathbb{R}^2)^n \to \overline{\mathbb{R}}, \qquad \tilde{E}_n(x^n; b^n) := \frac{1}{2n^2} \sum_{i=1}^{n} \sum_{\substack{j=1 \\ j \neq i}}^{n} b_i^n b_j^n V(x_i^n - x_j^n),$$

where V is the interaction potential in two dimensions given by

$$V : \mathbb{R}^2 \to (-\infty, \infty], \qquad V(r) := \frac{r_1^2}{|r|^2} - \log |r|. \tag{7}$$

The problem with this approach is that $\tilde{E}_n$ is unbounded from below. In fact, any small energy value can be obtained by taking two dislocations of opposite orientation close enough. Therefore, variational techniques are not suited for the upscaling of $\tilde{E}_n$.

We solve this problem by introducing an approximation to the energy $\tilde{E}_n$. The usual approximation in the literature consists of regularising the singularity in V. In contrast to choosing a specific regularisation, which is commonly done (for instance, the different regularisations used in [3, 7, 9, 22, 24]), we consider a large class of regularisations (specified below). For any such regularisation V_n of V, we consider the energy

$$E_n(\,\cdot\,; b^n) : (\mathbb{R}^2)^n \to \mathbb{R}, \qquad E_n(x^n; b^n) := \frac{1}{2n^2} \sum_{i=1}^{n} \sum_{j=1}^{n} b_i^n b_j^n V_n(x_i^n - x_j^n). \tag{8}$$

3.2 *Γ*-Convergence

We follow a similar program as in Sect. 2.2. To any given $x^n \in (\mathbb{R}^2)^n$ and $b^n \in \{-1, 1\}^n$, we associate the signed finite Borel measure $\kappa_n := \frac{1}{n} \sum_{i=1}^{n} b_i \delta_{x_i^n}$. In terms of κ_n, the energy in (8) reads

$$E_n(x^n; b^n) = \frac{1}{2} \int V_n * \kappa_n \, d\kappa_n,$$

which we denote by $E_n(\kappa_n)$. We use the following modified version of the narrow topology:

$$\kappa_n \xrightarrow{c} \kappa \quad :\Longleftrightarrow \quad \begin{cases} \kappa_n \rightharpoonup \kappa, \text{ and} \\ \bigcup_{n=1}^{\infty} \operatorname{supp} \kappa_n \text{ is bounded.} \end{cases} \tag{9}$$

The additional condition concerning the support of κ_n is artificial. It allows us to focus on the collective behaviour of the non-local dislocation interaction while side-stepping technical difficulties when dislocations spread infinitely far.

We motivate the Γ-convergence theorem below in terms of the main assumptions that we put on V_n. We observe [30, Proposition 5.3.1] that V can be split as

$$V = W * W + V_0$$

for some $W \in L^1(\mathbb{R}^2)$ and $V_0 \in C(\mathbb{R}^2)$. Then, we define

$$V_n := W_n * W_n + V_0^n \tag{10}$$

for some $W_n \to W$ in $L^1(\mathbb{R}^2)$ and $V_0^n \to V_0$ in $L^\infty_{\text{loc}}(\mathbb{R}^2)$ as $n \to \infty$, such that V_n converges uniformly to V in any annulus centred around 0. These are the core assumptions which we need on V_n. For the precise assumptions on V_n we refer to [30]. The main gain of the structure of V_n as in (10) is that the energy can be written as

$$E_n(\kappa_n) = \int_{\mathbb{R}^2} (W_n * \kappa_n)^2 + \int V_0^n * \kappa_n \, d\kappa_n,$$

which is crucial for our proof of the following theorem:

Theorem 3 (Γ-convergence [30]) *If $V_n(0)/n \to 0$ as $n \to \infty$, then E_n Γ-converges with respect to the modified narrow topology* (9) *to*

$$E : \left\{\kappa \in \mathcal{M}(\mathbb{R}^2) : \|\kappa\|_{TV} \leq 1\right\} \to [0, \infty], \qquad E(\kappa) := \int V * \kappa \, d\kappa. \tag{11}$$

Here, $\mathcal{M}(\mathbb{R}^2)$ is the space of finite Borel measures on $\mathbb{R}^2$. The condition $V_n(0)/n \to 0$ as $n \to \infty$ ensures that the energy contribution of the self-interactions (i.e. $i = j$ in (8)) vanishes in the limit. Regarding the case in which $V_n(0) \gtrsim n$, the only known Γ-convergence result of E_n is proved in [10, 14] for a specific regularisation V_n.

3.3 Convergence of Gradient Flows

A special feature of (edge) dislocations is that they move horizontally. Together with the linear drag law, their motion is described by the following gradient flow:

$$\begin{cases} \dfrac{dx}{dt}(t) = -nP\nabla E_n(x(t); b), & t \in (0, T], \\ x(0) = x_{\text{init}}, \end{cases} \tag{12}$$

where $P \in R^{2n \times 2n}$ is the diagonal projection matrix characterised by its diagonal given by $(1, 0, 1, 0, \ldots, 1, 0)$. As a consequence of the horizontal movement, the vertical coordinates of the dislocation positions can be considered as parameters.

Next we focus our attention on positive dislocations only, i.e. $b_i^n = 1$ for all $i = 1, \ldots, n$. Consequently, we characterise their positions by the probability measure $\mu_n := \frac{1}{n}\sum_{i=1}^n \delta_{x_i^n}$, and we use the restriction of the energy functionals E_n (8) and E (11) to $\mathscr{P}(\mathbb{R}^2)$ without changing notation.

Conjecture 1 (Convergence of gradient flows [30]) *Let $\mu_{init} \in \mathscr{P}(\mathbb{R}^2)$ be such that $E(\mu_{init}) < \infty$ and* supp μ_{init} *bounded. If $V_n(0)/n \to 0$, then for any recovery sequence μ_{init}^n (in the sense of* (5b)*) converging to μ_{init} as $n \to \infty$, it holds for the solutions $\mu_n(t)$ to* (12) *that there exists a solution $\mu(t)$ to the gradient flow of E such that, along a subsequence, $\mu_n(t) \overset{c}{\rightharpoonup} \mu(t)$ pointwise as $n \to \infty$ for all $t \in [0, T]$.*

Up to a few technical conjectures, the proof of Conjecture 1 is complete. It relies on the framework developed in [26, 28]. Other than Γ-convergence, an important requirement of this framework is that a similar liminf-inequality to (5a) holds for the *slope* of the energy, which is a generalisation for the length of the gradient for functionals defined on metric spaces. To prove this inequality, we require more regularity on $\partial_1 V_n$. Again, we refer to [30] for the details.

The gradient flow of E is characterised in terms of an energy-dissipation inequality, which we can formally write as

$$\frac{d}{dt}\mu = \partial_1(\mu \partial_1 V * \mu), \qquad \text{in } \mathscr{D}'(\mathbb{R}^2 \times (0, T)),$$

where ∂_1 denotes differentiation in the horizontal direction.

In future work, we aim to complete the proof of Conjecture 1, and extend it to mixed positive and negative dislocations. The challenge for proving this extension is to control the singularity in V, which may require us to put more restrictions on the regularisation V_n.

4 Implications for Plasticity Models

With the three theorems above and Conjecture 1 we have partially answered our main question in Sect. 1 about connecting discrete dislocation models with continuum dislocation models for several scenarios. Here, we elaborate on the implications of our results to existing dislocation density models.

The main result of [15] is that the local arrangement of dislocations, measured in terms of the parameter ℓ_n, can drastically change the pile-up profile of the dislocation density in equilibrium. Theorems 1 and 2 give a precise meaning to the extension of this statement to finite domains and to the dynamics of dislocation walls. As a result, it seems that an accurate two-dimensional model for the dislocation density should depend on more detailed information on the local arrangement of dislocations. Highly speculating, such a model may unite the currently available dislocation density models.

Theorem 3 and Conjecture 1 are closely related to the setting in [16, 17]. These theorems are consistent with the continuum model in [16], but only if the regularisation of the interaction potential converges *slowly* enough as $n \to \infty$. This is surprising, because the derivation of the continuum model in [16] does not rely on any regularisation. This suggests that the regularisation in the discrete model should converge, instead, *fast* enough to obtain the continuum gradient flow proposed in [16]. In the near future we aim to shed more light on this peculiar observation.

References

1. Acharya, A.: New inroads in an old subject: plasticity, from around the atomic to the macroscopic scale. J. Mech. Phys. Solids **58**(5), 766–778 (2010)
2. Alicandro, R., De Luca, L., Garroni, A., Ponsiglione, M.: Metastability and dynamics of discrete topological singularities in two dimensions: a Γ-convergence approach. Arch. Ration. Mech. Anal. **214**(1), 269–330 (2014)
3. Alvarez, O., Carlini, E., Hoch, P., Le Bouar, Y., Monneau, R.: Dislocation dynamics described by non-local hamilton-jacobi equations. Mater. Sci. Eng.: A **400**, 162–165 (2005)
4. Ambrosio, L., Gigli, N., Savaré, G.: Gradient Flows: In Metric Spaces and in the Space of Probability Measures. Birkhauser Verlag, New York (2008)
5. Ambrosio, L., Savaré, G., Zambotti, L.: Existence and stability for Fokker-Planck equations with log-concave reference measure. Probab. Theory Relat. Fields **145**, 517–564 (2009)
6. Braides, A., Gelli, M.S.: Continuum limits of discrete systems without convexity hypotheses. Math. Mech. Solids **7**(1), 41–66 (2002)
7. Cai, W., Arsenlis, A., Weinberger, C.R., Bulatov, V.V.: A non-singular continuum theory of dislocations. J. Mech. Phys. Solids **54**(3), 561–587 (2006)
8. Canizo, J.A., Carrillo, J.A., Rosado, J.: A well-posedness theory in measures for some kinetic models of collective motion. Math. Models Methods Appl. Sci. **21**(03), 515–539 (2011)
9. Cermelli, P., Leoni, G.: Renormalized energy and forces on dislocations. SIAM J. Math. Anal. **37**(4), 1131–1160 (2005)
10. De Luca, L., Garroni, A., Ponsiglione, M.: Γ-convergence analysis of systems of edge dislocations: the self energy regime. Arch. Ration. Mech. Anal. **206**(3), 885–910 (2012)
11. Focardi, M., Garroni, A.: A 1D macroscopic phase field model for dislocations and a second order Γ-limit. Multiscale Model. Simul. **6**(4), 1098–1124 (2007)

12. Forcadel, N., Imbert, C., Monneau, R.: Homogenization of the dislocation dynamics and of some particle systems with two-body interactions. Discrete Continuous Dyn. Syst. A **23**(3), 785–826 (2009)
13. Garroni, A., Müller, S.: A variational model for dislocations in the line tension limit. Arch. Rational Mech. Anal. **181**(3), 535–578 (2006)
14. Garroni, A., Leoni, G., Ponsiglione, M.: Gradient theory for plasticity via homogenization of discrete dislocations. J. Eur. Math. Soc. **12**(5), 1231–1266 (2010)
15. Geers, M.G.D., Peerlings, R.H.J., Peletier, M.A., Scardia, L.: Asymptotic behaviour of a pile-up of infinite walls of edge dislocations. Arch. Rational Mech. Anal. **209**, 495–539 (2013)
16. Groma, I., Balogh, P.: Investigation of dislocation pattern formation in a two-dimensional self-consistent field approximation. Acta Mater. **47**(13), 3647–3654 (1999)
17. Groma, I., Csikor, F.F., Zaiser, M.: Spatial correlations and higher-order gradient terms in a continuum description of dislocation dynamics. Acta Mater. **51**(5), 1271–1281 (2003)
18. Hudson, T.: Gamma-expansion for a 1D confined Lennard-Jones model with point defect. Netw. Heterogen. Media **8**(2), 501–527 (2013)
19. Hull, D., Bacon, D.J.: Introduction to Dislocations. Butterworth Heinemann, Oxford (2001)
20. Louat, N.: The distribution of dislocations in stacked linear arrays. Philos. Mag. **8**(91), 1219–1224 (1963)
21. Mora, M.G., Peletier, M.A., Scardia, L.: Convergence of interaction-driven evolutions of dislocations with Wasserstein dissipation and slip-plane confinement. arXiv:1409.4236 (2014)
22. Nabarro, F.R.N.: Dislocations in a simple cubic lattice. Proc. Phys. Soc. **59**(2), 256 (1947)
23. Oelschläger, K.: Large systems of interacting particles and the porous medium equation. J. Differ. Equ. **88**(2), 294–346 (1990)
24. Peierls, R.: The size of a dislocation. Proc. Phys. Soc. **52**(1), 34–37 (1940)
25. Roy, A., Peerlings, R.H.J., Geers, M.G.D., Kasyanyuk, Y.: Continuum modeling of dislocation interactions: why discreteness matters? Mater. Sci. Eng.: A **486**, 653–661 (2008)
26. Sandier, E., Serfaty, S.: Gamma-convergence of gradient flows with applications to Ginzburg-Landau. Commun. Pure Appl. Math. **57**, 1627–1672 (2004)
27. Scardia, L., Schlömerkemper, A., Zanini, C.: Boundary layer energies for nonconvex discrete systems. Math. Models Methods Appl. Sci. **21**(4), 777–817 (2011)
28. Serfaty, S.: Gamma-convergence of gradient flows on Hilbert and metric spaces and applications. Discret. Contin. Dyn. Syst. A **31**, 1427–1451 (2011)
29. Smith, E.: The spread of plasticity from stress concentrations. Proc. Roy. Soc. Lond. Ser. A. Math. Phys. Sci. **282**(1390), 422–432 (1964)
30. van Meurs, P.: Discrete-to-Continuum Limits of Interacting Dislocations. PhD thesis, Technische Universiteit Eindhoven (2015)
31. van Meurs, P., Muntean, A.: Upscaling of the dynamics of dislocation walls. Adv. Math. Sci. Appl. **24**(2), 401–414 (2014)
32. van Meurs, P., Muntean, A., Peletier, M.A.: Upscaling of dislocation walls in finite domains. Eur. J. Appl. Math. **25**(6), 749–781 (2014)

[illegible] Homogenization of the dislocation dynamics and of some particle systems with two-body interactions. Discrete Continuous Dyn. Syst. A 23 [illegible]

[illegible]

Phase Field Crack Growth Model with Hydrogen Embrittlement

Takeshi Takaishi

Abstract As an application of the phase field model for crack propagation in elastic body, chemical-diffuse crack growth model with the effect of the hydrogen embrittlement is considered. Numerical results show the difference of crack path between data with the effects and data without the effect. Temporal evolution of the normalized difference of phase field depict the time when start the difference of crack path.

1 Phase Field Approach for Crack Propagation

To know about fracture is very important to make our life safe. Most of the studies about fracture are focused on its threshold of breaking materials. Many numerical simulations are done with extended finite element method (XFEM) using suitable threshold value of breaking [1]. Since there are various crack propagation models in fracture mechanics, the phase field approximation of the crack has some advantages of mathematical and numerical treatment of this phenomena. In [2, 3], the author and Kimura introduced the phase field crack growth model which is based on the elastic-fracture energy by Francfort-Marigo [4] with the Ambrosio–Tortorelli's approximation [5]. This model describes the temporal evolution of crack using the phase field that shows the crack surface as the continuous function. This uses the similar derivation of the time-dependent Ginzburg–Landau equation which assumes the dissipative property of its free energy at nearly equilibrium state. There are many difficulties in computation of the crack growth, remeshing for new crack surface, blow-up of the stress on the crack tip, however, this model equation makes possible to compute the crack growth with branching, etc., by the well-known numerical methods. It is also remarked that adaptive mesh method, especially adaptive mesh FEM, is effective to calculate this phase field model of crack growth.

This model has also an advantage for extension to the complex crack growth. In this paper, crack growth model with chemical reaction which takes effect to the elastic

T. Takaishi (✉)
Hiroshima Kokusai Gakuin University, Aki-ku Nakano 6-20-1, Hiroshima, Japan
e-mail: t.takaishi@edu.hkg.ac.jp

H. Itou et al. (eds.), *Mathematical Analysis of Continuum Mechanics and Industrial Applications*, Mathematics for Industry 26,
DOI 10.1007/978-981-10-2633-1_3

properties and toughness is introduced. When chemical reaction proceeds, elasticity and toughness change. Basically, two method can be considered as followings:

(A) adding new descriptions in energy,
(B) adding new equation in this system.

In the next section, a phase field model with hydrogen embrittlement is considered with the type (B). In Sects. 3 and 4, some numerical results of it are shown and a simple method which can estimate the effect is proposed.

2 Phase Field Crack Growth Model with Hydrogen Embrittlement

Various factor changes the brittleness of the material, such as UV irradiation, rust, etc. Here, we focus on the hydrogen embrittlement of the steel material. hydrogen in the steel plays a important role in the brittle fracture of the steel. Hydrogen atom diffuses in the material and is trapped in it. Sometimes it makes steel weak for crack or break. The distribution of the hydrogen density is important for studying these phenomena (for example, [6, 7]).

Here, we consider the mode-III crack growth. Let $\Omega \subset \mathbf{R}^2$ be a bounded elastic body with crack. Let $u \in \mathbf{R}$ be an anti-plane displacement, $z \in [0, 1]$ be an phase field, The energy of F-M with A-T approximation is written as

$$E = \frac{1}{2}\int_\Omega (1-z)^2|\nabla z|^2 dx - \int_\Omega f(x)u(x)dx + \frac{1}{2}\int_\Omega \gamma(x)\left(\varepsilon|\nabla z|^2 + \frac{1}{\varepsilon}z^2\right)dx \tag{1}$$

where body force $f(x)$ and fracture toughness $\gamma(x) > 0$ is a given function at x, $\varepsilon > 0$ is a small regularization parameter for approximation (Fig. 1). The energy description introduced by F-M is based on Griffith's criterion of crack [8], and crack propagation phenomena are studied by minimizing this approximated energy (1) [9, 10]. Following the derivation of the phase field model in [11], the author and Kimura derived the temporal evolution model of the displacement and the crack shape [2].

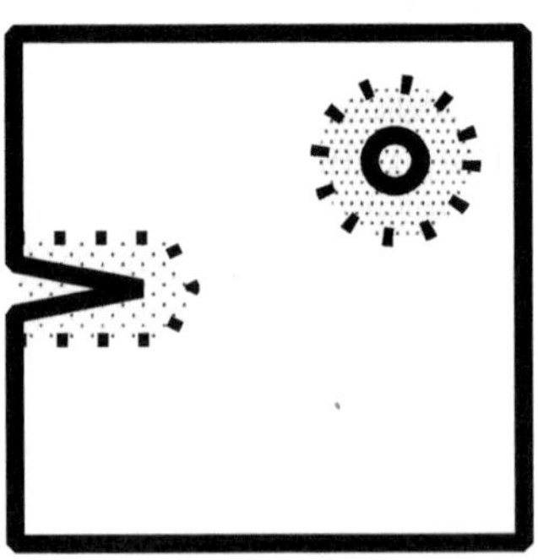

Fig. 1 Concept of hydrogen embrittlement. Hydrogen which is introduced to the crack surface and hole diffuses into the material

Here, let $w(x)$ be an normalized density of hydrogen. We assume that the density of hydrogen atom follows equation of diffusion with diffusion coefficient D_w, and it affects the toughness of material $\gamma(w)$. The phase field crack evolution model of [2] is extended for the hydrogen embrittlement as followings:

$$\begin{cases} \alpha_1 \dfrac{\partial u}{\partial t} = \mu \operatorname{div}\left((1-z)^2 \nabla u\right) + f(x,t) & x \in \Omega \\ \alpha_2 \dfrac{\partial z}{\partial t} = \left(\varepsilon \operatorname{div}(\gamma(w)\nabla z) - \dfrac{\gamma(w)}{\varepsilon} z + \mu |\nabla u|^2 (1-z)\right)_+ & x \in \Omega \\ \dfrac{\partial w}{\partial t} = \operatorname{div}(D_w \nabla w) & x \in \Omega \\ u = g(x,t) & x \in \Gamma_D \\ \dfrac{\partial u}{\partial n} = 0 & x \in \Gamma_N \\ \dfrac{\partial z}{\partial n} = 0 & x \in \Gamma \\ + \text{I.C.}, \end{cases} \tag{2}$$

where $(a)_+ = \max(a, 0)$. μ is one of Lamè's constant. $g(x,t)$ is a given function of displacement at Dirichlet boundary Γ_D. Neumann boundary is set at $\Gamma_N = \Gamma \backslash \Gamma_D$.

For treating the hydrogen embrittlement, three assumptions are set for this model: (i) diffusive property of hydrogen atom in the elastic material, (ii) fracture toughness is decreased by the increment of hydrogen density, (iii) hydrogen diffuses much faster at the crack surface. From (i), we introduce the diffusion equation for the density of hydrogen atom.

$$\frac{\partial w}{\partial t} = \operatorname{div}(D_w \nabla w) \tag{3}$$

From (ii), we simply assume that the fracture toughness of the material increases linearly with w from γ_0 to γ_1.

$$\gamma(w(x,t)) = \gamma_0(1-w) + \gamma_1 w \tag{4}$$

From (iii), we assume that the diffusion coefficient of hydrogen atom increases similar to the Arrhenius's mass diffusivity. Let E_a be an activation energy for diffusion, diffusion coefficient D_w can be written as $D \propto exp(-E_a/kT)$, where k is Boltzmann's constant, T is temperature. Here we assume that the activation energy can be estimated with subtract of the surface energy $E(z=1) - E(z=z_0)$ when the average of phase field z is z_0 in a small domain, and $E \sim z$. Then, we set the diffusion coefficient of the hydrogen atom, $D_w(z(x,t))$, as following:

$$D_w(z(x,t)) = D_{w_0} \exp(z \log(D_{w_1}/D_{w_0})) \tag{5}$$

where $D_w(0) = D_{w0}$, $D_w(1) = D_{w1}$.

3 Numerical Results

Here, we show some numerical results of this crack growth model with hydrogen embrittlement effect. Equations (2) in domain $\Omega = (-1, 1) \times (-1, 1)$ with a hole $(r = 0.1)$ is solved numerically by using FreeFem++ [12]. We set $\tau > 0$ as constant time step, $u^k(x)$, $z^k(x)$ and $w^k(x)$ are the approximated solution of u, z, and w, respectively, at $t = k\tau (k = 0, 1, 2, \dots)$. We make numerical simulation of single-line crack growth of the mode III type, and its numerical domain has a hole filled with hydrogen atom (Fig. 2). We set $N_{vertices} = 59379$. (u^k, z^k, w^k) is calculate from $(u^{k-1}, z^{k-1}, w^{k-1})$ with semi-implicit scheme for the first three equations of (2).

$$\begin{cases} \alpha_1 \dfrac{u^k - u^{k-1}}{\tau} = \mu \operatorname{div}\left((1 - z^{k-1})^2 \nabla u^k\right) \\ \alpha_2 \dfrac{\tilde{z}^k - z^{k-1}}{\tau} = \varepsilon \operatorname{div}\left(\gamma(x) \nabla \tilde{z}^k\right) - \dfrac{\gamma(x)}{\varepsilon} \tilde{z}^k + \mu |\nabla u^{k-1}|^2 (1 - \tilde{z}^k) \\ z^k = \max(\tilde{z}^k, z^{k-1}) \\ \dfrac{w^k - w^{k-1}}{\tau} = \operatorname{div}\left(D_w(z^{k-1}) \nabla w^k\right) \end{cases} \tag{6}$$

We solve (6) by FreeFem++ with P2-element. In the following simulations, we put $\varepsilon = 0.01, \mu = 1, \alpha_1 = 0.01, \alpha_2 = 1, \tau = 0.1$ in (6). $\gamma(w) = 0.5(1 - w) + 0.01w$ and $D_w(z) = 0.1\exp(z \log(100))$ is assumed in it. For diffusing hydrogen, we set $w = 1$ at the hole as a boundary condition. For Dirichlet boundary condition, $g(x, t) = \pm 0.04t$ on $\Gamma_D = \{(x_1, x_2) | x_1 \in (-1, 1), x_2 = \pm 1\}$. Initial crack is set as Fig. 2, single-line from left hand side.

Figure 3 shows numerical results of crack propagation without hydrogen effect. The crack propagation starts straight at $t \sim 37.5$, after that, initial crack grows attracting by the hole (upper right).

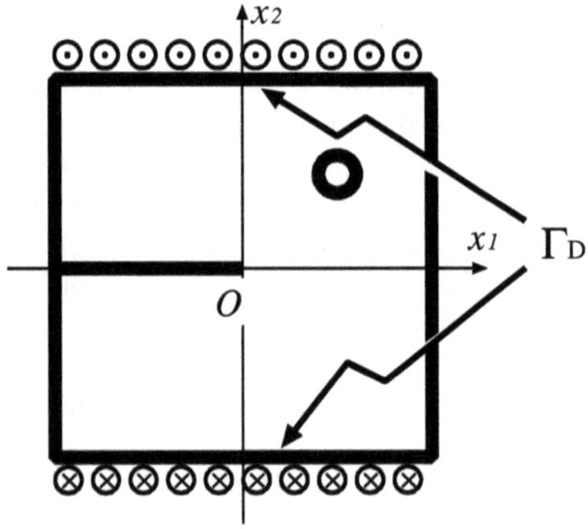

Fig. 2 Numerical domain has Dirichlet boundaries Γ_D at $x_2 = -1, 1$, initial crack at $x_1 < 0, x_2 = 0$, and a hole

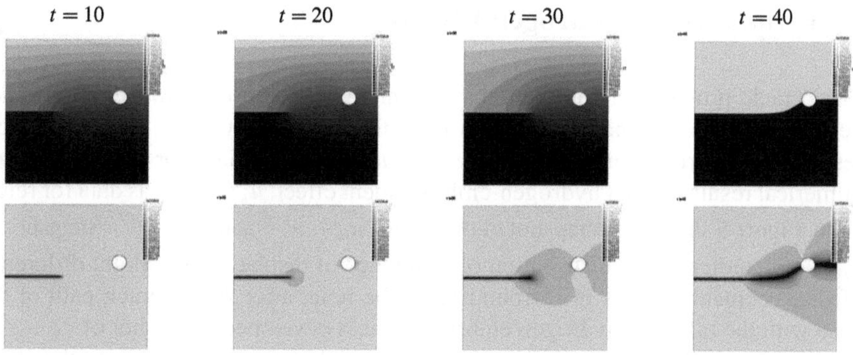

Fig. 3 Snap shots of numerical results without hydrogen effect. Temporal evolution of u (*upper*) shows the vertical displacement of material, and temporal evolution of z (*bottom*) shows the crack propagation

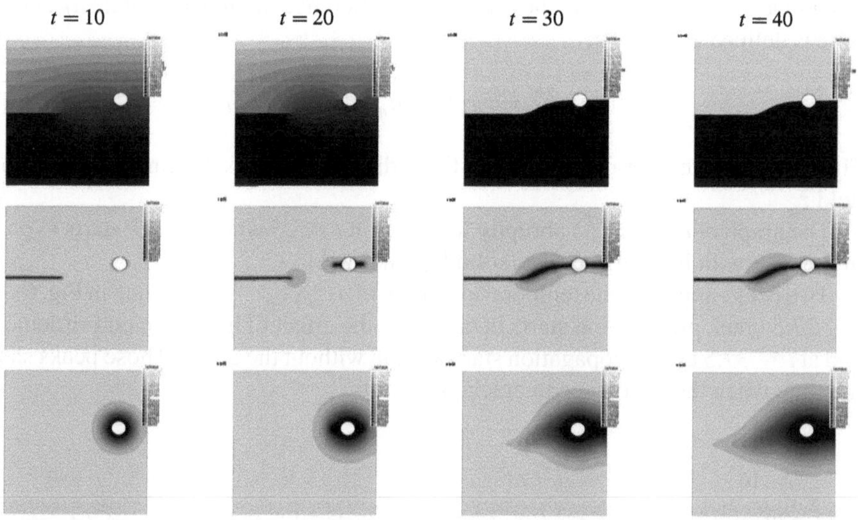

Fig. 4 Snap shots of numerical results without hydrogen effect. Temporal evolution of u (*upper*) shows the vertical displacement of material, temporal evolution of z (*middle*) shows the crack propagation, and temporal evolution of w (*bottom*) shows the diffusion of hydrogen

Figure 4 shows the snapshot of the numerical simulation of this model. There is a hole in the material where hydrogen diffuses from. First, the crack propagation starts from the hole at $t \sim 27.5$. The effect of diffusion and embrittlement of hydrogen accelerating the growth of crack.

4 Estimation of Hydrogen Effects

This crack propagation problem has no analytic solution, however, difference between data for estimation and data for reference is calculated. Here, numerical result with hydrogen-embrittlement effect, u_h, is set as data for estimation, and numerical result without hydrogen-embrittlement effect, u_h^{ref}, is set as data for reference. Figure 5 shows the snapshot of difference $u_h - u_h^{ref}$ and $z_h - z_h^{ref}$. Snapshot of $u_h - u_h^{ref}$ indicates the distribution of difference of displacement. Slight difference of displacement appears in surrounding of the hole, after that, a crack path of the case with the effect of hydrogen embrittlement is developed. Snapshot of $z_h - z_h^{ref}$ indicates the difference of crack path. First, a crack path of the case with the effect of hydrogen embrittlement appears at the edge of the hole, after that, the difference of crack path is developed.

For estimating these difference, a simple estimation method is proposed. Treating as an index of the difference, the normalized difference of numerical data u_h at each time is defined as follows:

$$d(u_h, u_h^{ref}) = ||u_h - u_h^{ref}||_{L^2}/||u_h^{ref}||_{L^2} \tag{7}$$

Figure 6 shows temporal evolution of these difference in displacement u and phase field z.

The graph of $d(u_h, u_h^{ref})$ abruptly increases at $t \sim 25$ when a crack starts to grow in the case with hydrogen effect (solid line in Fig. 6).

Two peaks appear in the temporal evolution of $d(z_h, z_h^{ref})$ (dotted line in Fig. 6). At $t \sim 27.5$ crack propagation starts in data with the effect of hydrogen embrittlement, and at $t \sim 37.5$ crack propagation starts in data without the effect. Those peaks show the time when the difference in crack path starts.

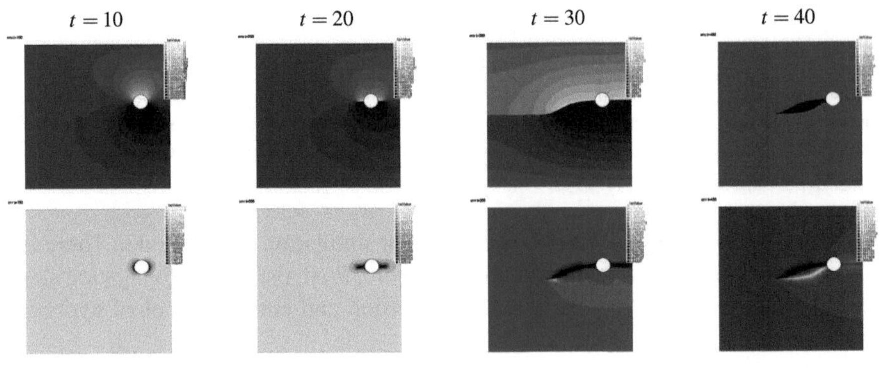

Fig. 5 $u_h - u_h^{ref}$ (*upper*) shows the difference of displacement u, and $z_h - z_h^{ref}$ (*bottom*) shows the difference of phase field (*crack shape*) z

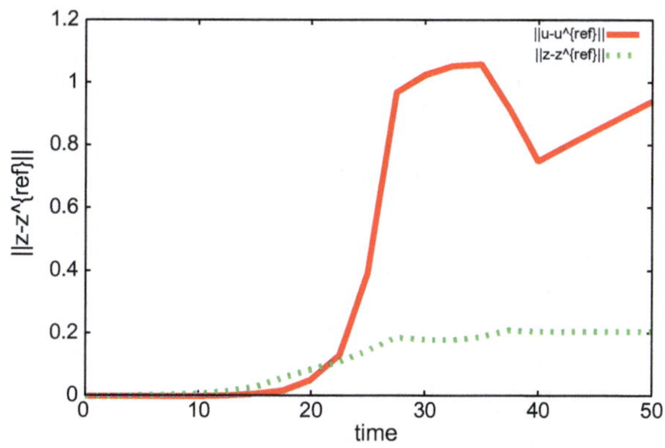

Fig. 6 *Solid line* shows temporal evolution of $d(u_h, u_h^{ref})$, and *dotted line* shows $d(z_h, z_h^{ref})$

5 Concluding Remarks

A complex model of the crack growth is proposed, and shows some numerical results with some simple assumptions of the hydrogen embrittlement and faster diffusivity on crack surface. In numerical results, this effect causes the difference of the path of crack, start time, and the speed of crack propagation. By estimating the temporal evolution of normalized difference between data, we can find the characteristic behavior of this effect.

This model is simply counting the effect of hydrogen diffusion, and the aggregation of hydrogen atom at the edge of crack tip is not considered in this model. However, some studies pointed out the importance of this aggregation phenomena for understanding the mechanism of hydrogen embrittlement. By taking into account the hydrogen aggregation effect, more precise model which may explain the mechanism of this embrittlement will be introduced. Not only in the case of hydrogen effect, embrittlement process of materials occurs due to many reason. Oxidization and UV degradation is very popular. In its simplicity, crack propagation model using the phase field approach is considered to be an effective method for these case.

Acknowledgments This work is in collaboration with M.Kimura.

References

1. Belytschko, T., Black, T.: Elastic crack growth in finite elements with minimal remeshing. Int. J. Numer. Methods Eng. **45**, 602–620 (1999)
2. Takaishi, T., Kimura, M.: Phase field model for mode III crack growth. Kybernetika **45**, 605–614 (2009)
3. Kimura, M., Takaishi, T.: A phase field approach to matehmatical modeling of crack propagation. A mathematical approach to research problems of science and technology mathematics for industry **5**, 161–170 (2014)

4. Francfort, G.A., Marigo, J.-J.: Revisiting brittle fracture as an energy minimization problem. J. Mech. Phys. Solids **46**, 1319–342 (1998)
5. Ambrosio, L., Tortorelli, V.M.: On the approximation of free discontinuity problems. Boll. Un. Mat. Ital. **7**, 105–123 (1992)
6. Troiano, A.R.: The role of Hydrogen and other interstitials in the mechanical behavier of metals. Trans. ASM **52**, 54–80 (1960)
7. Beachem, C.D.: A new model for Hydrogen-asisted cracking. Metall. Trans. **3**, 437–451 (1972)
8. Griffith, A.A.: The phenomenon of rupture and flow in solids. Philos. Trans. Roy. Soc. Lond. **A221**, 163–198 (1921)
9. Bourdin, B.: Numerical implementation of the variational formulation of brittle fracture. Interfaces Free Bound. **9**, 411–430 (2007)
10. Buliga, M.: Energy minimizing brittle crack propagation. J. Elast. **52**, 201–238 (1998/99)
11. Kobayashi, R.: Modeling and numerical simulations of dendritic crystal growth. Physica D **63**, 410–423 (1993)
12. Hecht, F.: New development in FreeFem++. J. Numer. Math. **20**, 251–265 (2012)

On Singularities in 2D Linearized Elasticity

Hiromichi Itou

Abstract The aim of this paper is to introduce some convergent expansion formulae of solutions of two-dimensional linearized elasticity equation so-called as Navier's equation around a crack tip and a tip of thin rigid inclusion, explicitly. In particular, three boundary value problems are treated; the first case is that a homogeneous and anisotropic body has a linear crack; the second case is that two different isotropic homogeneous bodies have an interfacial crack under the non-penetration condition and the frictional condition in Coulomb's law; the third case is that a homogeneous and isotropic body has a rigid line inclusion. Moreover, the paper intends to clarify how the order of singularities in expansions is determined, which depends on the geometry of the material, the governing equation, boundary conditions, material properties, and so on.

1 Introduction

Theory of partial differential equations has been thoroughly developed mainly in smooth domains. In non-smooth domains such as polyhedral or cracked domains, difficulties appear because domains have singular points. Then, it is important to analyze the precise behavior of the solution of the corresponding boundary value problems near singular points. This kind of analysis has possibility of application in various fields of science and engineering such as Fracture problems, Inverse problems (nondestructive evaluation) and so on. For example, by using the *Enclosure method* introduced by Ikehata, we considered reconstruction problems of a linear crack and a polygonal cavity in linearized elasticity from a single set of boundary data, refer to [5–7]. In this paper, we introduce some convergent expansion formulae of solutions of boundary value problems at a crack tip and a tip of thin rigid inclusion, like (1). This chapter is a sort of review of papers [5, 6, 8–10].

H. Itou (✉)
Department of Mathematics, Tokyo University of Science, Tokyo 162-8601, Japan
e-mail: h-itou@rs.tus.ac.jp

H. Itou et al. (eds.), *Mathematical Analysis of Continuum Mechanics and Industrial Applications*, Mathematics for Industry 26,
DOI 10.1007/978-981-10-2633-1_4

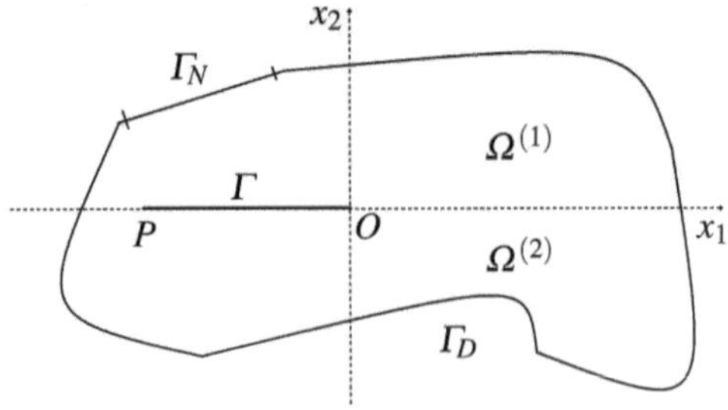

Fig. 1 An example of an illustration of the geometry

A brief outline of this paper is as follows: In Sect. 2, three problems which we deal with are stated. Section 3 is a main part of this paper and convergent series expansions for each problems are shown. Finally, in Sect. 4, we summarize the results and mention future works.

1.1 Domain

In the present paper, we restrict ourselves to two dimensional linearized elastostatic material denoted by $\Omega \subset \mathbf{R}^2$ assuming a bounded domain with Lipschitz boundary and divided into two Lipschitz domains $\Omega^{(1)} := \Omega \cap \{x_2 > 0\}$ and $\Omega^{(2)} = \Omega \cap \{x_2 < 0\}$ by the x_1-axis. We denote the interface of $\Omega^{(k)}$ $(k = 1, 2)$ by Γ'. Let Γ be a line segment $\boldsymbol{OP}$ on Γ' such that $\boldsymbol{P} \notin \partial\Omega$ and $\boldsymbol{O}$ is the origin of the coordinate system $\mathbf{x} = (x_1, x_2)$. Let Γ_N be an arbitrary nonempty open subset of $\partial\Omega^{(1)} \setminus \Gamma'$ such that $\Gamma_N \cap \overline{\Gamma'} = \emptyset$ and $\Gamma_D := \partial\Omega \setminus \Gamma_N$, see Fig. 1.

1.2 The Linearized Elasticity Equation

For the displacement vector $\mathbf{u} = (u_1, u_2)$ in Ω the linearized elasticity equation, called Navier's equation, for a homogeneous material consists of the constitutive law (Hooke's law) and the equilibrium conditions without any body forces;

$$\mathbf{A}\mathbf{u} := \sum_{j,k,l=1}^{2} \frac{\partial}{\partial x_j}\left(C_{ijkl}\frac{\partial \boldsymbol{u}_k}{\partial x_l}\right) = 0 \quad i = 1, 2.$$

Here $C = (C_{ijkl})_{i,j,k,l=1,2}$ is the constant elasticity tensor satisfying three symmetries: $C_{ijkl} = C_{jikl}$, $C_{ijkl} = C_{ijlk}$, $C_{ijkl} = C_{klij}$ and positive definiteness. As the result, in anisotropic materials one sees C has six independent components C_{1111}, C_{2222}, C_{1212}, C_{1122}, C_{1112}, C_{1222} and in isotropic materials C has two independent components, that is, we usually use *Lamé* constants $\lambda := C_{1122}$, $\mu := C_{1212}$ satisfying $\mu > 0$, $\lambda + \mu > 0$. We define $\kappa := \frac{\lambda+3\mu}{\lambda+\mu}$. Next, the stress vector on $\partial\Omega$ and Γ is denoted by $\boldsymbol{T}\mathbf{u} := \boldsymbol{\sigma}\mathbf{n}$, where $\boldsymbol{\sigma} = (\sigma_{ij})_{i,j=1,2}$ is the stress tensor and $\mathbf{n} = (n_1, n_2)$ is the unit outward normal. In the two dimensional case a rigid displacement can be written in the form $F(\mathbf{x})\mathbf{k} := (k_1 + k_0x_2, k_2 - k_0x_1)$ with an arbitrary constant vector $\mathbf{k} = (k_1, k_2, k_0)$. We denote the set of all rigid displacements on Γ by $\mathscr{R}(\Gamma)$. The superscripts (k) for $k = 1, 2$ refers to the values in $\Omega^{(k)}$, respectively, such as $\mathbf{u}^{(k)}$, $\mu^{(k)}$, etc.

2 Problems

Now, the following three cases are considered and the convergent series expansions of the solutions near $\boldsymbol{O}$ are derived. For them we set polar coordinate system (r, θ) with respect to the origin $\boldsymbol{O}, r \in (0, R), \theta \in (-\pi, \pi)$ with a sufficiently small $R > 0$. Let $B_R := B_R(\boldsymbol{O})$ be a disk of the radius $R > 0$ centered at $\boldsymbol{O}$ and fix some notations $B_R^{(1)} := B_R \cap \Omega^{(1)}$, $B_R^{(2)} := B_R \cap \Omega^{(2)}$.

2.1 Case 1

Assume that Ω is a homogeneous *anisotropic* elasticity and Γ is a linear crack. On Γ free traction condition is imposed. Then, we consider the boundary value problem; For given surface force $\boldsymbol{g} \in L^2(\Gamma_N)$, find $\boldsymbol{u}$ satisfying

$$(*)_1 \begin{cases} A\mathbf{u} = \mathbf{0} & \text{in } \Omega \setminus \overline{\Gamma}, \\ T\mathbf{u}^{(1)} = T\mathbf{u}^{(2)} = \mathbf{0} & \text{on } \Gamma, \\ T\mathbf{u} = \mathbf{g} & \text{on } \Gamma_N, \\ \mathbf{u} = \mathbf{0} & \text{on } \Gamma_D. \end{cases}$$

One sees existence of the unique weak solution in $H^1(\Omega \setminus \overline{\Gamma})$ by Lax-Milgram lemma. The derivation of series expansion of the solution of $(*)_1$ is based on complex analysis, especially, analytic continuation, Goursat–Kolosov–Muskhelishvili stress function [13], Lekhnitskii formalism [11], for the details refer to [6].

2.2 Case 2

Assume that Ω consists of two different homogeneous and isotropic elasticity $\Omega^{(1)}$ and $\Omega^{(2)}$ and Γ is a linear *interfacial* crack. On Γ *non-penetration* condition and *frictional* condition obeying Coulomb's law are imposed. Then, we consider the boundary value problem; For given surface force $\mathbf{g} \in L^2(\Gamma_N)$ and a small constant friction coefficient $f \in (0, 1)$, find $\mathbf{u}^{(k)} \in H^1(\Omega^{(k)})$ $(k = 1, 2)$ satisfying

$$(*)_2 \begin{cases} \mu^{(k)}\Delta\mathbf{u}^{(k)} + (\lambda^{(k)} + \mu^{(k)})\nabla(\nabla\cdot\mathbf{u}^{(k)}) = \mathbf{0} & \text{in } \Omega^{(k)}, \\ T\mathbf{u}^{(1)} = \mathbf{g} & \text{on } \Gamma_N, \\ \mathbf{u}^{(k)} = \mathbf{0} & \text{on } \Gamma_D, \\ [u_1] = [u_2] = [\sigma_{12}] = [\sigma_{22}] = 0 & \text{on } \Gamma' \setminus \overline{\Gamma}, \\ [\sigma_{22}] = 0,\ \sigma_{22}^{(k)} \le 0,\ [u_2] \ge 0,\ \sigma_{22}^{(k)}[u_2] = 0 & \text{on } \Gamma, \\ [\sigma_{12}] = 0,\ |\sigma_{12}^{(k)}| \le -f\sigma_{22}^{(k)},\ \sigma_{12}^{(k)}[u_1] + f\sigma_{22}^{(k)}|[u_1]| = 0 & \text{on } \Gamma. \end{cases}$$

We denote Jump of $\mathbf{u}$ at Γ by $[\mathbf{u}] := \mathbf{u}^{(1)} - \mathbf{u}^{(2)}$ on Γ. Conditions on Γ includes three states: open crack, slip state, slip state, as follows.

- $[u_2] > 0$ on Γ (open crack).
  In this case, they can be reduced to

$$\sigma_{12}^{(k)} = \sigma_{22}^{(k)} = 0 \quad \text{on} \quad \Gamma.$$

- $[u_1] = [u_2] = 0$ on Γ (stick state).
  In this case, they can be reduced to

$$[\sigma_{22}] = [\sigma_{12}] = 0, \quad \sigma_{22}^{(k)} \le 0, \quad |\sigma_{12}^{(k)}| \le -f\sigma_{22}^{(k)} \quad \text{on} \quad \Gamma.$$

- $[u_2] = 0$ and $[u_1] \neq 0$ on Γ (slip state).
  In this case, they can be reduced to

$$[\sigma_{22}] = [\sigma_{12}] = 0, \quad \sigma_{22}^{(k)} \le 0, \quad \sigma_{12}^{(k)} \pm f\sigma_{22}^{(k)} = 0 \quad \text{on} \quad \Gamma,$$

  where the upper sign "+" is taken for $[u_1] > 0$ on Γ and the lower sign "−" is taken for $[u_1] < 0$ on Γ.

In [10], we can show existence of the solution of $(*)_2$, however, uniqueness of the solution remains an open issue. Under assumption of no switches among 3 states near $\boldsymbol{O}$ on Γ, we derive a convergent expansion of the solution of $(*)_2$ in each of 3 states (open, stick, slip) by solving the Riemann–Hilbert problem for stress functions (e.g. [14]).

2.3 Case 3

Assume that Ω is a homogeneous and isotropic elasticity and Γ is a *rigid line inclusion* with delamination in a linear case. On the upper and lower surface of Γ free traction condition and *rigidity* condition are imposed, respectively. Then, we consider the boundary value problem; For given surface force $\mathbf{g} \in L^2(\Gamma_N)$, find $\mathbf{u} \in H^1(\Omega \setminus \overline{\Gamma})$ and $\rho_0 \in \mathscr{R}(\Gamma)$ satisfying

$$(*)_3 \begin{cases} A\mathbf{u} = \mathbf{0} & \text{in} \quad \Omega \setminus \overline{\Gamma}, \\ T\mathbf{u}^{(1)} = \mathbf{0},\ \mathbf{u}^{(2)} = \boldsymbol{\rho}_0 & \text{on} \quad \Gamma, \\ T\mathbf{u} = \mathbf{g} & \text{on} \quad \Gamma_N, \\ \mathbf{u} = \mathbf{0} & \text{on} \quad \Gamma_D, \\ \int_\Gamma [T\mathbf{u}]_\Gamma \cdot \boldsymbol{\rho}\, \mathrm{d}S_{\mathbf{x}} = \mathbf{0} & \text{for } \forall \boldsymbol{\rho} \in \mathscr{R}(\Gamma) \end{cases}$$

Define a convex set of admissible displacements

$$\mathscr{K} := \{\mathbf{v} \in H^1(\Omega \setminus \overline{\Gamma}) \mid \mathbf{v}|_{\Gamma_D} = \mathbf{0},\ \mathbf{v}^-|_\Gamma \in \mathscr{R}(\Gamma)\}.$$

Then, we can show that there exists a unique solution $\mathbf{u} \in \mathscr{K}$ of $(*)_3$ through corresponding minimization problem of the potential energy and we obtain a convergent series expansion of the solution of $(*)_3$ near the tip of rigid line inclusion $\boldsymbol{O}$ by solving corresponding a system of Riemann–Hilbert problems for stress functions [9].

3 Convergent Series Expansions

For respective three cases defined above, we derive *convergent* series expansions of the solution near $\boldsymbol{O}$ like

$$\mathbf{u}(r,\theta) = \sum_{n=0}^{\infty} A_n r^{\alpha_n} \boldsymbol{\Phi}_n(\theta) + B_n r^{\alpha_n} \boldsymbol{\Psi}_n(\theta). \tag{1}$$

Here α_n is called the order of singularity, $\boldsymbol{\Phi}_n(\theta)$ and $\boldsymbol{\Psi}_n(\theta)$ are vector of angular function, which are given explicitly for respective cases. In particular, the coefficients of the leading terms in the expansion are very important in fracture mechanics, called the stress intensity factors, to predict the stress state near the tip of a crack, see [2].

3.1 Case 1

Let $S = (S_{ijkl})_{i,j,k,l=1,2} := C^{-1}$ be the compliance tensor and $\mu_1, \mu_2 \in \mathbf{C}$ be roots of the following characteristic equation:

$$S_{1111}t^4 - 2S_{1112}t^3 + (2S_{1122} + S_{1212})t^2 - 2S_{2212}t + S_{2222} = 0.$$

Proposition 1 ([6]) *In the case $\mu_1 \neq \mu_2$, there exist $A_n, B_n \in \mathbf{R}$ $(n = 0, 1, 2, \ldots)$ and $\mathbf{k} \in \mathbf{R}^3$ such that*

$$\mathbf{u}(r,\theta) = \sum_{n=0}^{\infty} r^{\frac{n}{2}} \mathbf{Re}\left[A_n \frac{i^{n-1}\psi_{1,n}(\theta)}{\mu_1 - \mu_2} + B_n \frac{i^{n-1}\psi_{2,n}(\theta)}{\mu_1 - \mu_2} \right] + F(\mathbf{x})\mathbf{k},$$

where

$$\psi_{1,n}(\theta) = \begin{pmatrix} \mu_1 p(\mu_2)(\cos\theta + \mu_2 \sin\theta)^{\frac{n}{2}} - \mu_2 p(\mu_1)(\cos\theta + \mu_1 \sin\theta)^{\frac{n}{2}} \\ \mu_1 q(\mu_2)(\cos\theta + \mu_2 \sin\theta)^{\frac{n}{2}} - \mu_2 q(\mu_1)(\cos\theta + \mu_1 \sin\theta)^{\frac{n}{2}} \end{pmatrix}$$

$$\psi_{2,n}(\theta) = \begin{pmatrix} -p(\mu_1)(\cos\theta + \mu_1 \sin\theta)^{\frac{n}{2}} + p(\mu_2)(\cos\theta + \mu_2 \sin\theta)^{\frac{n}{2}} \\ -q(\mu_1)(\cos\theta + \mu_1 \sin\theta)^{\frac{n}{2}} + q(\mu_2)(\cos\theta + \mu_2 \sin\theta)^{\frac{n}{2}} \end{pmatrix}$$

and $p(\mu_k) := S_{1111}\mu_k^2 + S_{1122} - S_{1112}\mu_k$, $q(\mu_k) := S_{1122}\mu_k + \frac{S_{2222}}{\mu_k} - S_{2212}$. *The series is convergent, absolutely in* $H^1(B_R \setminus \overline{\Gamma})$ *and uniformly on compact sets in* $B_R \setminus \overline{\Gamma}$.

Proposition 2 ([6]) *In the case* $\mu_1 = \mu_2 =: \mu_0 = \alpha_0 + i\beta_0$, *there exist* $A_n, B_n \in \mathbf{R}$ $(n = 0, 1, 2, \ldots)$ *and* $\mathbf{k} \in \mathbf{R}^3$ *such that*

$$\mathbf{u}(r,\theta) = \sum_{n=0}^{\infty} r^{\frac{n}{2}} \left\{ A_n \mathbf{Re}\left[\tilde{\psi}_{1,n}(\theta)\right] + B_n \mathbf{Im}\left[\tilde{\psi}_{2,n}(\theta)\right]\right\} + F(\mathbf{x})\mathbf{k},$$

where

$$\tilde{\psi}_{1,n}(\theta) = \begin{pmatrix} p(\mu_0)\left\{\frac{n}{2}\Theta^{\frac{n}{2}-2} - \left(\frac{n}{2} + (-1)^n\right)\Theta^{\frac{n}{2}} + \overline{\Theta}^{\frac{n}{2}}\right\} + 4S_{1111}\beta_0^2\Theta^{\frac{n}{2}} \\ q(\mu_0)\left\{\frac{n}{2}\Theta^{\frac{n}{2}-2} - \left(\frac{n}{2} + (-1)^n\right)\Theta^{\frac{n}{2}} + \overline{\Theta}^{\frac{n}{2}}\right\} + S_{2222}\frac{4\beta_0^2}{|\mu_0|^2}\overline{\mu_0}\Theta^{\frac{n}{2}} \end{pmatrix}$$

$$\tilde{\psi}_{2,n}(\theta) = \begin{pmatrix} p(\mu_0)\left\{-\frac{n}{2}\Theta^{\frac{n}{2}-2} + \left(\frac{n}{2} - (-1)^n\right)\Theta^{\frac{n}{2}} + \overline{\Theta}^{\frac{n}{2}}\right\} - 4S_{1111}\beta_0^2\Theta^{\frac{n}{2}} \\ q(\mu_0)\left\{-\frac{n}{2}\Theta^{\frac{n}{2}-2} + \left(\frac{n}{2} - (-1)^n\right)\Theta^{\frac{n}{2}} + \overline{\Theta}^{\frac{n}{2}}\right\} - S_{2222}\frac{4\beta_0^2}{|\mu_0|^2}\overline{\mu_0}\Theta^{\frac{n}{2}} \end{pmatrix}$$

with $\Theta = \cos\theta + \mu_0 \sin\theta$. *The series is convergent, absolutely in* $H^1(B_R \setminus \overline{\Gamma})$ *and uniformly on compact sets in* $B_R \setminus \overline{\Gamma}$. *Especially,* $\mu_0 = i$ *(i.e.,* $\alpha_0 = 0$, $\beta_0 = 1$*) implies the material is isotropic.*

3.2 Case 2

Assume no switches among three states (open, stick, slip) near $\boldsymbol{O}$ on Γ. Let

$$m_1 := \frac{\kappa^{(1)}}{\mu^{(1)}} + \frac{1}{\mu^{(2)}}, \quad m_2 := \frac{\kappa^{(2)}}{\mu^{(2)}} + \frac{1}{\mu^{(1)}}.$$

By employing the Dundurs parameter [3]

$$\beta := \frac{\mu^{(2)}(\kappa^{(1)} - 1) - \mu^{(1)}(\kappa^{(2)} - 1)}{\mu^{(2)}(\kappa^{(1)} + 1) + \mu^{(1)}(\kappa^{(2)} + 1)} = \frac{m_1 - m_2}{m_1 + m_2},$$

we set ϵ as $\epsilon := -\frac{1}{2\pi}\ln\left(\frac{1-\beta}{1+\beta}\right)$. We see that β varies from $-1/2$ to $1/2$ and vanishes for identical materials or special materials.

3.2.1 Open Crack

Proposition 3 ([10]) *For open crack case there exist complex numbers $\hat{a}_n$ satisfying the condition*

$$\sum_{n=0}^{\infty}(-1)^n r^{\frac{1}{2}+n}\mathbf{Re}\left[\hat{a}_n r^{-i\epsilon}\right] > 0, \tag{2}$$

$\hat{b}_n$ and a constant vector $\mathbf{c} \in \mathbf{R}^3$ such that for $k = 1, 2$

$$\mathbf{u}^{(k)}(r,\theta) = \sum_{n=0}^{\infty}\frac{e^{(-1)^k\epsilon\pi}r^{\frac{1}{2}+n}}{2\mu^{(k)}}\left\{\mathbf{Re}\left[\hat{a}_n r^{-i\epsilon}\right]\boldsymbol{P}^{(k)}_{1,n}(\theta) - \mathbf{Im}\left[\hat{a}_n r^{-i\epsilon}\right]\boldsymbol{Q}^{(k)}_{1,n}(\theta)\right\}$$
$$+\sum_{n=0}^{\infty}r^{n+1}d_k\left\{\mathbf{Re}\left[\hat{b}_n\right]\boldsymbol{R}^{(k)}_{1,n}(\theta) - \mathbf{Im}\left[\hat{b}_n\right]\boldsymbol{S}^{(k)}_{1,n}(\theta)\right\} + \boldsymbol{F}(\mathbf{x})\mathbf{c},$$

where $d_1 = \kappa^{(2)} + 1$, $d_2 = \kappa^{(1)} + 1$,

$$\boldsymbol{P}^{(k)}_{1,n}(\theta) = e^{\epsilon\theta}\begin{pmatrix}\left(\kappa^{(k)} + n + \frac{1}{2} - e^{-2\epsilon(\theta+(-1)^k\pi)}\right)\cos\left(n+\frac{1}{2}\right)\theta \\ +\epsilon\left(\sin\left(n+\frac{1}{2}\right)\theta - \sin\left(n-\frac{3}{2}\right)\theta\right) - \left(n+\frac{1}{2}\right)\cos\left(n-\frac{3}{2}\right)\theta \\ \left(\kappa^{(k)} - n - \frac{1}{2} + e^{-2\epsilon(\theta+(-1)^k\pi)}\right)\sin\left(n+\frac{1}{2}\right)\theta \\ +\epsilon\left(\cos\left(n+\frac{1}{2}\right)\theta - \cos\left(n-\frac{3}{2}\right)\theta\right) + \left(n+\frac{1}{2}\right)\sin\left(n-\frac{3}{2}\right)\theta\end{pmatrix},$$

$$\boldsymbol{Q}^{(k)}_{1,n}(\theta) = e^{\epsilon\theta}\begin{pmatrix}\left(\kappa^{(k)} + n + \frac{1}{2} + e^{-2\epsilon(\theta+(-1)^k\pi)}\right)\sin\left(n+\frac{1}{2}\right)\theta \\ -\epsilon\left(\cos\left(n+\frac{1}{2}\right)\theta - \cos\left(n-\frac{3}{2}\right)\theta\right) - \left(n+\frac{1}{2}\right)\sin\left(n-\frac{3}{2}\right)\theta \\ \left(-\kappa^{(k)} + n + \frac{1}{2} + e^{-2\epsilon(\theta+(-1)^k\pi)}\right)\cos\left(n+\frac{1}{2}\right)\theta \\ +\epsilon\left(\sin\left(n+\frac{1}{2}\right)\theta - \sin\left(n-\frac{3}{2}\right)\theta\right) - \left(n+\frac{1}{2}\right)\cos\left(n-\frac{3}{2}\right)\theta\end{pmatrix},$$

$$\boldsymbol{R}^{(k)}_{1,n}(\theta) = \begin{pmatrix}\kappa^{(k)}\cos(n+1)\theta - (n+1)\cos(n-1)\theta + (n+2)\cos(n+1)\theta \\ \kappa^{(k)}\sin(n+1)\theta + (n+1)\sin(n-1)\theta - (n+2)\sin(n+1)\theta\end{pmatrix},$$

$$\boldsymbol{S}^{(k)}_{1,n}(\theta) = \begin{pmatrix}\kappa^{(k)}\sin(n+1)\theta - (n+1)\sin(n-1)\theta + n\sin(n+1)\theta \\ -\kappa^{(k)}\cos(n+1)\theta - (n+1)\cos(n-1)\theta + n\cos(n+1)\theta\end{pmatrix}.$$

The series are convergent, absolutely in $H^1(B^{(k)}_R)$ and uniformly on compact sets in $B^{(k)}_R$ for $k = 1, 2$, respectively. For $n \geq 0$, $\hat{a}_n$ and $\hat{b}_n$ satisfy

$$|\hat{a}_n| \le c\frac{1}{\sqrt{2n+1}}\, R^{-(n+\frac{1}{2})}\|\nabla \mathbf{u}\|_{L^2(B_R)},$$
$$|\hat{b}_n| \le c\frac{1}{\sqrt{n+1}}\, R^{-(n+1)}\|\nabla \mathbf{u}\|_{L^2(B_R)}.$$

Note here that one can evaluate coefficients of the expansion in Proposition 4, especially stress intensity factors. In the case of homogeneous material which follows $\epsilon = 0$, the condition (2) implies $\mathbf{Re}\left[\hat{a}_0\right] \ge 0$.

3.2.2 Stick State

Proposition 4 ([10]) *For stick state case there exist complex numbers c_n, b_n satisfying the conditions (3)–(6) and a constant vector $\mathbf{c} \in \mathbf{R}^3$ such that for $k = 1, 2$*

$$\mathbf{u}^{(k)}(r,\theta) = \sum_{n=0}^{\infty} \frac{r^{n+1}}{2\mu^{(k)} m_k (n+1)} \left\{\mathbf{Re}\,[c_n]\, \boldsymbol{P}^{(k)}_{2a,n}(\theta) - \mathbf{Im}\,[c_n]\, \boldsymbol{Q}^{(k)}_{2a,n}(\theta)\right\}$$
$$+ \sum_{n=0}^{\infty} \frac{r^{n+1}}{\mu^{(1)}\mu^{(2)}(n+1)} \left\{\mathbf{Re}\,[b_n]\, \boldsymbol{R}^{(k)}_{2a,n}(\theta) - \mathbf{Im}\,[b_n]\, \boldsymbol{S}^{(k)}_{2a,n}(\theta)\right\} + \boldsymbol{F}(\mathbf{x})\mathbf{c},$$

where

$$\boldsymbol{P}^{(k)}_{2a,n}(\theta) = \begin{pmatrix} (\kappa^{(k)} + n + 1 - e^{(-1)^{k+1}2\pi\epsilon}) \cos{(n+1)\theta} - (n+1)\cos{(n-1)\theta} \\ (\kappa^{(k)} - n - 1 + e^{(-1)^{k+1}2\pi\epsilon}) \sin{(n+1)\theta} + (n+1)\sin{(n-1)\theta} \end{pmatrix},$$

$$\boldsymbol{Q}^{(k)}_{2a,n}(\theta) = \begin{pmatrix} (\kappa^{(k)} + n + 1 + e^{(-1)^{k+1}2\pi\epsilon}) \sin{(n+1)\theta} - (n+1)\sin{(n-1)\theta} \\ (-\kappa^{(k)} + n + 1 + e^{(-1)^{k+1}2\pi\epsilon}) \cos{(n+1)\theta} - (n+1)\cos{(n-1)\theta} \end{pmatrix},$$

$$\boldsymbol{R}^{(k)}_{2a,n}(\theta) = \begin{pmatrix} \left(\frac{\kappa^{(1)}}{m_1} + \frac{\kappa^{(2)}}{m_2} + \frac{n+1}{m_k}\right) \cos{(n+1)\theta} - \frac{n+1}{m_k}\cos{(n-1)\theta} \\ \left((-1)^{k+1}\frac{\kappa^{(1)}}{m_1} + (-1)^k\frac{\kappa^{(2)}}{m_2} - \frac{n+1}{m_k}\right) \sin{(n+1)\theta} + \frac{n+1}{m_k}\sin{(n-1)\theta} \end{pmatrix},$$

$$\boldsymbol{S}^{(k)}_{2a,n}(\theta) = \begin{pmatrix} \left((-1)^{k+1}\frac{\kappa^{(1)}}{m_1} + (-1)^k\frac{\kappa^{(2)}}{m_2} + \frac{n+1}{m_k}\right) \sin{(n+1)\theta} - \frac{n+1}{m_k}\sin{(n-1)\theta} \\ \left(-\frac{\kappa^{(1)}}{m_1} - \frac{\kappa^{(2)}}{m_2} + \frac{n+1}{m_k}\right) \cos{(n+1)\theta} - \frac{n+1}{m_k}\cos{(n-1)\theta} \end{pmatrix}.$$

The series are convergent, absolutely in $H^1(B_R^{(k)})$ and uniformly on compact sets in $B_R^{(k)}$ for $k = 1, 2$, respectively. For $n \ge 0$, c_n and b_n satisfy

$$|c_n| \le c\sqrt{n+1}\, R^{-(n+1)}\|\nabla \mathbf{u}\|_{L^2(B_R)},$$
$$|b_n| \le c\sqrt{n+1}\, R^{-(n+1)}\|\nabla \mathbf{u}\|_{L^2(B_R)}.$$

One can see that the condition $\sigma_{22}^{(k)} \leq 0$ on Γ is equivalent to a condition on $B_R \cap \Gamma$

$$\sum_{n=0}^{\infty} r^n(-1)^n \left\{(m_1 + m_2)\,\mathbf{Re}\,[c_n] - \frac{2(\kappa^{(1)}\kappa^{(2)} - 1)}{\mu^{(1)}\mu^{(2)}}\mathbf{Re}\,[b_n]\right\} \leq 0. \tag{3}$$

Moreover, since (3) is valid as r tends to 0, *one sees*

$$(m_1 + m_2)\,\mathbf{Re}\,[c_0] - \frac{2(\kappa^{(1)}\kappa^{(2)} - 1)}{\mu^{(1)}\mu^{(2)}}\mathbf{Re}\,[b_0] \leq 0. \tag{4}$$

On the other hand, one knows that the condition $|\sigma_{12}^{(k)}| \leq -f\sigma_{22}^{(k)}$ on Γ is reduced to a condition on $B_R \cap \Gamma$

$$\begin{aligned} &\left|\sum_{n=0}^{\infty} r^n(-1)^{n+1} \left\{(m_1 + m_2)\,\mathbf{Im}\,[c_n] - \frac{2(\kappa^{(1)}\kappa^{(2)} - 1)}{\mu^{(1)}\mu^{(2)}}\mathbf{Im}\,[b_n]\right\}\right| \\ &\leq -f\sum_{n=0}^{\infty} r^n(-1)^n \left\{(m_1 + m_2)\,\mathbf{Re}\,[c_n] - \frac{2(\kappa^{(1)}\kappa^{(2)} - 1)}{\mu^{(1)}\mu^{(2)}}\mathbf{Re}\,[b_n]\right\} \end{aligned} \tag{5}$$

and also we have

$$\begin{aligned} &\left|(m_1 + m_2)\,\mathbf{Im}\,[c_0] - \frac{2(\kappa^{(1)}\kappa^{(2)} - 1)}{\mu^{(1)}\mu^{(2)}}\mathbf{Im}\,[b_0]\right| \\ &\leq -f\left\{(m_1 + m_2)\,\mathbf{Re}\,[c_0] - \frac{2(\kappa^{(1)}\kappa^{(2)} - 1)}{\mu^{(1)}\mu^{(2)}}\mathbf{Re}\,[b_0]\right\} \end{aligned} \tag{6}$$

3.2.3 Slip State

In this case, we have

$$\cot(\pi\check{\gamma}) = \frac{1 + \cos(2\pi\check{\gamma})}{\sin(2\pi\check{\gamma})} = \mp f\beta, \tag{7}$$

where the upper sign "−" is for the case $[u_1] > 0$ on Γ and the lower sign "+" is for the case $[u_1] < 0$ on Γ. Therefore, since given f is assumed to be less than 1, β varies from $-1/2$ to $1/2$, we can uniquely choose $\check{\gamma} \in \mathbf{R}$ satisfying (7) and $0 < \check{\gamma} < 1$. In fact, a possibility of a case $\frac{1}{2} < \check{\gamma} < 1$ is precluded by inequality conditions on Γ, for the detail see Remark 1. According to [1], it is shown in that $\check{\gamma}$ cannot be larger than $\frac{1}{2}$ by excluding an inconsistent situation of a backward propagation of the crack, see also [3]. Moreover, note that $\check{\gamma} = \frac{1}{2}$ if and only if $\beta = 0$ which includes identical materials.

Proposition 5 ([10]) *For slip state case, there exist complex numbers $\check{a}_n$, b_n, $\check{c}_n$ satisfying the condition (8) and a constant vector $\mathbf{c} \in \mathbf{R}^3$ such that for $k = 1, 2$*

$$\begin{aligned}\mathbf{u}^{(k)}(r,\theta) = &\sum_{n=0}^{\infty} \frac{e^{(-1)^k \epsilon\pi} r^{n+1-\check{\gamma}}}{2\mu^{(k)}(n+1-\check{\gamma})} \left\{ -\mathbf{Im}\left[\check{a}_n\right] \boldsymbol{Q}^{(k)}_{2b,n}(\theta) \right\} \\ &+ \sum_{n=0}^{\infty} \frac{r^{n+1}}{\mu^{(1)}\mu^{(2)}(n+1)} \mathbf{Re}\left[b_n\right] \left(\boldsymbol{R}^{(k)}_{2a,n}(\theta) \pm \frac{f(\kappa^{(1)}\kappa^{(2)}-1)}{\mu^{(k)} m_k (m_1+m_2)} \boldsymbol{Q}^{(k)}_{2a,n}(\theta) \right) \\ &- \sum_{n=0}^{\infty} \frac{r^{n+1}\tilde{d}_k}{\mu^{(1)}\mu^{(2)}(m_1+m_2)(n+1)} \mathbf{Im}\left[b_n\right] \boldsymbol{S}^{(k)}_{1,n}(\theta) \\ &+ \sum_{n=0}^{\infty} \frac{r^{n+1}}{2\mu^{(k)} m_k (n+1)} \mathbf{Re}\left[\check{c}_n\right] \left\{ \boldsymbol{P}^{(k)}_{2a,n}(\theta) \mp f \boldsymbol{Q}^{(k)}_{2a,n}(\theta) \right\} + \boldsymbol{F}(\mathbf{x})\mathbf{c},\end{aligned}$$

where the upper and lower signs are taken when (10) is positive and negative, respectively,

$$\boldsymbol{Q}^{(k)}_{2b,n}(\theta) = \begin{pmatrix} (\kappa^{(k)} + n + 1 - \check{\gamma} + e^{(-1)^{k+1}2\pi\epsilon}) \sin\left(n+1-\check{\gamma}\right)\theta - (n+1-\check{\gamma})\sin\left(n-1-\check{\gamma}\right)\theta \\ (-\kappa^{(k)} + n + 1 - \check{\gamma} + e^{(-1)^{k+1}2\pi\epsilon}) \cos\left(n+1-\check{\gamma}\right)\theta - (n+1-\check{\gamma})\cos\left(n-1-\check{\gamma}\right)\theta \end{pmatrix}.$$

The series are convergent, absolutely in $H^1(B_R^{(k)})$ and uniformly on compact sets in $B_R^{(k)}$ for $k = 1, 2$, respectively. For $n \geq 0$, $\check{a}_n$, b_n and $\check{c}_n$ satisfy

$$\begin{aligned}|\check{a}_n| &\leq c\sqrt{n+1-\check{\gamma}}\; R^{-(n+1-\check{\gamma})} \|\nabla \mathbf{u}\|_{L^2(B_R)}, \\ |b_n| &\leq c\sqrt{n+1}\; R^{-(n+1)} \|\nabla \mathbf{u}\|_{L^2(B_R)}, \\ |\check{c}_n| &\leq c\sqrt{n+1}\; R^{-(n+1)} \|\nabla \mathbf{u}\|_{L^2(B_R)}.\end{aligned}$$

One can see that the condition $\sigma_{22}^{(k)} \leq 0$ on Γ is equivalent to a condition on $B_R \cap \Gamma$

$$\begin{aligned}&\sum_{n=0}^{\infty} (-1)^n \left\{ r^{n-\check{\gamma}} \left(e^{-\epsilon\pi} - e^{\epsilon\pi}\right) \sin\check{\gamma}\pi\, \mathbf{Im}\left[\check{a}_n\right] - r^n \frac{2(\kappa^{(1)}\kappa^{(2)}-1)}{\mu^{(1)}\mu^{(2)} m_1 m_2} \mathbf{Re}\left[b_n\right] \right\} \\ &+ \sum_{n=0}^{\infty} (-1)^n r^n \frac{1}{2}\left(\frac{1}{m_1} + \frac{1}{m_2}\right) \mathbf{Re}\left[\check{c}_n\right] \leq 0.\end{aligned} \tag{8}$$

It follows from this and $\check{\gamma} > 0$ that

$$\left(e^{-\epsilon\pi} - e^{\epsilon\pi}\right) \mathbf{Im}\left[\check{a}_0\right] = \frac{-2\beta}{\sqrt{1-\beta^2}} \mathbf{Im}\left[\check{a}_0\right] \leq 0. \tag{9}$$

Furthermore, we obtain

$$[u_1] = \sum_{n=0}^{\infty} \frac{(-1)^{n+1} r^{n+1-\check{\gamma}}}{n+1-\check{\gamma}} \sqrt{m_1 m_2} \sin \check{\gamma}\pi \, \mathbf{Im}\left[\check{a}_n\right] \quad \text{on} \quad B_R \cap \Gamma. \tag{10}$$

Remark 1 Note here that $\check{\gamma}$ can vary from 0 to 1 which implies a possibility of the stronger singularity of the stress at the crack tip than inverse square root. However, this case can be precluded by the following reason. Now, we assume $0 < \check{\gamma} < 1$ and $\beta \neq 0$. If $[u_1] > 0$ on $B_\rho \cap \Gamma$, then $\cot \check{\gamma}\pi = -f\beta$ and $\mathbf{Im}\left[\check{a}_0\right] \leq 0$ from (10). Combining with (9), one sees that $\beta < 0$ or $\mathbf{Im}\left[\check{a}_0\right] = 0$ which means the singular term of the expansion disappears. $\beta < 0$ follows $-f\beta > 0$ and thus we conclude $0 < \check{\gamma} < \frac{1}{2}$.

Similarly, if $[u_1] < 0$ on $B_\rho \cap \Gamma$, then $\cot \check{\gamma}\pi = f\beta$ and $\mathbf{Im}\left[\check{a}_0\right] \geq 0$ from (10). Combining with (9), one sees that $\beta > 0$ or $\mathbf{Im}\left[\check{a}_0\right] = 0$ and thus we conclude $0 < \check{\gamma} < \frac{1}{2}$. In the case $\beta = 0$, one knows $\check{\gamma} = \frac{1}{2}$ by (7) and from (10) we have $\mathbf{Im}\left[\check{a}_0\right] \leq 0$ for $[u_1] > 0$ on $B_R \cap \Gamma$ and $\mathbf{Im}\left[\check{a}_0\right] \geq 0$ for $[u_1] < 0$ on $B_R \cap \Gamma$.

3.3 Case 3

Proposition 6 ([9]) *There exists $c_n \in \mathbf{C}$ and $\mathbf{k} \in \mathbf{R}^3$ such that*

$$\mathbf{u}(r,\theta) = \sum_{n=1}^{\infty} \frac{1}{2\mu} r^{\frac{2n-1}{4}} \mathbf{Re}[c_n r^{-\tilde{\epsilon} i}] \boldsymbol{R}_n(\theta) e^{\tilde{\epsilon}\theta} - \frac{1}{2\mu} r^{\frac{2n-1}{4}} \mathbf{Im}[c_n r^{-\tilde{\epsilon} i}] \boldsymbol{S}_n(\theta) e^{\tilde{\epsilon}\theta} + F(\mathbf{x})\mathbf{k},$$

where

$$\boldsymbol{R}_n(\theta) = \begin{pmatrix} \left(\kappa + \frac{2n-1}{4}\right)\cos\frac{2n-1}{4}\theta - \left((-1)^n\sqrt{\kappa}e^{-2\tilde{\epsilon}\theta} - \tilde{\epsilon}\right)\sin\frac{2n-1}{4}\theta \\ -\frac{2n-1}{4}\cos\frac{2n-9}{4}\theta - \tilde{\epsilon}\sin\frac{2n-9}{4}\theta \\ \left(\kappa - \frac{2n-1}{4}\right)\sin\frac{2n-1}{4}\theta - \left((-1)^n\sqrt{\kappa}e^{-2\tilde{\epsilon}\theta} - \tilde{\epsilon}\right)\cos\frac{2n-1}{4}\theta \\ +\frac{2n-1}{4}\sin\frac{2n-9}{4}\theta - \tilde{\epsilon}\cos\frac{2n-9}{4}\theta \end{pmatrix},$$

$$\boldsymbol{S}_n(\theta) = \begin{pmatrix} \left(\kappa + \frac{2n-1}{4}\right)\sin\frac{2n-1}{4}\theta - \left((-1)^n\sqrt{\kappa}e^{-2\tilde{\epsilon}\theta} + \tilde{\epsilon}\right)\cos\frac{2n-1}{4}\theta \\ -\frac{2n-1}{4}\sin\frac{2n-9}{4}\theta + \tilde{\epsilon}\cos\frac{2n-9}{4}\theta \\ -\left(\kappa - \frac{2n-1}{4}\right)\cos\frac{2n-1}{4}\theta + \left((-1)^n\sqrt{\kappa}e^{-2\tilde{\epsilon}\theta} + \tilde{\epsilon}\right)\sin\frac{2n-1}{4}\theta \\ -\frac{2n-1}{4}\cos\frac{2n-9}{4}\theta - \tilde{\epsilon}\sin\frac{2n-9}{4}\theta \end{pmatrix}$$

with $\tilde{\epsilon} := \frac{\log\kappa}{4\pi}$. The series are convergent, absolutely in $H^1(B_R)$ and uniformly on compact sets in B_R. Moreover, for $n \geq 1$, c_n satisfies

$$|c_n| \leq c \frac{1}{\sqrt{2n-1}} R^{-\frac{n}{2}+\frac{1}{4}} \|\nabla \mathbf{u}\|_{L^2(B_R)}.$$

Table 1 Summarized results

Case	Ref.	Singular point	Material	Boundary condition on Γ	Order of singularity $n \in \mathbf{N}$
1	[5, 6]	A linear crack tip	Homogeneous, anisotropic	Traction-free	$\frac{n}{2}$
2	[10]	a Tip of a linear interfacial crack	Two different homogeneous and isotropic media	Non-penetration, Coulomb's friction	$n+1-\alpha$ open crack: $\alpha = \frac{1}{2} + i\epsilon$, stick state: $\alpha = 0$, slip state: $\cot(\pi\alpha) = \mp f\beta$
3	[9]	A tip of a rigid line inclusion	Homogeneous, isotropic	Upper and lower surface: traction-free, rigidity	$\frac{2n-1}{4} - \tilde{\epsilon}i$

4 Conclusion

In this paper, we introduced the complete asymptotic expansions of the displacement near the tip of a linear crack and a line rigid inclusion, written in Propositions 1–6. Then, it is shown how the order of singularity is determined for some cases (briefly summarized as Table 1), further, coefficients in expansions are estimated. We also get the exact forms with respect to the distance to $\boldsymbol{O}$ as well as the explicit expression of the angular functions around $\boldsymbol{O}$.

Especially, by virtue of the results, it enables us to have an a priori regularity of the solution near $\boldsymbol{O}$. Indeed, in Case 1, open crack in Case 2 and Case 3, it implies $\mathbf{u} \notin H^{3/2}(B_\rho \setminus \overline{\Gamma})$ generally, the solution is smooth in the stick state in Case 2, and for general different shear moduli materials, i.e., $\beta \neq 0$, $\mathbf{u} \in H^{3/2}(B_R \setminus \overline{\Gamma})$ in the slip state in Case 2.

As for the future works, it is expected to apply the result in frictional crack problems to the case of a dynamic crack. It has been treated in a case characterized by steady-state propagation, allowing use of the Galilean transformation, and a self-similar problem, i.e., scaling of the spatial coordinates, with a scaling factor increasing in proportion to time, see [2]. However, more general cases should be considered, for example, self-similar crack problems between bimaterials and crack propagation speed is not constant, etc. This kind of problems is very important in seismology because earthquake is dynamic phenomenon for faulting with friction and the inner structure of the earth is very complex. By finding analytical solution, it may be possible to know how the rupture velocity is depending on material coefficients.

Acknowledgements H. Itou was partially supported by Grant-in-Aid for Scientific Research (C) (No. 26400178) of Japan Society for the Promotion of Science. I would like to thank M. Ikehata (Hiroshima University), V.A. Kovtunenko (University of Graz), A. Tani (Keio University), A.M. Khludnev and E.M. Rudoy (Lavrentyev Institute of Hydrodynamics of the Russian Academy of Sciences) for the cooperation and fruitful discussions.

References

1. Audoly, B.: Asymptotic study of the interfacial crack with friction. J. Mech. Phys. Solids **48**, 1851–1864 (2000)
2. Broberg, K.B.: Cracks and Fracture. Academic Press, San Diego (1999)
3. Dundurs, J., Comninou, M.: Some consequences of the inequality conditions in contact and crack problems. J. Elast. **9**, 71–82 (1979)
4. Fichera, G.: Existence theorems in elasticity. In: Truesdell, C. (ed.) Mechanics of Solids, vol. II, pp. 347–389. Springer, Berlin (1984)
5. Ikehata, M., Itou, H.: Reconstruction of a linear crack in an isotropic elastic body from a single set of measured data. Inverse Prob. **23**, 589–607 (2007)
6. Ikehata, M., Itou, H.: An inverse problem for a linear crack in an anisotropic elastic body and the enclosure method. Inverse Prob. **24**, 025005(21pp) (2008)
7. Ikehata, M., Itou, H.: Extracting the support function of a cavity in an isotropic elastic body from a single set of boundary data. Inverse Prob. **25**, 105005(21pp) (2009)
8. Itou, H.: On convergent expansions of solutions of the linearized elasticity equation near singular points. AIP Conf. Proc. **1389**, 465–468 (2011)
9. Itou, H., Khludnev, A.M., Rudoy, E.M., Tani, A.: Asymptotic behaviour at a tip of a rigid line inclusion in linearized elasticity. Z. Angew. Math. Mech. **92**, 716–730 (2012)
10. Itou, H., Kovtunenko, V.A., Tani, A.: The interface crack with Coulomb friction between two bonded dissimilar elastic media. Appl. Math. **56**, 69–97 (2011)
11. Lekhnitskii, S.G.: Anisotropic Plates. Gordon and Breach Science Publishers, New York (1968)
12. Maz'ya, V., Nazarov, S., Plamenevskii, B.: Asymptotic Theory of Elliptic Boundary Value Problems in Singularly Perturbed Domains, vol. II. Birkhauser, Basel (2000)
13. Muskhelishvili, N.I.: Singular Integral Equations. Noordhoff, Groningen (1972)
14. Rice, J.R.: Elastic fracture mechanics concepts for interfacial cracks. J. Appl. Mech. **55** (1988), 98–103

Part II
Shape Optimization

Two-Parameter Topological Expansion of Helmholtz Problems with Inhomogeneity

Victor A. Kovtunenko

Abstract For forward and inverse Helmholtz problems with inhomogeneity in the 2-D setting, high-order topological analysis is provided based on singular perturbations and variational methods. When diminishing the inhomogeneity, the two-parameter asymptotic result is proved rigorously with respect to the size of inhomogeneity and its refractive index. In particular, for a fixed refractive index this implies the topological derivative. For identifying an unknown inhomogeneity put in a test domain, variation of a complex refractive index leads to the zero-order necessary optimality condition of minimum of the objective function. This condition is realized as an imaging function for finding center of the inhomogeneity.

1 Introduction

Forward and inverse Helmholtz problems with inhomogeneity are considered in 2-D spatial setting of the problem. By this, the inhomogeneity is characterized by a complex refractive index, see [22, Chap. 6]. This formulation is motivated by applications to nondestructive testing of scattering media with acoustic, elastic, and electromagnetic waves. We provide rigorously two-parameter topological analysis based on singular perturbations and variational methods.

The classic analysis of Helmholtz problems is based on the theory of potential operators, see [8, 10, 11, 22, 35]. Specifically for inverse scattering problems, the direct (noniterative) methods of factorization [2], sampling and probe [30], enclosure [15, 16], MUSIC [32], as well as iterative methods [18] are established. For the general mathematical theory of inverse problems we refer to [22, 27].

V.A. Kovtunenko (✉)
Institute for Mathematics and Scientific Computing, Karl-Franzens University of Graz,
NAWI Graz, Heinrichstr. 36, 8010 Graz, Austria
e-mail: victor.kovtunenko@uni-graz.at

V.A. Kovtunenko
Lavrentyev Institute of Hydrodynamics, Siberian Division
of the Russian Academy of Sciences, 630090 Novosibirsk, Russia

H. Itou et al. (eds.), *Mathematical Analysis of Continuum Mechanics and Industrial Applications*, Mathematics for Industry 26,
DOI 10.1007/978-981-10-2633-1_5

Within shape and topology optimization (see e.g. [1, 13, 28]), the theory of identification of small inhomogeneities was developed in [7, 12]. Newly, the concept of topological derivatives was introduced in [34] and adapted to inhomogeneous problems in, e.g., [3, 5, 6, 9]. It proceeds with high-order topological expansions in [14, 23]. The topological analysis is realized by adapting methods of singular perturbations which we refer to [17, 19, 21, 29] and references therein.

Within topological analysis of Helmholtz problems in inhomogeneous medium, inhomogeneities are described by arbitrary geometries and variable refractive indexes. An admissible geometry implies the triple of implicit geometric variables of the shape ω, the center x_0, and the size ε. This geometric description is treated within variational theory. We provide the underlying forward and inverse Helmholtz problems with minimax variational principles according to the Fenchel–Legendre duality.

A positive refractive index α is determined in the field of complex numbers. Varying the refractive index describes two important limit cases. First, $|\alpha| \searrow +0$ describes the homogeneous Neumann problem, and, second, $|\alpha| \nearrow +\infty$ corresponds to the homogeneous Dirichlet problem in the perforated domain outside inhomogeneity. The former case implies the sound-hard obstacle, and the latter one corresponds to the sound-soft obstacle in acoustics, which were treated previously in [23].

For the perturbation analysis, asymptotic arguments of small size $\varepsilon \searrow +0$ are employed. The problem of singular perturbations is supported by construction of a boundary layer. It is described by the transmission problem for the Laplace operator stated in the exterior domain, where weighted Sobolev spaces are useful. As the result, we get a two-parameter asymptotic expansion with respect to ε and α.

From the perspective of topology optimization, the topology variables $(\omega, x_0, \varepsilon, \alpha)$ enter the objective implicitly through the solution of a geometry-dependent Helmholtz problem. This fact makes difficult to find optimality conditions based on conventional directional derivatives of the objective.

Following [23, 24] in the present contribution, we show that an unknown parameter of the boundary impedance is well suitable for the purpose of variation of trial geometries. Passing $\mathrm{Im}(\alpha) \nearrow +\infty$ in the two-parameter expansion gets a zero-order necessary optimality condition which is derivative-free. For the objective minimizing misfit from boundary measurements, this optimality condition gives an imaging function identifying the optimal center x_0.

2 Forward Helmholtz Problem in Inhomogeneous Medium

Let $\Omega \subset \mathbf{R}^2$ be a reference domain with the Lipschitz boundary $\partial\Omega$ and the unit normal vector $n = (n_1, n_2)^\top$ which is outward to Ω, where the upper $^\top$ denotes transposition swapping columns and rows. We assume that the boundary $\partial\Omega = \overline{\Gamma_N} \cup \overline{\Gamma_D}$ consists of two disjoint parts Γ_N and Γ_D associated with the Neumann and the Dirichlet conditions, respectively.

Let ω be a compact set in $\mathbf{R}^2$ with the Lipschitz boundary $\partial\omega$ and the normal vector $n = (n_1, n_2)^\top$ outward to ω. We assume that $0 \in \omega$ and $\omega \subset B_1(0)$ such that the unit ball $B_1(0)$ is the minimum enclosing ball centered at origin 0. Then the set G_ω of such shapes ω is invariant to translations and isotropic scaling.

The reason of such a construction is to separate the near field $B_1(0)$ containing ω from the far field $\mathbf{R}^2 \setminus \overline{B_1(0)}$.

Rescaling $\omega \in G_\omega$ by a size parameter $\varepsilon > 0$ produces the geometric object

$$\omega_\varepsilon(x_0) = \left\{x \in \mathbf{R}^2 : \tfrac{x-x_0}{\varepsilon} \in \omega\right\} \subset B_\varepsilon(x_0)$$

posed at a center $x_0 \in \mathbf{R}^2$. In the reference domain Ω, admissible geometries $G := G_\omega \times G_\varepsilon \times G_x$ consist of triples: the shape $\omega \in G_\omega$, the size $\varepsilon \in G_\varepsilon \subset \mathbf{R}_+$, and the center $x_0 \in G_x \subset \Omega$ which satisfy the geometric condition:

$$\omega_\varepsilon(x_0) \subset B_\varepsilon(x_0) \subset \Omega.$$

We distinguish two faces of the interface $\partial\omega_\varepsilon(x_0)$, respectively: the boundary of the inhomogeneity $\omega_\varepsilon(x_0)$ called the negative face $\partial\omega_\varepsilon(x_0)^-$ (having the outward normal vector n), and the part of the boundary of the perforated domain $\Omega \setminus \overline{\omega_\varepsilon(x_0)}$ called the positive face $\partial\omega_\varepsilon(x_0)^+$ (having the inward normal vector n). The respective jump across $\partial\omega_\varepsilon(x_0)$ is denoted by

$$[\![u]\!] = u|_{\partial\omega_\varepsilon(x_0)^+} - u|_{\partial\omega_\varepsilon(x_0)^-}.$$

Let $\alpha \in \mathbf{C}_+$ be a complex refractive index $\alpha = \mathrm{Re}(\alpha) + \mathrm{i}\mathrm{Im}(\alpha)$ with positive the real $\mathrm{Re}(\alpha) \in \mathbf{R}_+$ and the imaginary $\mathrm{Im}(\alpha) \in \mathbf{R}_+$ parts. It characterizes the inhomogeneity $\omega_\varepsilon(x_0)$ with the help of the complex-valued function $\chi^\alpha_{\omega_\varepsilon(x_0)} : \mathbf{R}^2 \mapsto \mathbf{C}_+$:

$$\chi^\alpha_{\omega_\varepsilon(x_0)}(x) = \begin{cases} 1 & \text{for } x \in \mathbf{R}^2 \setminus \overline{\omega_\varepsilon(x_0)} \\ \alpha & \text{for } x \in \omega_\varepsilon(x_0) \end{cases}$$

which is piecewise-constant with the jump $[\![\chi^\alpha_{\omega_\varepsilon(x_0)}]\!] = 1 - \alpha$ across the interface $\partial\omega_\varepsilon(x_0)$. We note that $\alpha = 1$ implies homogeneity since $\chi^1_{\omega_\varepsilon(x_0)} \equiv 1$ in $\mathbf{R}^2$.

The topology variables $(\omega, \varepsilon, x_0, \alpha)$ will be used for the sake of variation of inhomogeneities for inverse problems in Sect. 3.

For forward problems in Sect. 2, the geometry $(\omega, \varepsilon, x_0) \in G$ of inhomogeneity and its refractive index $\alpha \in \mathbf{C}_+$ are fixed. In the following, we will mark the dependence of functions on the size ε and the refractive index α for the reason of subsequent two-parameter asymptotic analysis.

Given the Neumann $g \in L^2(\Gamma_N; \mathbf{C})$ and the Dirichlet $h \in H^{1/2}(\Gamma_D; \mathbf{C})$ complex data and the real wave number $k \in \mathbf{R}_+$, the *Helmholtz problem in inhomogeneous medium* is stated for the wave potential $u^{(\varepsilon,\alpha)}(x)$ satisfying for $x \in \overline{\Omega}$ the system:

$$-[\Delta + k^2]u^{(\varepsilon,\alpha)} = 0 \text{ in } \Omega \setminus \overline{\omega_\varepsilon(x_0)}, \quad -[\mathrm{Re}(\alpha)\Delta + \alpha k^2]u^{(\varepsilon,\alpha)} = 0 \text{ in } \omega_\varepsilon(x_0), \quad (1)$$

$$[\![u^{(\varepsilon,\alpha)}]\!] = 0, \quad \frac{\partial u^{(\varepsilon,\alpha)}}{\partial n}\Big|_{\partial\omega_\varepsilon(x_0)^+} - \mathrm{Re}(\alpha)\frac{\partial u^{(\varepsilon,\alpha)}}{\partial n}\Big|_{\partial\omega_\varepsilon(x_0)^-} = 0 \quad \text{on } \partial\omega_\varepsilon(x_0), \quad (2)$$

$$\frac{\partial u^{(\varepsilon,\alpha)}}{\partial n} = g \quad \text{on } \Gamma_N, \quad (3)$$

$$u^{(\varepsilon,\alpha)} = h \quad \text{on } \Gamma_D. \quad (4)$$

Here and in what follows, Δ is the Laplace operator, the notation $\frac{\partial}{\partial n} := n \cdot \nabla = n^\top \nabla$ stands for the normal derivative at the boundary, the dot $\cdot$ means the inner product of vectors, and ∇ is the gradient.

For the strong solution of (1)–(4), a weak formulation can be derived by standard variational arguments: multiplying Eq. (1) with a smooth test function u and subsequent integration by parts over Ω due to transmission conditions (2) and boundary conditions (3) and (4). The weak solution to (1)–(4) is described by the following variational problem: Find $u^{(\varepsilon,\alpha)} \in H^1(\Omega; \mathbf{C})$ such that (4) holds and

$$\int_\Omega \big(\chi^{\mathrm{Re}(\alpha)}_{\omega_\varepsilon(x_0)} \nabla u^{(\varepsilon,\alpha)} \cdot \nabla\overline{u} - \chi^{\alpha}_{\omega_\varepsilon(x_0)} k^2 u^{(\varepsilon,\alpha)}\overline{u}\big)\, dx = \int_{\Gamma_N} g\overline{u}\, dS_x$$

$$\text{for all test-functions } u \in H^1(\Omega; \mathbf{C}) \text{ such that } u = 0 \text{ on } \Gamma_D. \quad (5)$$

Here the discontinuous piecewise-constant function $\chi^{\mathrm{Re}(\alpha)}_{\omega_\varepsilon(x_0)} = 1$ in $\mathbf{R}^2 \setminus \overline{\omega_\varepsilon(x_0)}$ and $\chi^{\mathrm{Re}(\alpha)}_{\omega_\varepsilon(x_0)} = \mathrm{Re}(\alpha)$ in $\omega_\varepsilon(x_0)$, the usual notation $\overline{u} = \mathrm{Re}(u) - \mathrm{i}\mathrm{Im}(u)$ implies the complex conjugate of $u = \mathrm{Re}(u) + \mathrm{i}\mathrm{Im}(u)$, and i stands for the imaginary unit.

Avoiding eigenvalues, the unique variational solution of (5) can be argued by the Fredholm alternative using Rellich's lemma, see [22, Theorem 6.9].

In return, Green's formulas hold for every function $\tilde{u} \in H^1(\Omega; \mathbf{C})$ such that $\Delta\tilde{u} \in L^2(\Omega; \mathbf{C})$ and all test-functions $u \in H^1(\Omega; \mathbf{C})$ such that $u = 0$ on Γ_D:

$$\int_{\Omega\setminus\omega_\varepsilon(x_0)} (\nabla\tilde{u} \cdot \nabla\overline{u} + \overline{u}\,\Delta\tilde{u})\, dx = \langle \tfrac{\partial\tilde{u}}{\partial n}, \overline{u}\rangle_{\Gamma_N} - \langle \tfrac{\partial\tilde{u}}{\partial n}, \overline{u}\rangle_{\partial\omega_\varepsilon(x_0)^+}, \quad (6)$$

$$\int_{\omega_\varepsilon(x_0)} (\nabla\tilde{u} \cdot \nabla\overline{u} + \overline{u}\,\Delta\tilde{u})\, dx = \langle \tfrac{\partial\tilde{u}}{\partial n}, \overline{u}\rangle_{\partial\omega_\varepsilon(x_0)^-}. \quad (7)$$

Here $\langle \frac{\partial\tilde{u}}{\partial n}, \overline{u}\rangle_{\partial\omega_\varepsilon(x_0)^\pm}$ implies the duality between $u \in H^{1/2}(\partial\omega_\varepsilon(x_0)^\pm; \mathbf{C})$ and $\frac{\partial\tilde{u}}{\partial n} \in H^{-1/2}(\partial\omega_\varepsilon(x_0)^\pm; \mathbf{C})$, while $\langle \frac{\partial\tilde{u}}{\partial n}, \overline{u}\rangle_{\Gamma_N}$ stands for the duality pairing between $u \in H^{1/2}_{00}(\Gamma_N; \mathbf{C})$ and $\frac{\partial\tilde{u}}{\partial n} \in H^{-1/2}(\Gamma_N; \mathbf{C})$ in the Lions–Magenes dual spaces, see, e.g., [19, Sect. 1.4] for detail. Summing up (6) and (7) multiplied by $\mathrm{Re}(\alpha)$ provides the Green's formula in inhomogeneous medium:

$$\int_\Omega \big(\chi^{\mathrm{Re}(\alpha)}_{\omega_\varepsilon(x_0)} \nabla\tilde{u} \cdot \nabla\overline{u} + \chi^{\mathrm{Re}(\alpha)}_{\omega_\varepsilon(x_0)} \overline{u}\,\Delta\tilde{u}\big)\, dx = \langle \tfrac{\partial\tilde{u}}{\partial n}, \overline{u}\rangle_{\Gamma_N} - \langle [\![\chi^{\mathrm{Re}(\alpha)}_{\omega_\varepsilon(x_0)} \tfrac{\partial\tilde{u}}{\partial n}]\!], \overline{u}\rangle_{\partial\omega_\varepsilon(x_0)}. \quad (8)$$

With the help of (8) with $\tilde{u} = u^{(\varepsilon,\alpha)}$, we get from (5):

$$\int_\Omega \big([\chi^{\mathrm{Re}(\alpha)}_{\omega_\varepsilon(x_0)} \Delta + \chi^{\alpha}_{\omega_\varepsilon(x_0)} k^2] u^{(\varepsilon,\alpha)}\big)\overline{u}\, dx = \langle \tfrac{\partial u^{(\varepsilon,\alpha)}}{\partial n} - g, \overline{u}\rangle_{\Gamma_N} - \langle [\![\chi^{\mathrm{Re}(\alpha)}_{\omega_\varepsilon(x_0)} \tfrac{\partial u^{(\varepsilon,\alpha)}}{\partial n}]\!], \overline{u}\rangle_{\partial\omega_\varepsilon(x_0)}$$

and derive (1)–(3) by the fundamental lemma of calculus of variations when varying the test function u such that first $u = 0$ on $\Gamma_N \cup \partial\omega_\varepsilon(x_0)$ and then $u = 0$ on Γ_N.

Moreover, the following variational principle holds:

Theorem 1 *The variational equation (5) implies the first-order necessary optimality condition for the minimax variational problem:*

$$\mathscr{P}(u^{(\varepsilon,\alpha)}) = \min_{\mathrm{Re}(v)} \max_{\mathrm{Im}(v)} \mathscr{P}(v) \text{ over } v \in H^1(\Omega; \mathbf{C}) \text{ such that } v = h \text{ on } \Gamma_D \tag{9}$$

with the following Lagrangian $\mathscr{P} : H^1(\Omega; \mathbf{C}) \mapsto \mathbf{R}$,

$$\mathscr{P}(v) = \mathrm{Re}\Big\{\tfrac{1}{2}\int_\Omega \big(\chi^{\mathrm{Re}(\alpha)}_{\omega_\varepsilon(x_0)} \nabla v \cdot \nabla v - \chi^{\alpha}_{\omega_\varepsilon(x_0)} k^2 v^2\big)\, dx - \int_{\Gamma_N} g v \, dS_x\Big\}.$$

Proof If we rewrite $\mathscr{P}$ component-wisely for $v = \mathrm{Re}(v) + \mathrm{i}\mathrm{Im}(v)$ as

$$\begin{aligned}\mathscr{P}(v) = {} & \tfrac{1}{2}\int_\Omega \big\{\chi^{\mathrm{Re}(\alpha)}_{\omega_\varepsilon(x_0)} \big(|\nabla(\mathrm{Re}(v))|^2 - |\nabla(\mathrm{Im}(v))|^2\big) \\ & -\chi^{\alpha}_{\omega_\varepsilon(x_0)} k^2 \big(\mathrm{Re}(v)^2 - \mathrm{Im}(v)^2\big)\big\}\, dx - \int_{\Gamma_N} \big(\mathrm{Re}(g)\mathrm{Re}(v) - \mathrm{Im}(g)\mathrm{Im}(v)\big)\, dS_x\end{aligned}$$

and differentiate it with respect to $\mathrm{Re}(v)$ and $\mathrm{Im}(v)$, then the necessary optimality condition for (9) implies two variational inequalities

$$\big\langle \tfrac{\partial}{\partial \mathrm{Re}(v)} \mathscr{P}(u^{(\varepsilon,\alpha)}), \mathrm{Re}(v - u^{(\varepsilon,\alpha)})\big\rangle \ge 0, \quad \big\langle \tfrac{\partial}{\partial \mathrm{Im}(v)} \mathscr{P}(u^{(\varepsilon,\alpha)}), \mathrm{Im}(v - u^{(\varepsilon,\alpha)})\big\rangle \le 0$$

holding for all $v \in H^1(\Omega; \mathbf{C})$ such that $v = h$ on Γ_D. Inserting here $v = u^{(\varepsilon,\alpha)} \pm \tilde{u}$ with $\tilde{u} \in H^1(\Omega; \mathbf{C})$ such that $\tilde{u} = 0$ on Γ_D results in two variational equations:

$$\begin{aligned}\int_\Omega \big(\chi^{\mathrm{Re}(\alpha)}_{\omega_\varepsilon(x_0)} \nabla\mathrm{Re}(u^{(\varepsilon,\alpha)}) \cdot \nabla\mathrm{Re}(\tilde{u}) - \chi^{\alpha}_{\omega_\varepsilon(x_0)} k^2 \mathrm{Re}(u^{(\varepsilon,\alpha)})\mathrm{Re}(\tilde{u})\big)\, dx \\ = \int_{\Gamma_N} \mathrm{Re}(g)\mathrm{Re}(\tilde{u})\, dS_x,\\ \int_\Omega \big(\chi^{\mathrm{Re}(\alpha)}_{\omega_\varepsilon(x_0)} \nabla\mathrm{Im}(u^{(\varepsilon,\alpha)}) \cdot \nabla\mathrm{Im}(\tilde{u}) - \chi^{\alpha}_{\omega_\varepsilon(x_0)} k^2 \mathrm{Im}(u^{(\varepsilon,\alpha)})\mathrm{Im}(\tilde{u})\big)\, dx \\ = \int_{\Gamma_N} \mathrm{Im}(g)\mathrm{Im}(\tilde{u})\, dS_x.\end{aligned}$$

The summation of these equations for the test function $\tilde{u} = u$:

$$\int_\Omega \{\chi^{\mathrm{Re}(\alpha)}_{\omega_\varepsilon(x_0)} (\nabla \mathrm{Re}(u^{(\varepsilon,\alpha)}) \cdot \nabla \mathrm{Re}(u) + \nabla \mathrm{Im}(u^{(\varepsilon,\alpha)}) \cdot \nabla \mathrm{Im}(u))$$
$$- \chi^{\alpha}_{\omega_\varepsilon(x_0)} k^2 \big(\mathrm{Re}(u^{(\varepsilon,\alpha)})\mathrm{Re}(u) + \mathrm{Im}(u^{(\varepsilon,\alpha)})\mathrm{Im}(u)\big)\big\}\, dx$$
$$= \int_{\Gamma_N} \big(\mathrm{Re}(g)\mathrm{Re}(u) + \mathrm{Im}(g)\mathrm{Im}(u)\big)\, dS_x,$$

with arbitrary $u \in H^1(\Omega; \mathbf{C})$ such that $u = 0$ on Γ_D, and for $\tilde{u} = \mathrm{i}u$:

$$\int_\Omega \{\chi^{\mathrm{Re}(\alpha)}_{\omega_\varepsilon(x_0)} (\nabla \mathrm{Im}(u^{(\varepsilon,\alpha)}) \cdot \nabla \mathrm{Re}(u) - \nabla \mathrm{Re}(u^{(\varepsilon,\alpha)}) \cdot \nabla \mathrm{Im}(u))$$
$$- \chi^{\alpha}_{\omega_\varepsilon(x_0)} k^2 \big(\mathrm{Im}(u^{(\varepsilon,\alpha)})\mathrm{Re}(u) - \mathrm{Re}(u^{(\varepsilon,\alpha)})\mathrm{Im}(u)\big)\big\}\, dx$$
$$= \int_{\Gamma_N} \big(\mathrm{Im}(g)\mathrm{Re}(u) - \mathrm{Re}(g)\mathrm{Im}(u)\big)\, dS_x,$$

constitutes respectively the real and imaginary parts of (5) and completes the proof. □

Now we gain insight into the two-parameter dependence on (ε, α).

On the one hand, for fixed $\varepsilon \in \mathbf{R}_+$, if $\mathrm{Im}(\alpha) \searrow +0$ and $\mathrm{Re}(\alpha) \searrow +0$, then the Helmholtz problem (1)–(4) decouples into the mixed problem in the perforated domain with homogeneous Neumann condition at the interface (sound-hard):

$$-[\Delta + k^2]u^{(\varepsilon,0)} = 0 \quad \text{in } \Omega \setminus \overline{\omega_\varepsilon(x_0)},$$
$$\frac{\partial u^{(\varepsilon,0)}}{\partial n} = 0 \quad \text{on } \partial\omega_\varepsilon(x_0)^+,$$
$$\frac{\partial u^{(\varepsilon,0)}}{\partial n} = g \quad \text{on } \Gamma_N, \quad u^{(\varepsilon,0)} = h \quad \text{on } \Gamma_D,$$

and the Dirichlet problem inside the inhomogeneity:

$$-[\Delta + k^2]u^{(\varepsilon,0)} = 0 \text{ in } \omega_\varepsilon(x_0),$$
$$u^{(\varepsilon,0)}\big|_{\partial\omega_\varepsilon(x_0)^-} = u^{(\varepsilon,0)}\big|_{\partial\omega_\varepsilon(x_0)^+} \quad \text{on } \partial\omega_\varepsilon(x_0)^-.$$

If $\mathrm{Re}(\alpha) \nearrow +\infty$ or $\mathrm{Im}(\alpha) \nearrow +\infty$, then $u^{(\varepsilon,\infty)} \equiv 0$ trivially inside $\omega_\varepsilon(x_0)$, and (1)–(4) turns into the mixed problem in the perforated domain with homogeneous Dirichlet boundary condition at the interface (sound-soft):

$$\begin{cases} -[\Delta + k^2]u^{(\varepsilon,\infty)} = 0 \quad \text{in } \Omega \setminus \overline{\omega_\varepsilon(x_0)}, \\ u^{(\varepsilon,\infty)}\big|_{\partial\omega_\varepsilon(x_0)^+} = 0 \quad \text{on } \partial\omega_\varepsilon(x_0)^+, \\ \frac{\partial u^{(\varepsilon,\infty)}}{\partial n} = g \quad \text{on } \Gamma_N, \quad u^{(\varepsilon,\infty)} = h \quad \text{on } \Gamma_D. \end{cases} \tag{10}$$

These limit cases of boundary conditions were treated separately in [23].

On the other hand, if $\varepsilon \searrow +0$, then we derive the *background Helmholtz problem in homogeneous medium* for the wave potential $u^0(x)$:

$$-[\Delta + k^2]u^0 = 0 \quad \text{in } \Omega, \tag{11}$$

$$\tfrac{\partial u^0}{\partial n} = g \quad \text{on } \Gamma_N, \tag{12}$$

$$u^0 = h \quad \text{on } \Gamma_D. \tag{13}$$

In the weak form, the system (11)–(13) is described by the following variational problem: Find $u^0 \in H^1(\Omega; \mathbf{C})$ such that (13) holds and

$$\int_\Omega (\nabla u^0 \cdot \nabla \overline{u} - k^2 u^0 \overline{u})\, dx = \int_{\Gamma_N} g\overline{u}\, dS_x \text{ for all } u \in H^1(\Omega; \mathbf{C}) : u = 0 \text{ on } \Gamma_D. \tag{14}$$

For an inhomogeneity $\omega_\varepsilon(x_0)$, applying Green's formulas (6) and (7) to the variational equation (14) it can be restated equivalently in the inhomogeneous medium similarly to (5):

$$\int_\Omega \big(\chi^{\mathrm{Re}(\alpha)}_{\omega_\varepsilon(x_0)} \nabla u^0 \cdot \nabla \overline{u} - \chi^{\alpha}_{\omega_\varepsilon(x_0)} k^2 u^0 \overline{u}\big)\, dx = \int_{\Gamma_N} g\overline{u}\, dS_x$$
$$-(1 - \mathrm{Re}(\alpha)) \int_{\partial\omega_\varepsilon(x_0)} \tfrac{\partial u^0}{\partial n} \overline{u}\, dS_x - \mathrm{i}\mathrm{Im}(\alpha) \int_{\omega_\varepsilon(x_0)} k^2 u^0 \overline{u}\, dx. \tag{15}$$

In Sect. 2.1 we start with an inner asymptotic expansion of u^0 in the near-field, Sect. 2.2 proceeds with an outer asymptotic expansion in the far-filed, and in Sect. 2.3 an uniform asymptotic expansion of $u^{(\varepsilon,\alpha)} - u^0$ is matched in Ω.

2.1 *Inner Asymptotic Expansion in the Near-Field*

We start with some preliminaries of asymptotic analysis.

We introduce the local polar coordinate system associated to the center $x_0 \in \Omega$ and determined by the polar radius $\rho \in (0, R)$ and the polar angle $\theta \in (-\pi, \pi]$ such that $x - x_0 = \rho\widehat{x}$ with

$$\rho := |x - x_0|, \quad \widehat{x} := \tfrac{x - x_0}{|x - x_0|} = (\cos\theta, \sin\theta)^\top, \quad \widehat{x}' = (-\sin\theta, \cos\theta)^\top, \tag{16}$$

where $R > 0$ can be chosen such that the ball of radius R and center x_0 lies in the reference domain, i.e., $B_R(x_0) \subset \Omega$.

For indexes $m \geq 0$, Bessel functions of the first kind $J_m(k\rho)$ and the second kind $Y_m(k\rho)$ are two linearly independent solutions to the Bessel equation:

$$\big(\rho (u^0_m)'_\rho\big)'_\rho + \rho\big(k^2 - \tfrac{m^2}{\rho^2}\big) u^0_m = 0 \quad \text{for } \rho \mapsto u^0_m : \mathbf{R}_+ \mapsto \mathbf{R}. \tag{17}$$

The functions J_0, J_1, and Y_0 yield the following expansions as $k\rho \searrow +0$:

$$J_0(k\rho) = 1 + a_0(k\rho),\ a_0(k\rho) = -\tfrac{(k\rho)^2}{4} + \mathrm{O}((k\rho)^4), \tag{18}$$

$$J_1(k\rho) = \tfrac{1}{2}(k\rho + a_1(k\rho)),\ a_1(k\rho) = -\tfrac{(k\rho)^3}{8} + \mathrm{O}((k\rho)^5), \tag{19}$$

$$Y_0(k\rho) = \tfrac{2}{\pi}\big(\ln \tfrac{k\rho}{2} + \gamma\big) J_0(k\rho) + a_2(k\rho),\ a_2(k\rho) = \mathrm{O}((k\rho)^2), \tag{20}$$

where $\gamma > 0$ is the Euler constant.

Using (16)–(20) we prove the following truncated Fourier series.

Theorem 2 *The solution u^0 of the background problem (14) yields the first-order asymptotic representation in the near-field $B_R(x_0) \subset \Omega$:*

$$u^0(x) = u^0(x_0) J_0(k\rho) + U_0^0(x), \tag{21}$$

$$U_0^0(x) = \tfrac{2}{k} J_1(k\rho) \nabla u^0(x_0) \cdot \widehat{x} + U_1^0(x), \tag{22}$$

with the residuals $U_0^0, U_1^0 \in H^1(B_R(x_0); \mathbf{C})$ such that

$$\int_{-\pi}^{\pi} U_0^0\, d\theta = 0, \quad \int_{-\pi}^{\pi} U_1^0\, d\theta = \int_{-\pi}^{\pi} U_1^0 \widehat{x}\, d\theta = 0, \tag{23}$$

$$U_0^0 = \mathrm{O}(\rho), \quad U_1^0 = \mathrm{O}(\rho^2). \tag{24}$$

Moreover, the following formula for the gradient holds in $B_R(x_0)$:

$$\begin{aligned}
\nabla u^0(x) =&\ \nabla u^0(x_0) + b_u^0(x) + \nabla U_1^0(x),\\
b_u^0(x) :=&\ \big(u^0(x_0) k a_0'(k\rho) + a_1'(k\rho) \nabla u^0(x_0) \cdot \widehat{x}\big)\widehat{x} + \tfrac{a_1(k\rho)}{k\rho}(\nabla u^0(x_0) \cdot \widehat{x}')\widehat{x}', \quad (25)\\
\nabla U_0^0 =&\ \mathrm{O}(1), \quad \nabla U_1^0 = \mathrm{O}(\rho), \quad b_u^0 = \mathrm{O}(\rho). \quad (26)
\end{aligned}$$

Proof In a ball $B_\delta(x_0)$ of radius $\delta \in (0, R)$ we decompose the solution u^0 as

$$u^0(x) = u_0^0(\rho) + U_0^0(x) \text{ with } u_0^0(\rho) := \tfrac{1}{2\pi} \int_{-\pi}^{\pi} u^0\, d\theta, \quad U_0^0 := u^0 - u_0^0, \tag{27}$$

then the residual function U_0^0 has the zero average as written in (23).

Using this decomposition, we substitute a smooth cut-off function $\eta(\rho)$ supported in $B_\delta(x_0)$ as the test function $u = \eta$ in (14) and integrate it by parts providing

$$\begin{aligned}
0 &= \int_{B_\delta(x_0)} (\nabla u^0 \cdot \nabla \eta - k^2 u^0 \eta)\, dx\\
&= \int_0^\delta \Big\{ \tfrac{\partial}{\partial \rho}\Big(\int_{-\pi}^{\pi} (u_0^0 + U_0^0)\, d\theta \Big) \eta' - k^2 \int_{-\pi}^{\pi} (u_0^0 + U_0^0)\, d\theta\, \eta \Big\} \rho d\rho\\
&= 2\pi \int_0^\delta ((u_0^0)'_\rho \eta' - k^2 u_0^0 \eta)\, \rho d\rho = -2\pi \int_0^\delta \Big(\big(\rho (u_0^0)'_\rho\big)'_\rho + \rho k^2 u_0^0 \Big) \eta\, d\rho
\end{aligned}$$

for all η. This results in the Bessel equation (17) for u_0^0 as $m = 0$, which has the general solution of the form

$$u_0^0(\rho) = K_0^0 J_0(k\rho) + S_0^0 Y_0(k\rho) \quad \text{with} \quad K_0^0, S_0^0 \in \mathbf{C}.$$

But the Neumann function $Y_0(k\rho) = \mathrm{O}(|\ln \rho|)$ in (20) disagrees the fact that $u_0^0 \in H^1((0,\delta);\mathbf{C})$ in (27), hence parameter $S_0^0 = 0$ and the radial function is

$$u_0^0(\rho) = K_0^0 J_0(k\rho) \quad \text{with} \quad K_0^0 \in \mathbf{C}. \tag{28}$$

Next, we decompose the residual U_0^0 in the manner of (27) as

$$U_0^0(x) = u_1^0(\rho)\cdot \widehat{x} + U_1^0(x), \quad u_1^0(\rho) := \tfrac{1}{\pi}\int_{-\pi}^{\pi} u^0 \widehat{x}\, d\theta, \quad U_1^0 := U_0^0 - u_1^0 \cdot \widehat{x}, \tag{29}$$

with the residual function U_1^0 satisfying the equality in (23).

Inserting $u^0 = u_0^0 + u_1^0 \cdot \widehat{x} + U_1^0$ into (14) with the test vector-function $u = \widehat{x}\eta(\rho)$ supported in $B_\delta(x_0)$ and recalling the trigonometric calculus:

$$\widehat{x}' = (-\widehat{x}_2, \widehat{x}_1)^\top, \quad \int_{-\pi}^{\pi} \widehat{x}\, d\theta = 0, \quad \int_{-\pi}^{\pi} \widehat{x}_i \widehat{x}_j\, d\theta = \begin{cases} \pi & \text{for } i = j, \\ 0 & \text{otherwise}, \end{cases} \quad i, j = 1, 2,$$

after integration by parts the variational equation succeeds in

$$\begin{aligned} 0 = \int_{B_\delta(x_0)} \big(\nabla u^0 \cdot \nabla(\widehat{x}\eta) - k^2 u^0 \widehat{x}\eta\big)\, dx = \int_0^\delta \Big\{ \tfrac{\partial}{\partial\rho}\Big(\int_{-\pi}^{\pi} (u_0^0 + u_1^0 \cdot \widehat{x} + U_1^0)\widehat{x}\, d\theta\Big)\eta' \\ + \tfrac{\eta}{\rho^2} \int_{-\pi}^{\pi} \Big(u_1^0 \cdot \tfrac{\partial \widehat{x}}{\partial\theta} + \tfrac{\partial U_1^0}{\partial\theta}\Big)\tfrac{\partial \widehat{x}}{\partial\theta}\, d\theta - k^2 \eta \int_{-\pi}^{\pi} (u_0^0 + u_1^0 \cdot \widehat{x} + U_1^0)\, d\theta \Big\} \rho d\rho \\ = \pi \int_0^\delta \big((u_1^0)'_\rho \eta' + \tfrac{\eta}{\rho^2} u_1^0 - k^2 u_1^0 \eta\big)\rho d\rho = -\pi \int_0^\delta \Big(\big(\rho (u_1^0)'_\rho\big)'_\rho + \rho\big(k^2 - \tfrac{1}{\rho^2}\big)u_1^0\Big)\eta d\rho \end{aligned}$$

which follows the Bessel equation (17) for u_1^0 as $m = 1$ possessing the solution

$$u_1^0(\rho) = K_1^0 J_1(k\rho) \quad \text{with} \quad K_1^0 \in \mathbf{C}^2 \tag{30}$$

since the Neumann function $Y_1(k\rho)$ is singular as $\rho \searrow +0$.

In order to justify the asymptotic rate in (24), we apply the following Wirtinger inequality hold for U_1^0 due to (23):

$$\int_{-\pi}^{\pi} |U_1^0|^2\, d\theta \le \tfrac{1}{4} \int_{-\pi}^{\pi} \big|\tfrac{\partial U_1^0}{\partial\theta}\big|^2\, d\theta. \tag{31}$$

Indeed, decomposing the residual in the Fourier series:

$$U_1^0 = c_0^1 + \sum_{m=1}^{\infty} c_m^1 \cdot \widehat{x}^m, \quad c_0^1 := \tfrac{1}{2\pi} \int_{-\pi}^{\pi} U_1^0 \, d\theta \in \mathbf{C}, \quad c_m^1 := \tfrac{1}{\pi} \int_{-\pi}^{\pi} U_1^0 \widehat{x}^m \, d\theta \in \mathbf{C}^2,$$

where $\widehat{x}^m := (\cos(m\theta), \sin(m\theta))^\top$, and its derivative

$$\tfrac{\partial U_1^0}{\partial \theta} = \sum_{m=1}^{\infty} c_m^1 \cdot (\widehat{x}^m)', \quad \text{where } (\widehat{x}^m)' = m(-\sin(m\theta), \cos(m\theta))^\top,$$

this leads to the following Parseval identities

$$\tfrac{1}{2\pi} \int_{-\pi}^{\pi} |U_1^0|^2 \, d\theta = |c_0^1|^2 + \tfrac{1}{2} \sum_{m=1}^{\infty} |c_m^1|^2, \quad \tfrac{1}{2\pi} \int_{-\pi}^{\pi} |\tfrac{\partial U_1^0}{\partial \theta}|^2 \, d\theta = \tfrac{1}{2} \sum_{m=1}^{\infty} m^2 |c_m^1|^2.$$

Conditions in (23) imply $c_0^1 = c_1^1 = 0$ that allows us to derive (31):

$$\int_{-\pi}^{\pi} |U_1^0|^2 \, d\theta = \pi \sum_{m=2}^{\infty} |c_m^1|^2 \le \tfrac{\pi}{2^2} \sum_{m=2}^{\infty} m^2 |c_m^1|^2 = \tfrac{1}{4} \int_{-\pi}^{\pi} |\tfrac{\partial U_1^0}{\partial \theta}|^2 \, d\theta.$$

Considering the residual in the energy norm, after integration by parts and using the Helmholtz equation $-[\Delta + k^2]U_1^0 = 0$ hold in $B_\delta(x_0)$, we have

$$0 \le I(\delta) := \int_{B_\delta(x_0)} |\nabla U_1^0|^2 \, dx = \int_{B_\delta(x_0)} k^2 |U_1^0|^2 \, dx + \int_{\partial B_\delta(x_0)} \tfrac{\partial U_1^0}{\partial \rho} \overline{U_1^0} \, dS_x. \quad (32)$$

Due to Wirtinger's inequality (31) the boundary integral in $I(\delta)$ can be estimated as

$$\int_{\partial B_\delta(x_0)} \tfrac{\partial U_1^0}{\partial \rho} \overline{U_1^0} \, dS_x \le \int_{-\pi}^{\pi} \Big(\tfrac{\delta}{4} |\tfrac{\partial U_1^0}{\partial \rho}|^2 + \tfrac{1}{\delta} |U_1^0|^2 \Big) \delta d\theta$$
$$\le \int_{-\pi}^{\pi} \Big(\tfrac{\delta}{4} |\tfrac{\partial U_1^0}{\partial \rho}|^2 + \tfrac{1}{4\delta} |\tfrac{\partial U_1^0}{\partial \theta}|^2 \Big) \delta d\theta = \tfrac{\delta}{4} \int_{\partial B_\delta(x_0)} |\nabla U_1^0|^2 \, dS_x.$$

Applying to (32) the Poincare inequality:

$$\int_{B_\delta(x_0)} |U_1^0|^2 \, dx \le \tfrac{(2\delta)^2}{\pi^2} \int_{B_\delta(x_0)} |\nabla U_1^0|^2 \, dx \quad (33)$$

and the co-area formula:

$$\tfrac{d}{d\delta} \int_{B_\delta(x_0)} |\nabla U_1^0|^2 \, dx = \int_{\partial B_\delta(x_0)} |\nabla U_1^0|^2 \, dS_x,$$

we obtain the differential inequality

$$\big(1-k^2\tfrac{4\delta^2}{\pi^2}\big)I(\delta)\le\tfrac{\delta}{4}\tfrac{d}{d\delta}I(\delta).$$

Its integration over $\delta\in(r,R)$ leads to Grönwall's inequality

$$0\le I(r)\le\big(\tfrac{r}{R}\big)^4\exp\big(\tfrac{8k^2}{\pi^2}(R^2-r^2)\big)I(R)=\mathrm{O}(r^4). \tag{34}$$

Due to the fundamental theorem of calculus and using homogeneity arguments, the function oscillation in $\mathbf{R}^2$ can be estimated from above (see e.g., [20]) as

$$\sup_{x,y\in B_\delta(x_0)}|U_1^0(x)-U_1^0(y)|^2\le C\int_{B_\delta(x_0)}(|\nabla U_1^0|^2+\delta^2|\Delta U_1^0|^2)\,dx$$

$$=C\int_{B_\delta(x_0)}\big(|\nabla U_1^0|^2+\delta^2k^2|U_1^0|^2\big)\,dx\le C_1I(\delta)\text{ with }C_1:=\big(1+\tfrac{4k^2}{\pi^2}\big)C>0, \tag{35}$$

where we have used $-[\Delta+k^2]U_1^0=0$ in $B_r(x_0)$, the notation $I(\delta)$ from (32), and the Poincare inequality (33). For $x=x_0+\rho\widehat{x}$ and $y=x_0$ we have $U_1^0(x_0)=0$ due to the zero average in (23), then (35) with $\delta=\rho$ provides the point-wise estimate

$$|U_1^0(x_0+\rho\widehat{x})|^2\le C_1I(\rho)=\mathrm{O}(\rho^4),$$

which justifies the asymptotic rate in (24).

Due to (24), passing $\rho\searrow+0$ in (28) we specify the constant $K_0^0=u^0(x_0)$. After differentiation of u^0 in (21) and (22) according to (16), (18), and (19), this provides formulas (25) and (26) for the gradient ∇u^0 in $B_R(x_0)$. Here the constant $K_1^0=\frac{2}{k}\nabla u^0(x_0)$ in (30) is specified when $\rho\searrow+0$, and the asymptotic rate of $\nabla U_1^0=\mathrm{O}(\rho)$ in (24) can be argued by (34). The proof is complete. □

In the next section, we proceed with outer asymptotic expansion in the far-field with respect to the geometry ω expressed by a boundary layer.

2.2 *Outer Asymptotic Expansion in the Far-Field*

We introduce the weighted Sobolev space (see [4, 31, 33]):

$$\begin{aligned}H_\mu^1(\mathbf{R}^2;\mathbf{C})&=\big\{v:\ \tfrac{v}{\mu}\in L^2(\mathbf{R}^2;\mathbf{C}),\ \nabla v\in L^2(\mathbf{R}^2;\mathbf{C}^2)\big\}\text{ with the weight}\\ \mu(y)&=\mathrm{O}(|y|\ln|y|)\quad\text{in }\mathbf{R}^2\setminus\overline{B_2(0)},\quad \mu(y)=\mathrm{O}(1)\quad\text{in }B_2(0),\end{aligned}$$

suggested by the weighted Poincare inequality in exterior domains

$$\int_{\mathbf{R}^2\setminus B_2(0)} \Big(\frac{v}{|y|\ln|y|}\Big)^2 dy \le 4\int_{\mathbf{R}^2\setminus B_2(0)} |\nabla v|^2 dy \quad \text{if} \int_{\partial B_2(0)} v\, dS_x = 0.$$

We note that constant functions are allowed in H^1_μ.

Excluding constants implying polynomials of degree zero $\mathbf{P}_0$, a boundary layer depending on $1-\mathrm{Re}(\alpha)$ is described by the real-valued *exterior transmission problem*: Find vector-function $w(y) = (w_1, w_2)^\top \in H^1_\mu(\mathbf{R}^2;\mathbf{R}^2)/\mathbf{P}_0^2$ such that

$$\int_{\mathbf{R}^2} \chi_\omega^{\mathrm{Re}(\alpha)} Dw\nabla v\, dy = (1-\mathrm{Re}(\alpha))\int_{\partial\omega} nv\, dS_y \quad \text{for all } v \in H^1_\mu(\mathbf{R}^2;\mathbf{R}), \tag{36}$$

where $(Dw)_{ij} = \frac{\partial w_i}{\partial y_j}$ for $i, j = 1, 2$ denotes the derivative matrix, and recalling that $n = (n_1, n_2)^\top$ is the normal vector outward to ω.

From the result of [4], it follows existence of the solution to the variational equation (36) which implies the boundary value problem:

$$-\Delta w = 0 \quad \text{in } \mathbf{R}^2\setminus\partial\omega, \tag{37}$$

$$[\![w]\!] = 0, \quad (Dw)n|_{\partial\omega^+} - \mathrm{Re}(\alpha)(Dw)n|_{\partial\omega^-} = -(1-\mathrm{Re}(\alpha))n \quad \text{on } \partial\omega, \tag{38}$$

$$w = \mathrm{O}\big(\tfrac{1}{|y|}\big) \quad \text{as } |y| \nearrow \infty. \tag{39}$$

If $\mathrm{Re}(\alpha) > \alpha_0 > 0$, then from (36) it follows that $\|w\|_{H^1_\mu(\mathbf{R}^2;\mathbf{R}^2)} = \mathrm{O}(|1-\mathrm{Re}(\alpha)|)$.

Rescaling $y = \frac{x-x_0}{\varepsilon}$, we reduce the exterior problem to Ω next.

Theorem 3 *The rescaled solution $w^\varepsilon(x) := w(\frac{x-x_0}{\varepsilon}) \in H^1(\Omega;\mathbf{R}^2)$ of (36) satisfies the variational equation:*

$$\int_\Omega \chi_{\omega_\varepsilon(x_0)}^{\mathrm{Re}(\alpha)} Dw^\varepsilon \nabla u\, dx = \int_{\Gamma_N} (Dw^\varepsilon n)u\, dS_x + \frac{1-\mathrm{Re}(\alpha)}{\varepsilon}\int_{\partial\omega_\varepsilon(x_0)} nu\, dS_x$$
$$\textit{for all test-functions } u \in H^1(\Omega;\mathbf{R}) \textit{ such that } u = 0 \textit{ on } \Gamma_D. \tag{40}$$

The solution has the far-field representation in the truncated Fourier series

$$w^\varepsilon(x) = \frac{\varepsilon}{\rho}\frac{1}{2\pi} A_{(\omega,\mathrm{Re}(\alpha))}\widehat{x} + W^\varepsilon(x) \quad \textit{for } x \in \mathbf{R}^2\setminus\overline{B_\varepsilon(x_0)}, \tag{41}$$

with the residual vector-function $W^\varepsilon = (W_1^\varepsilon, W_2^\varepsilon)^\top$ such that

$$\int_{-\pi}^{\pi} W^\varepsilon\, d\theta = \int_{-\pi}^{\pi} W^\varepsilon\,\widehat{x}\, d\theta = 0, \tag{42}$$

$$W^\varepsilon = \mathrm{O}\big((\tfrac{\varepsilon}{\rho})^2\big), \quad DW^\varepsilon = \mathrm{O}\big(\tfrac{\varepsilon^2}{\rho^3}\big) \quad \textit{for } \rho > \varepsilon, \theta \in (-\pi,\pi]. \tag{43}$$

Entries of the 2-by-2 real matrix $A_{(\omega,\mathrm{Re}(\alpha))}$ have the implicit expression:

$$(A_{(\omega,\mathrm{Re}(\alpha))})_{ij} = (1-\mathrm{Re}(\alpha))\big(\delta_{ij}\,|\omega| + \int_{\partial\omega} w_i n_j\, dS_y\big), \quad i,j=1,2. \tag{44}$$

If $\mathrm{Re}(\alpha) \in [0,1)$ *and* $|\omega|>0$*, then* $A_{(\omega,\mathrm{Re}(\alpha))} \in \mathrm{Spd}(\mathbf{R}^{2\times 2})$*, i.e., symmetric positive definite. For ellipsoidal shapes* ω*, the matrix has the explicit expression*

$$A_{(\omega,\mathrm{Re}(\alpha))} = \Theta(\phi) A_{(\omega',\mathrm{Re}(\alpha))} \Theta(\phi)^\top, \quad \Theta(\phi) := \begin{pmatrix} \cos\phi & -\sin\phi \\ \sin\phi & \cos\phi \end{pmatrix}, \tag{45}$$

$$A_{(\omega',\mathrm{Re}(\alpha))} = \pi(a+b)\begin{pmatrix} \frac{(1-\mathrm{Re}(\alpha))ab}{a+b\mathrm{Re}(\alpha)} & 0 \\ 0 & \frac{(1-\mathrm{Re}(\alpha))ab}{a\mathrm{Re}(\alpha)+b} \end{pmatrix}, \tag{46}$$

with the ellipse major $a=1$ *and minor* $b \in (0,1]$ *semi-axes, where the major axis has an angle of* $\phi \in (-\frac{\pi}{2}, \frac{\pi}{2})$ *with the* y_1*-axis counted in the anticlockwise direction.*

Proof The proof is based on the techniques from [12, Lemma 3.2] and [25].

The local coordinate system (16) in the stretched variables y implies the polar radius $|y| \in \mathbf{R}_+$ and the polar angle $\theta \in (-\pi,\pi]$ such that

$$y = \tfrac{x-x_0}{\varepsilon} = |y|\widehat{x}, \quad |y| = \tfrac{\varrho}{\varepsilon}, \quad \widehat{x} = (\cos\theta, \sin\theta)^\top. \tag{47}$$

We apply the coordinate transformation (47) and differential calculus:

$$\tfrac{\partial}{\partial y} = \varepsilon\tfrac{\partial}{\partial x}, \quad dy = \tfrac{1}{\varepsilon^2}dx, \quad dS_y = \tfrac{1}{\varepsilon}dS_x$$

to Eqs. (37) and (38) and derive the following relations for $w^\varepsilon(x) = w(\frac{x-x_0}{\varepsilon})$:

$$-\Delta w^\varepsilon = 0 \quad \text{in } \mathbf{R}^2 \setminus \partial\omega_\varepsilon(x_0), \tag{48}$$

$$[\![w^\varepsilon]\!] = 0, \quad [\![\chi^{\mathrm{Re}(\alpha)}_{\omega_\varepsilon(x_0)} Dw^\varepsilon n]\!] = -\tfrac{1-\mathrm{Re}(\alpha)}{\varepsilon} n \quad \text{on } \partial\omega_\varepsilon(x_0). \tag{49}$$

Using Green's formulas (6) and (7), it follows variational formulation (40) written in the bounded domain.

We split $y \in \mathbf{R}^2$ in the far-field $\mathbf{R}^2 \setminus \overline{B_1(0)}$ and the near-field $B_1(0)$.

In the far-field, using the notation $\widehat{x}^m = (\cos(m\theta), \sin(m\theta))^\top$, the harmonic vector-valued function w solving (37) and (39) admits the Fourier series as

$$w(y) = \sum_{m=1}^{\infty} \tfrac{1}{|y|^m} C_m \widehat{x}^m \quad \text{with coefficient matrices } C_m \in \mathbf{R}^{2\times 2} \text{ for } y \in \mathbf{R}^2 \setminus \overline{B_1(0)}.$$

This formula implies the outer asymptotic expansion with $A_{(\omega,\mathrm{Re}(\alpha))} := 2\pi C_1$:

$$w(y) = \tfrac{1}{2\pi} A_{(\omega,\mathrm{Re}(\alpha))} \tfrac{\widehat{x}}{|y|} + W(y) \quad \text{for } y \in \mathbf{R}^2 \setminus \overline{B_1(0)}, \tag{50}$$
$$\int_{-\pi}^{\pi} W\, d\theta = \int_{-\pi}^{\pi} W\widehat{x}\, d\theta = 0, \quad W = \mathrm{O}\big((\tfrac{1}{|y|})^2\big), \; DW = \mathrm{O}\big((\tfrac{1}{|y|})^3\big)$$

and turns into (41)–(43) for $W^\varepsilon(x) := W(\frac{x-x_0}{\varepsilon})$ after rescaling.

In the near-field, we apply the second Green's formula hold for $i, j = 1, 2$:

$$0 = \int_{B_1(0)} \chi_{\omega_1(0)}^{\mathrm{Re}(\alpha)} \{\Delta w_i y_j - w_i \Delta y_j\}\, dy = \int_{\partial B_1(0)} \{\tfrac{\partial w_i}{\partial |y|} y_j - w_i \tfrac{\partial y_j}{\partial |y|}\}\, dS_y$$
$$- \int_{\partial\omega} \{[\tfrac{\partial w_i}{\partial n}|_{\partial\omega^+} - \mathrm{Re}(\alpha)\tfrac{\partial w_i}{\partial n}|_{\partial\omega^-}] y_j - (1-\mathrm{Re}(\alpha)) w_i \tfrac{\partial y_j}{\partial n}\}\, dS_y,$$

and substitute here the transmission condition (38) and $\frac{\partial y_j}{\partial n} = n_j$ to derive that

$$- \int_{\partial B_1(0)} \{\tfrac{\partial w_i}{\partial |y|} - w_i\} \widehat{x}_j\, dS_y = (1-\mathrm{Re}(\alpha)) \int_{\partial\omega} \{n_i y_j + w_i n_j\}\, dS_y. \tag{51}$$

We apply to (51) the divergence theorem with Kronecker's delta

$$\int_{\partial\omega} n_i y_j\, dS_y = \int_{\omega} \tfrac{\partial y_j}{\partial y_i}\, dy = \delta_{ij}\, |\omega|, \quad \text{where } |\omega| := \int_{\omega} dy, \quad \delta_{ij} = \begin{cases} 1 \text{ if } i = j, \\ 0 \text{ if } i \neq j, \end{cases}$$

and substitute (50) to calculate the integral over $\partial B_1(0)$ as

$$- \int_{\partial B_1(0)} \{\tfrac{\partial w_i}{\partial |y|} - w_i\} \widehat{x}_j\, dS_y = \tfrac{1}{\pi} \int_{-\pi}^{\pi} (A_{(\omega,\mathrm{Re}(\alpha))})_{ik} \widehat{x}_k \widehat{x}_j\, d\theta = (A_{(\omega,\mathrm{Re}(\alpha))})_{ij},$$

which together with (51) follows formula (44).

In order to prove the symmetry of $A_{(\omega,\mathrm{Re}(\alpha))}$, we insert $v = w_j$ into (36) written component-wisely for $i, j = 1, 2$ as

$$\int_{\mathbf{R}^2} \chi_{\omega}^{\mathrm{Re}(\alpha)} \nabla w_i^\top \nabla w_j\, dy = (1-\mathrm{Re}(\alpha)) \int_{\partial\omega} n_i w_j\, dS_y = (1-\mathrm{Re}(\alpha)) \int_{\partial\omega} n_j w_i\, dS_y.$$

Henceforth, $(A_{(\omega,\mathrm{Re}(\alpha))})_{ij} = (A_{(\omega,\mathrm{Re}(\alpha))})_{ji}$ follows due to the matrix entries expression by (44). For arbitrary $z \in \mathbf{R}^2$, from (36) we have

$$0 \le \int_{\mathbf{R}^2} \chi_{\omega}^{\mathrm{Re}(\alpha)} |\nabla(z_1 w_1 + z_2 w_2)|^2\, dy = (1-\mathrm{Re}(\alpha)) \int_{\partial\omega} (n \cdot z)(w \cdot z)\, dS_y.$$

Multiplying (44) with $z_i z_j$ and summing the result over $i, j = 1, 2$ implies

$$z^\top A_{(\omega,\mathrm{Re}(\alpha))} z = (1 - \mathrm{Re}(\alpha))\Big\{|z|^2|\omega| + \int_{\partial\omega} (n \cdot z)(w \cdot z)\, dS_y\Big\} \ge (1 - \mathrm{Re}(\alpha))|z|^2|\omega|,$$

which is positive, hence $A_{(\omega,\mathrm{Re}(\alpha))} \in \mathrm{Spd}(\mathbf{R}^{2\times 2})$, if $1 - \mathrm{Re}(\alpha) > 0$ and $|\omega| > 0$.

An explicit representation of $A_{(\omega,\mathrm{Re}(\alpha))}$ will be established for ellipses ω.

Let an ellipse $\omega' \subset B_1(0)$ have the major $a = 1$ and the minor $b \in (0, 1]$ semi-axes written with respect to y'-coordinates as

$$\omega' = \{y' \in \mathbf{R}^2 : (\tfrac{y'_1}{a})^2 + (\tfrac{y'_2}{b})^2 < 1\}, \quad a = 1.$$

Let the reference ellipse ω written in y-coordinates have an angle of $\phi \in (-\frac{\pi}{2}, \frac{\pi}{2})$ to the major y_1-axis counted in the anticlockwise direction:

$$\omega = \big\{y \in \mathbf{R}^2 : (\tfrac{y_1 \cos\phi + y_2 \sin\phi}{a})^2 + (\tfrac{-y_1 \sin\phi + y_2 \cos\phi}{b})^2 < 1\big\}, \quad a = 1.$$

This implies that $y' \in \omega'$ when $\Theta^\top y \in \omega$ with the orthogonal matrix $\Theta(\phi)$ given in (45). Therefore, we prove first formula (46) for ω' and then transform $y' = \Theta^\top y$.

We introduce the elliptic coordinates $r \in \mathbf{R}_+$ and $\psi \in (-\pi, \pi]$ such that

$$y'_1 = c\cosh(r)\cos\psi, \quad y'_2 = c\sinh(r)\sin\psi, \quad c = \sqrt{a^2 - b^2}, \quad a = 1, \tag{52}$$

and c is the linear eccentricity. Setting the distance $r_0 \in \mathbf{R}_+$ implicitly by

$$a = c\cosh(r_0) \quad b = c\sinh(r_0), \tag{53}$$

the geometry ω' can be restated as

$$\omega' = \{r < r_0, \psi \in (-\pi, \pi]\}, \quad \mathbf{R}^2 \setminus \overline{\omega'} = \{r > r_0, \psi \in (-\pi, \pi]\}.$$

From (52) it follows the differential calculus in elliptic coordinates:

$$\begin{cases} \frac{\partial}{\partial y'_1} = \frac{1}{\kappa^2(r,\psi)}\big(c\cosh(r)\cos\psi \frac{\partial}{\partial r} - c\cosh(r)\sin\psi \frac{\partial}{\partial \psi}\big), \\ \frac{\partial}{\partial y'_2} = \frac{1}{\kappa^2(r,\psi)}\big(c\cosh(r)\sin\psi \frac{\partial}{\partial r} + c\sinh(r)\cos\psi \frac{\partial}{\partial \psi}\big), \end{cases}$$

$$dy' = \kappa^2(r, \psi) dr d\psi, \quad \kappa(r, \psi) = c\sqrt{\sinh^2(r) + \sin^2\psi}$$

involving the scale factor $\kappa(r, \psi)$. In particular, at the ellipse boundary $\partial\omega'$ as $r = r_0$, using (52) with constant ψ and (53), for the normal vector n' we get expressions

$$\begin{aligned} n' &= \tfrac{1}{\kappa(r_0,\psi)}(b\cos\psi, a\sin\psi)^\top, \quad \tfrac{\partial}{\partial n'} = \tfrac{1}{\kappa(r_0,\psi)}\tfrac{\partial}{\partial r}, \\ dS_{y'} &= \kappa(r_0, \psi) d\psi, \quad \kappa(r_0, \psi) = \sqrt{a^2\cos^2\psi + b^2\sin^2\psi}. \end{aligned}$$

Applying transformation (52) and the differential calculus, the exterior problem (37)–(39) for $w'(r,\psi)$ associated to ω' can be rewritten in elliptic coordinates as

$$-\tfrac{1}{\kappa^2(r,\psi)}\big[\tfrac{\partial^2}{\partial r^2}+\tfrac{\partial^2}{\partial \psi^2}\big]w' = 0 \text{ for } r \neq r_0, \tag{54}$$

$$[\![w']\!] = 0,\ \tfrac{1}{\kappa(r_0,\psi)}\big(\tfrac{\partial w'}{\partial r}\big|_{r=r_0^+} - \mathrm{Re}(\alpha)\tfrac{\partial w'}{\partial r}\big|_{r=r_0^-}\big) = -(1-\mathrm{Re}(\alpha))n' \text{ as } r = r_0, \tag{55}$$

$$w' = o(1) \text{ as } r \nearrow \infty, \tag{56}$$

where $r_0^\pm$ denotes the one-sided limit corresponding to $(\partial\omega')^\pm$, respectively. Due to (54) and (56) the harmonic vector-function w' admits the Fourier series:

$$w' = \sum_{m=1}^{\infty} C'_m R'_m \begin{pmatrix}\cos(m\psi)\\ \sin(m\psi)\end{pmatrix},\quad R'_m = \begin{cases} e^{m(r_0-r)}I & \text{for } r > r_0,\\ \begin{pmatrix}\frac{\cosh(mr)}{\cosh(mr_0)} & 0\\ 0 & \frac{\sinh(mr)}{\sinh(mr_0)}\end{pmatrix} & \text{for } r < r_0,\end{cases}$$

where I stands for the identity 2-by-2 matrix. Then w' satisfies $w'(0,\psi) = w'(0,-\psi)$, $\frac{\partial w'}{\partial r}(0,\psi) = -\frac{\partial w'}{\partial r}(0,-\psi)$ as $r=0$ for $\psi \in (-\pi,\pi]$, and the first jump condition $w'|_{r=r_0^+} - w'|_{r=r_0^-} = 0$ in (55). To find the unknown coefficient matrices $C'_m \in \mathbf{R}^{2\times 2}$, we substitute the Fourier series into the second jump condition in (55) implying that

$$\sum_{m=1}^{\infty} \tfrac{mC'_m}{\kappa(r_0,\psi)}\Big(I + \mathrm{Re}(\alpha)\begin{pmatrix}\tanh(mr_0) & 0\\ 0 & \frac{1}{\tanh(mr_0)}\end{pmatrix}\Big)\begin{pmatrix}\cos(m\psi)\\ \sin(m\psi)\end{pmatrix} = \tfrac{1-\mathrm{Re}(\alpha)}{\kappa(r_0,\psi)}\begin{pmatrix}b\cos\psi\\ a\sin\psi\end{pmatrix}.$$

Henceforth, $C'_m = 0$ for all $m \geq 2$ and $C'_1 = (1-\mathrm{Re}(\alpha))\begin{pmatrix}\frac{b}{1+\mathrm{Re}(\alpha)ba^{-1}} & 0\\ 0 & \frac{a}{1+\mathrm{Re}(\alpha)ab^{-1}}\end{pmatrix}$ due to $\tanh(r_0) = ba^{-1}$ resulting in the following analytic solution to (54)–(56):

$$w' = \begin{cases} \big(\frac{(1-\mathrm{Re}(\alpha))ab}{a+b\mathrm{Re}(\alpha)}e^{r_0-r}\cos\psi,\ \frac{(1-\mathrm{Re}(\alpha))ab}{a\mathrm{Re}(\alpha)+b}e^{r_0-r}\sin\psi\big)^\top & \text{for } r > r_0,\\ \big(\frac{(1-\mathrm{Re}(\alpha))ab}{a+b\mathrm{Re}(\alpha)}\frac{\cosh(r)}{\cosh(r_0)}\cos\psi,\ \frac{(1-\mathrm{Re}(\alpha))ab}{a\mathrm{Re}(\alpha)+b}\frac{\sinh(r)}{\sinh(r_0)}\sin\psi\big)^\top & \text{for } r < r_0.\end{cases} \tag{57}$$

The matrix $A_{(\omega',\mathrm{Re}(\alpha))}$ is calculated analytically after substitution of (57) in the representation formula (44) that implies the following two vectors for $j = 1, 2$:

$$(A_{(\omega',\mathrm{Re}(\alpha))})_{(\cdot,j)} = (1-\mathrm{Re}(\alpha))\big(\delta_{(\cdot,j)}|\omega'| + I_{(\cdot,j)}\big),\quad \text{where}$$

$$I_{(\cdot,j)} := \int_{\partial\omega'} (w'_1 n'_j, w'_2 n'_j)^\top\, dS_{y'} = \begin{cases} \int_{-\pi}^{\pi} \big(\frac{(1-\mathrm{Re}(\alpha))ab^2}{a+b\mathrm{Re}(\alpha)}\cos^2\psi, 0\big)^\top d\psi & \text{for } j = 1,\\ \int_{-\pi}^{\pi} \big(0, \frac{(1-\mathrm{Re}(\alpha))a^2 b}{a\mathrm{Re}(\alpha)+b}\sin^2\psi\big)^\top d\psi & \text{for } j = 2.\end{cases}$$

Using $\int_{-\pi}^{\pi}\cos^2\psi\, d\psi = \int_{-\pi}^{\pi}\sin^2\psi\, d\psi = \pi$ and $|\omega'| = \pi ab$ we arrive at (46).

The transformation formula (45) is justified by rotation $y = \Theta y'$ applied to the variational equation (36) which, after the left multiplication with $\Theta^\top$, results in

$$\int_{\mathbf{R}^2} \chi_{\omega'}^{\mathrm{Re}(\alpha)} D_{y'}(\Theta^\top w(\Theta y'))\nabla_{y'} v\, dy' = (1-\mathrm{Re}(\alpha)) \int_{\partial\omega'} \Theta^\top n v\, dS_{y'} \;\forall v \in H^1_\mu(\mathbf{R}^2;\mathbf{R}).$$

Since $\Theta^\top n = n'$ this proves $w'(y') = \Theta^\top w(\Theta y')$, then from (50) it follows

$$\tfrac{1}{2\pi} A_{(\omega',\mathrm{Re}(\alpha))} \tfrac{y'}{|y'|^2} + W' = w'(y') = \Theta^\top w(\Theta y') = \tfrac{\Theta^\top}{2\pi} A_{(\omega,\mathrm{Re}(\alpha))} \tfrac{\Theta y'}{|\Theta y'|^2} + \Theta^\top W(\Theta y'),$$

which implies $A_{(\omega',\mathrm{Re}(\alpha))} = \Theta^\top A_{(\omega,\mathrm{Re}(\alpha))}\Theta$, hence (45), and completes the proof. □

2.3 Uniform Asymptotic Expansion of the Solution

The boundary layer w^ε depending on $1 - \mathrm{Re}(\alpha)$ will express a leading asymptotic term in the uniform expansion over Ω. Moreover, for its refinement we need the auxiliary Helmholtz problem: Find $u^1 \in H^1(\Omega; \mathbf{C}^2)$ such that $u^1 = 0$ at Γ_D and

$$\int_\Omega \big(\chi_{\omega_\varepsilon(x_0)}^{\mathrm{Re}(\alpha)} D u^1 \nabla \overline{u} - \chi_{\omega_\varepsilon(x_0)}^{\alpha} k^2 u^1 \overline{u}\big)\, dx = \tfrac{k^2}{\varepsilon\sqrt{|\ln\varepsilon|}} \int_{\Omega\setminus B_\varepsilon(x_0)} w^\varepsilon \overline{u}\, dx \tag{58}$$

for all test-functions $u \in H^1(\Omega; \mathbf{C})$ such that $u = 0$ on Γ_D.

We show that the solution u^1 to problem (58) is of the order $\mathrm{O}(|1 - \mathrm{Re}(\alpha)|)$ since its right-hand side can be estimated uniformly with respect to ε for fixed α. Indeed, we inscribe Ω in a ball $B_R(x_0)$ of radius $R > 0$, then due to (41)–(43) it holds

$$\int_{\Omega\setminus B_\varepsilon(x_0)} |w^\varepsilon|^2\, dx \le \int_{B_R(x_0)\setminus B_\varepsilon(x_0)} |w^\varepsilon|^2\, dx = \int_{-\pi}^{\pi}\int_\varepsilon^R \Big((\tfrac{\varepsilon}{\rho})^2 |\tfrac{1}{2\pi} A_{(\omega,\mathrm{Re}(\alpha))}\widehat{x}|^2$$
$$+ |W^\varepsilon|^2\Big)\rho\, d\rho\, d\theta = \mathrm{O}\big(|1-\mathrm{Re}(\alpha)|^2 \varepsilon^2 |\ln\varepsilon|\big),$$

because $A_{(\omega,\mathrm{Re}(\alpha))} = \mathrm{O}(|1 - \mathrm{Re}(\alpha)|)$ due to expression (44).

Theorem 4 *The inhomogeneous solution $u^{(\varepsilon,\alpha)}$ of (5), the background solution u^0 of (14), the rescaled solution $w^\varepsilon(x) = w(\frac{x-x_0}{\varepsilon})$ of (36) together with its refinement u^1 from (58) compose the residual*

$$q^\varepsilon := u^{(\varepsilon,\alpha)} - u^0 - \varepsilon \nabla u^0(x_0)\cdot w^1, \quad w^1 := w^\varepsilon + \varepsilon\sqrt{|\ln\varepsilon|}\,u^1, \tag{59}$$

which satisfies the two-parameter asymptotic relation

$$\Big|\int_\Omega (\chi^{\mathrm{Re}(\alpha)}_{\omega_\varepsilon(x_0)} \nabla q^\varepsilon \cdot \nabla\overline{u} - \chi^{\alpha}_{\omega_\varepsilon(x_0)} k^2 q^\varepsilon \overline{u})\, dx\Big| \le \underline{C}(\alpha,\varepsilon)\|u\|_{H^1(\Omega;\mathbf{C})} \tag{60}$$
$$\textit{where } 0 \le \underline{C}(\alpha,\varepsilon) = \mathrm{O}\big(|1-\alpha|\varepsilon^2 + |\alpha||1-\mathrm{Re}(\alpha)|\varepsilon^3\big).$$

If for all $\varepsilon \in (0,\varepsilon_0)$ with $\varepsilon_0 > 0$, $\mathrm{Re}(\alpha) > \alpha_0 > 0$ and $|\alpha| < \alpha_1$ with $\alpha_1 > \alpha_0 > 0$, and for $u, v \in H^1(\Omega;\mathbf{C})$ such that $u = v = 0$ on Γ_D, the inf-sup condition holds:

$$\Big|\int_\Omega (\chi^{\mathrm{Re}(\alpha)}_{\omega_\varepsilon(x_0)} \nabla v \cdot \nabla\overline{u} - \chi^{\alpha}_{\omega_\varepsilon(x_0)} k^2 v \overline{u})\, dx\Big| \ge \underline{c}\|u\|_{H^1(\Omega;\mathbf{C})}\|v\|_{H^1(\Omega;\mathbf{C})}, \quad \underline{c} > 0, \tag{61}$$

then it follows from (60) and (61) the residual error estimate

$$\|q^\varepsilon\|_{H^1(\Omega;\mathbf{C})} \le \underline{c}^{-1}\underline{C}(\alpha,\varepsilon). \tag{62}$$

Proof We subtract from (5) Eqs. (15) and (40) multiplied with $\varepsilon\nabla u^0(x_0)$, and (58) multiplied with $\varepsilon^2\sqrt{|\ln\varepsilon|}\nabla u^0(x_0)$. Using the expansion (25) of ∇u^0 at $\partial\omega_\varepsilon(x_0)$, the differential identity $\nabla(\nabla u^0(x_0)\cdot w^1) = \nabla u^0(x_0)^\top D w^1$, and the notation of q^ε and w^1 introduced in (59) we obtain the variational equation

$$\begin{aligned}\int_\Omega (\chi^{\mathrm{Re}(\alpha)}_{\omega_\varepsilon(x_0)} \nabla q^\varepsilon \cdot \nabla\overline{u} - \chi^{\alpha}_{\omega_\varepsilon(x_0)} k^2 q^\varepsilon \overline{u})\, dx &= -\varepsilon \int_{\Gamma_N} (\nabla u^0(x_0)^\top D w^\varepsilon n)\overline{u}\, dS_x \\ +(1-\mathrm{Re}(\alpha))\int_{\partial\omega_\varepsilon(x_0)} n\cdot(b^0_u + \nabla U^0_1)\,\overline{u}\, dS_x &+ \mathrm{i}\mathrm{Im}(\alpha)\int_{\omega_\varepsilon(x_0)} k^2 u^0\overline{u}\, dx \\ &+ \varepsilon\int_{B_\varepsilon(x_0)} \chi^{\alpha}_{\omega_\varepsilon(x_0)} k^2(\nabla u^0(x_0)\cdot w^\varepsilon)\overline{u}\, dx \end{aligned}\tag{63}$$

for all test-functions $u \in H^1(\Omega;\mathbf{C})$ such that $u = 0$ on Γ_D. Applying to the right-hand side of (63) the Cauchy–Schwarz inequality and trace theorems, with the help of the representations (21), (25), and (41), we get $w^\varepsilon = \mathrm{O}(|1-\alpha|\varepsilon)$ at Γ_N, while $w^\varepsilon = \mathrm{O}(|1-\alpha|)$ in $B_\varepsilon(x_0)$, and $b^0_u + \nabla U^0_1 = \mathrm{O}(\varepsilon)$ at $\partial\omega_\varepsilon(x_0)$, hence derive the upper bound in (60).

Now let the inf-sup condition (61) hold.

For a smooth cutoff function η_{Γ_D} supported in a neighborhood of the Dirichlet boundary Γ_D such that $\eta_{\Gamma_D} = 1$ at Γ_D, the lifting function is set

$$Q^\varepsilon := q^\varepsilon + R^\varepsilon, \quad R^\varepsilon := \varepsilon(\nabla u^0(x_0)\cdot w^\varepsilon)\eta_{\Gamma_D}, \quad Q^\varepsilon = 0 \text{ on } \Gamma_D,$$

and $R^\varepsilon = \mathrm{O}(|1-\mathrm{Re}(\alpha)|\varepsilon^2)$ can be estimated due to (41). Therefore, inserting $q^\varepsilon = Q^\varepsilon - R^\varepsilon$ into (60) and using (61) we derive (62). The proof is complete. □

As the consequence, from (63) we infer the boundary value problem for q^ε:

$$-[\chi^{\mathrm{Re}(\alpha)}_{\omega_\varepsilon(x_0)} \Delta + \chi^{\alpha}_{\omega_\varepsilon(x_0)} k^2] q^\varepsilon = \begin{cases} 0 & \text{in } \Omega \setminus B_\varepsilon(x_0) \\ F := \varepsilon \chi^{\alpha}_{\omega_\varepsilon(x_0)} k^2 \nabla u^0(x_0) \cdot w^\varepsilon & \text{in } B_\varepsilon(x_0) \setminus \omega_\varepsilon(x_0) \\ F + \mathrm{i}\mathrm{Im}(\alpha) k^2 u^0 & \text{in } \omega_\varepsilon(x_0), \end{cases} \tag{64}$$

$$[\![q^\varepsilon]\!] = 0, \quad [\chi^{\mathrm{Re}(\alpha)}_{\omega_\varepsilon(x_0)} \tfrac{\partial q^\varepsilon}{\partial n}] = -(1 - \mathrm{Re}(\alpha)) n \cdot (b^0_u + \nabla U^0_1) \quad \text{on } \partial\omega_\varepsilon(x_0), \tag{65}$$

$$\tfrac{\partial q^\varepsilon}{\partial n} = -\varepsilon \nabla u^0(x_0)^\top D w^\varepsilon n \quad \text{on } \Gamma_N, \quad q^\varepsilon = -\varepsilon \nabla u^0(x_0) \cdot w^1 \quad \text{on } \Gamma_D. \tag{66}$$

Theorem 4 will be applied for the two-parameter asymptotic expansion of a geometry-dependent objective function as described in the following section.

3 Inverse Helmholtz Problem in Inhomogeneous Medium

In the inverse setting of the Helmholtz problem in inhomogeneous medium, the geometry variables $(\omega^\star, \varepsilon^\star, x^\star) \in G$ and the refractive index $\alpha^\star \in \mathbf{C}_+$ of an unknown inhomogeneity $\omega^\star_{\varepsilon^\star}(x^\star)$ being tested in the reference domain Ω are to be identified and reconstructed from the known boundary measurement $u^\star \in L^2(\Gamma_N; \mathbf{C})$.

For variation of the topology, a trial inhomogeneity $\omega_\varepsilon(x_0)$ with admissible $(\omega, \varepsilon, x_0) \in G$ and $\alpha \in \mathbf{C}_+$ is put in Ω. For such trial variables we find a family of solutions $u^{(\varepsilon,\alpha)}$ to the forward Helmholtz problem (5) which determines the objective function of the misfit at the boundary

$$J : G \times \mathbf{C}_+ \mapsto \mathbf{R}_+, \quad J(\omega, \varepsilon, x_0, \alpha) := \tfrac{1}{2} \int_{\Gamma_N} |u^{(\varepsilon,\alpha)} - u^\star|^2 \, dS_x. \tag{67}$$

The objective (67) forces the state-constrained *topology optimization problem*: Find $(\omega^\star, \varepsilon^\star, x^\star, \alpha^\star) \in G \times \mathbf{C}_+$ such that

$$J(\omega^\star, \varepsilon^\star, x^\star, \alpha^\star) = \min_{(\omega,\varepsilon,x_0,\alpha) \in G \times \mathbf{C}_+} J(\omega, \varepsilon, x_0, \alpha) \quad \text{subject to (5)}. \tag{68}$$

If the test variables $(\omega^\star, \varepsilon^\star, x^\star, \alpha^\star) \in G \times \mathbf{C}_+$ are feasible, then the trivial minimum in (68) is attained at the solution $u^{(\varepsilon^\star,\alpha^\star)}$ of (5) because $u^{(\varepsilon^\star,\alpha^\star)} = u^\star$ at Γ_N in (67). Uniqueness of the minimum is open. Therefore,

Theorem 5 *For feasible measurement $u^\star = u^{(\varepsilon^\star,\alpha^\star)}$ at Γ_N, a solution to the inverse Helmholtz problem in inhomogeneous medium exists implying the trivial minimum in (68).*

While $u^{(\varepsilon,\alpha)}$ in (67) implies the primal state variable, a dual state variable $v^{(\varepsilon,\alpha)}$ associates a Fenchel–Legendre duality corresponding to the variational principle:

$$\mathscr{L}(u^{(\varepsilon,\alpha)}, v^{(\varepsilon,\alpha)}) = \min_{\mathrm{Re}(u),\mathrm{Re}(v)} \max_{\mathrm{Im}(u),\mathrm{Im}(v)} \mathscr{L}(u, v) \tag{69}$$

$$\text{over } u, v \in H^1(\Omega; \mathbf{C}) \text{ such that } u = h, v = 0 \text{ on } \Gamma_D,$$

where the Lagrangian $\mathscr{L}$ has the form (compare with $\mathscr{P}$ in (9)):

$$\mathscr{L}(u, v) := \mathrm{Re}\Big\{\int_\Omega (\chi^{\mathrm{Re}(\alpha)}_{\omega_\varepsilon(x_0)} \nabla u \cdot \nabla v - \chi^{\alpha}_{\omega_\varepsilon(x_0)} k^2 uv)\, dx - \int_{\Gamma_N} gv\, dS_x + \tfrac{1}{2}\int_{\Gamma_N} (u - u^\star)^2\, dS_x\Big\}. \tag{70}$$

Theorem 6 *The first-order necessary optimality conditions for the minimax problem (69) imply the primal problem (5) and the dual Helmholtz problem in inhomogeneous medium: Find $v^{(\varepsilon,\alpha)} \in H^1(\Omega; \mathbf{C})$ such that $v^{(\varepsilon,\alpha)} = 0$ at Γ_D and*

$$\int_\Omega (\chi^{\mathrm{Re}(\alpha)}_{\omega_\varepsilon(x_0)} \nabla v^{(\varepsilon,\alpha)} \cdot \nabla \overline{u} - \chi^{\alpha}_{\omega_\varepsilon(x_0)} k^2 v^{(\varepsilon,\alpha)} \overline{u})\, dx = -\int_{\Gamma_N} (u^{(\varepsilon,\alpha)} - u^\star)\overline{u}\, dS_x \tag{71}$$

for all test-functions $u \in H^1(\Omega; \mathbf{C})$ such that $u = 0$ on Γ_D.

Proof Applying variational calculus in the manner of Theorem 1, the first order optimality condition for (69) necessitates four variational inequalities:

$$\Big\langle \frac{\partial \mathscr{L}(u^{(\varepsilon,\alpha)}, v^{(\varepsilon,\alpha)})}{\partial \mathrm{Re}(v)}, \mathrm{Re}(v - v^{(\varepsilon,\alpha)})\Big\rangle \ge 0, \quad \Big\langle \frac{\partial \mathscr{L}(u^{(\varepsilon,\alpha)}, v^{(\varepsilon,\alpha)})}{\partial \mathrm{Im}(v)}, \mathrm{Im}(v - v^{(\varepsilon,\alpha)})\Big\rangle \le 0,$$
$$\Big\langle \frac{\partial \mathscr{L}(u^{(\varepsilon,\alpha)}, v^{(\varepsilon,\alpha)})}{\partial \mathrm{Re}(u)}, \mathrm{Re}(u - u^{(\varepsilon,\alpha)})\Big\rangle \ge 0, \quad \Big\langle \frac{\partial \mathscr{L}(u^{(\varepsilon,\alpha)}, v^{(\varepsilon,\alpha)})}{\partial \mathrm{Im}(u)}, \mathrm{Im}(u - u^{(\varepsilon,\alpha)})\Big\rangle \le 0,$$

holding for all $u, v \in H^1(\Omega; \mathbf{C})$ such that $u = h$, $v = 0$ on Γ_D. Inserting here $v = v^{(\varepsilon,\alpha)} \pm \widetilde{v}$ and $u = u^{(\varepsilon,\alpha)} \pm \widetilde{u}$ with $\widetilde{v}, \widetilde{u} \in H^1(\Omega; \mathbf{C})$ such that $\widetilde{v} = \widetilde{u} = 0$ on Γ_D we get four variational equations:

$$\Big\langle \frac{\partial \mathscr{L}(u^{(\varepsilon,\alpha)}, v^{(\varepsilon,\alpha)})}{\partial \mathrm{Re}(v)}, \mathrm{Re}(\widetilde{v})\Big\rangle = 0, \quad \Big\langle \frac{\partial \mathscr{L}(u^{(\varepsilon,\alpha)}, v^{(\varepsilon,\alpha)})}{\partial \mathrm{Im}(v)}, \mathrm{Im}(\widetilde{v})\Big\rangle = 0,$$
$$\Big\langle \frac{\partial \mathscr{L}(u^{(\varepsilon,\alpha)}, v^{(\varepsilon,\alpha)})}{\partial \mathrm{Re}(u)}, \mathrm{Re}(\widetilde{u})\Big\rangle = 0, \quad \Big\langle \frac{\partial \mathscr{L}(u^{(\varepsilon,\alpha)}, v^{(\varepsilon,\alpha)})}{\partial \mathrm{Im}(u)}, \mathrm{Im}(\widetilde{u})\Big\rangle = 0.$$

The summation of the first and the second equations for $\widetilde{v} = u$ and $\widetilde{v} = \mathrm{i}u$ constitutes the real and imaginary parts of (5), while the third and the fourth equations for $\widetilde{u} = u$ and $\widetilde{u} = \mathrm{i}u$ contribute to (71), respectively. This completes the proof. □

We note that the dual problem (71) is analogous to the primal problem (5) and differs by the boundary data given at $\partial\Omega$. Therefore, the asymptotic results stated for $u^{(\varepsilon,\alpha)}$ remains true also for $v^{(\varepsilon,\alpha)}$.

In particular, if $\varepsilon = 0$, then (71) turns into the dual background problem stated in the reference domain Ω as follows (compare to (14)): Find $v^0 \in H^1(\Omega; \mathbf{C})$ such that $v^0 = 0$ at Γ_D and

$$\int_{\Omega} (\nabla v^0 \cdot \nabla \overline{u} - k^2 v^0 \overline{u})\, dx = - \int_{\Gamma_N} (u^0 - u^\star)\overline{u}\, dS_x \tag{72}$$

for all test-functions $u \in H^1(\Omega; \mathbf{C})$ such that $u = 0$ on Γ_D.

which implies the weak solution to (cf. (11)–(13)):

$$-[\Delta + k^2]v^0 = 0 \quad \text{in } \Omega, \tag{73}$$

$$\frac{\partial v^0}{\partial n} = -(u^0 - u^\star) \quad \text{on } \Gamma_N, \tag{74}$$

$$v^0 = 0 \quad \text{on } \Gamma_D. \tag{75}$$

Theorem 2 applied to the problem (73)–(75) provides the near-field representation in $B_R(x_0) \subset \Omega$:

$$v^0(x) = v^0(x_0)J_0(k\rho) + V_0^0(x), \quad V_0^0(x) = \tfrac{2}{k}J_1(k\rho)\nabla v^0(x_0)\cdot\widehat{x} + V_1^0(x), \tag{76}$$

with the residuals $V_0^0, V_1^0 \in H^1(B_R(x_0); \mathbf{C})$ such that

$$\int_{-\pi}^{\pi} V_0^0\, d\theta = 0, \quad \int_{-\pi}^{\pi} V_1^0\, d\theta = \int_{-\pi}^{\pi} V_1^0 \widehat{x}\, d\theta = 0, \tag{77}$$

$$V_0^0 = \mathrm{O}(\rho), \quad V_1^0 = \mathrm{O}(\rho^2), \tag{78}$$

and similar to (25) and (26) representation of the gradient in $B_R(x_0)$:

$$\begin{aligned}\nabla v^0(x) =& \ \nabla v^0(x_0) + b_v^0(x) + \nabla V_1^0(x),\\ b_v^0(x) :=& \ (v^0(x_0)ka_0'(k\rho) + a_1'(k\rho)\nabla v^0(x_0)\cdot\widehat{x})\widehat{x} + \tfrac{a_1(k\rho)}{k\rho}(\nabla v^0(x_0)\cdot\widehat{x}')\widehat{x}',\end{aligned} \tag{79}$$

$$\nabla V_0^0 = \mathrm{O}(1), \quad \nabla V_1^0 = \mathrm{O}(\rho), \quad b_v^0 = \mathrm{O}(\rho). \tag{80}$$

Since the topological variables $(\omega, \varepsilon, x_0, \alpha)$ enter the objective J in fully implicit way through the solution $u^{(\varepsilon,\alpha)}$ of the state problem, further we get an explicit expansion of J applying asymptotic arguments as the size $\varepsilon \searrow +0$.

3.1 High-Order Topological Expansion of the Objective

We decompose $u^{(\varepsilon,\alpha)} - u^\star = u^{(\varepsilon,\alpha)} - u^0 + u^0 - u^\star$ and using the Neumann boundary condition (74) we express the objective in (67) equivalently as

$$J(\omega, \varepsilon, x_0, \alpha) = J_0 + \mathrm{Re}\{\mathscr{I}(u^{(\varepsilon,\alpha)} - u^0, v^0)\} + J_{(\varepsilon,\alpha)}, \tag{81}$$

with the following integral terms:

$$J_0 := \frac{1}{2}\int_{\Gamma_N} |u^0 - u^\star|^2\, dS_x, \quad J_{(\varepsilon,\alpha)} := \frac{1}{2}\int_{\Gamma_N} |u^{(\varepsilon,\alpha)} - u^0|^2\, dS_x \tag{82}$$

$$\mathscr{I}(u^{(\varepsilon,\alpha)} - u^0, v^0) := -\int_{\Gamma_N} (u^{(\varepsilon,\alpha)} - u^0)\frac{\partial \overline{v^0}}{\partial n}\, dS_x. \tag{83}$$

From (81)–(83) we infer the asymptotic result below.

Theorem 7 *The objective in (67) admits the two-parameter topological expansion*

$$\begin{aligned} J(\omega,\varepsilon,x_0,\alpha) = \; & J_0 + \mathrm{Re}\big\{\varepsilon^2 J_1(\omega,x_0,\alpha) + J_2^{(\varepsilon,\alpha)} + J_3^{(\varepsilon,\alpha)} + J_4^{(\varepsilon,\alpha)}\big\} \\ & + \mathrm{O}\big(|1-\mathrm{Re}(\alpha)|^2\varepsilon^4|\ln\varepsilon|\big), \end{aligned} \tag{84}$$

with the first-order asymptotic term implying the topological derivative:

$$J_1(\omega,x_0,\alpha) := -\nabla u^0(x_0)^\top A_{(\omega,\mathrm{Re}(\alpha))}\nabla\overline{v^0}(x_0) + (1-\alpha)k^2|\omega|u^0(x_0)\overline{v^0}(x_0) \tag{85}$$

and the high-order asymptotic terms expressed by the following formulas:

$$\begin{aligned} J_2^{(\varepsilon,\alpha)} &:= \varepsilon^3\sqrt{|\ln\varepsilon|}\int_{-\pi}^{\pi} \nabla u^0(x_0)\cdot\big(\tfrac{\partial u^1}{\partial\rho}\overline{v^0}(x_0) - u^1(\nabla\overline{v^0}(x_0)\cdot\widehat{x})\big)\, d\theta \\ &= \mathrm{O}\big(|1-\mathrm{Re}(\alpha)|\varepsilon^3\sqrt{|\ln\varepsilon|}\big), \end{aligned} \tag{86}$$

$$\begin{aligned} J_3^{(\varepsilon,\alpha)} &:= -\int_{\omega_\varepsilon(x_0)} \big((1-\mathrm{Re}(\alpha))\nabla q^\varepsilon\cdot\nabla\overline{v^0} - (1-\alpha)k^2 q^\varepsilon\overline{v^0}\big)\, dx \\ &-(1-\mathrm{Re}(\alpha))\int_{\partial\omega_\varepsilon(x_0)} n\cdot(b_u^0 + \nabla U_1^0)\overline{V_0^0}\, dS_x - \mathrm{i}\mathrm{Im}(\alpha)\int_{\omega_\varepsilon(x_0)} k^2(U_0^0\overline{v^0}(x_0) \\ &+ u^0(x_0)\overline{V_0^0})\, dx + \varepsilon^2\int_{-\pi}^{\pi} \nabla u^0(x_0)\cdot\big(\tfrac{\partial W^\varepsilon}{\partial\rho}\overline{V_1^0} - W^\varepsilon\tfrac{\partial\overline{V_1^0}}{\partial\rho}\big)d\theta \\ &-\varepsilon\int_{B_\varepsilon(x_0)} \chi^\alpha_{\omega_\varepsilon(x_0)} k^2\nabla u^0(x_0)\cdot w^\varepsilon\overline{v^0}\, dx = \mathrm{O}\{(|1-\mathrm{Re}(\alpha)|(|1-\alpha|+\mathrm{Re}(\alpha)))\varepsilon^3\}, \end{aligned} \tag{87}$$

$$\begin{aligned} J_4^{(\varepsilon,\alpha)} &:= \varepsilon^3\sqrt{|\ln\varepsilon|}\int_{-\pi}^{\pi} \nabla u^0(x_0)\cdot\big\{\tfrac{\partial u^1}{\partial\rho}\varepsilon(\nabla\overline{v^0}(x_0)\cdot\widehat{x}) \\ &-u^1\big(\overline{v^0}(x_0)ka_0' + \tfrac{\partial\overline{V_1^0}}{\partial\rho}\big)\big\}d\theta = \mathrm{O}\big(|1-\mathrm{Re}(\alpha)|\varepsilon^4\sqrt{|\ln\varepsilon|}\big). \end{aligned} \tag{88}$$

Proof Due to $u^1 = \mathrm{O}(|1-\mathrm{Re}(\alpha)|)$ in (59), the asymptotic rate of $J_{(\varepsilon,\alpha)}$ in (82) is

$$J_{(\varepsilon,\alpha)} = \mathrm{O}\big(|1-\mathrm{Re}(\alpha)|^2\varepsilon^4|\ln\varepsilon|\big). \tag{89}$$

To verify (84), the boundary integral $\mathscr{I}$ in (83) should be expanded up to the order in (89). For this task we employ the second Green's formula in $\Omega\setminus\overline{B_\varepsilon(x_0)}$ and rewrite

$$\mathscr{I}(u^{(\varepsilon,\alpha)} - u^0, v^0) = \int_{\Omega\setminus B_\varepsilon(x_0)} \big(\Delta(u^{(\varepsilon,\alpha)} - u^0)\overline{v^0} - (u^{(\varepsilon,\alpha)} - u^0)\Delta\overline{v^0}\big)\, dx + \int_{\partial B_\varepsilon(x_0)} \big(\tfrac{\partial(u^{(\varepsilon,\alpha)}-u^0)}{\partial\rho}\overline{v^0} - (u^{(\varepsilon,\alpha)} - u^0)\tfrac{\partial\overline{v^0}}{\partial\rho}\big)\, dS_x,$$

where the domain integral vanishes due to Helmholtz equations (1), (11), and (73).

We note that $\mathscr{I}$ is an invariant integral which can be written over arbitrary Lipschitz boundary $\partial\mathscr{O}$ of a domain $\mathscr{O}$ such that $\omega_\varepsilon(x_0) \subseteq \mathscr{O} \subset \Omega$. The integral over the circle $\partial B_\varepsilon(x_0)$ is advantageous when calculated analytically by substituting the uniform expansion (59) for $u^{(\varepsilon,\alpha)} - u^0$ and the inner expansion (76) for v^0.

The plan is to decompose $\mathscr{I}(u^{(\varepsilon,\alpha)} - u^0, v^0) = \mathscr{I}(q^\varepsilon, v^0) + \mathscr{I}(u^{(\varepsilon,\alpha)} - u^0 - q^\varepsilon, v^0)$ with the residual q^ε from Theorem 4 and calculate two integrals separately.

First, applying to $\mathscr{I}(q^\varepsilon, v^0)$ the second Green's formula in $B_\varepsilon(x_0) \setminus \overline{\omega_\varepsilon(x_0)}$ due to the Helmholtz equations (64) and the jump condition (65) for q^ε, we get

$$\mathscr{I}(q^\varepsilon, v^0) = \int_{\partial B_\varepsilon(x_0)} \big(\tfrac{\partial q^\varepsilon}{\partial\rho}\overline{v^0} - q^\varepsilon\tfrac{\partial\overline{v^0}}{\partial\rho}\big)\, dS_x = \int_{\partial\omega_\varepsilon(x_0)^+} \big(\tfrac{\partial q^\varepsilon}{\partial n}\overline{v^0} - q^\varepsilon\tfrac{\partial\overline{v^0}}{\partial n}\big)\, dS_x$$
$$-\varepsilon \int_{B_\varepsilon(x_0)\setminus\omega_\varepsilon(x_0)} k^2(\nabla u^0(x_0)\cdot w^\varepsilon)\overline{v^0}\, dx = -\varepsilon \int_{B_\varepsilon(x_0)\setminus\omega_\varepsilon(x_0)} k^2(\nabla u^0(x_0)\cdot w^\varepsilon)\overline{v^0}\, dx$$
$$+ \int_{\partial\omega_\varepsilon(x_0)^-} \Big(\big(\mathrm{Re}(\alpha)\tfrac{\partial q^\varepsilon}{\partial n} - (1-\mathrm{Re}(\alpha))n\cdot(b_u^0 + \nabla U_1^0)\big)\overline{v^0} - q^\varepsilon\tfrac{\partial\overline{v^0}}{\partial n}\Big)\, dS_x.$$

Further integration by part in $\omega_\varepsilon(x_0)$ of this expression gives

$$\mathscr{I}(q^\varepsilon, v^0) = -\int_{\omega_\varepsilon(x_0)} \big((1-\mathrm{Re}(\alpha))\nabla q^\varepsilon\cdot\nabla\overline{v^0} - (1-\alpha)k^2 q^\varepsilon\overline{v^0}\big)\, dx$$
$$-\mathrm{i}\,\mathrm{Im}(\alpha)\int_{\omega_\varepsilon(x_0)} k^2 u^0\overline{v^0}\, dx - (1-\mathrm{Re}(\alpha))\int_{\partial\omega_\varepsilon(x_0)} n\cdot(b_u^0 + \nabla U_1^0)\overline{v^0}\, dS_x$$
$$-\varepsilon\int_{B_\varepsilon(x_0)} \chi^\alpha_{\omega_\varepsilon(x_0)} k^2(\nabla u^0(x_0)\cdot w^\varepsilon)\overline{v^0}\, dx.$$

Expansions (21) for u^0 and (76) for v^0 in the second integral here proceed

$$\int_{\omega_\varepsilon(x_0)} u^0\overline{v^0}\, dx = \varepsilon^2|\omega|u^0(x_0)\overline{v^0}(x_0) + \int_{\omega_\varepsilon(x_0)} \big(U_0^0\overline{v^0}(x_0) + u^0(x_0)\overline{V_0^0} + U_0^0\overline{V_0^0}\big)\, dx,$$

while in the third integral, applying the divergence theorem, this leads to

$$\int_{\partial\omega_\varepsilon(x_0)} n\cdot(b_u^0 + \nabla U_1^0)\overline{v^0}\, dS_x = \int_{\partial\omega_\varepsilon(x_0)} n\cdot(b_u^0 + \nabla U_1^0)\overline{V_0^0}\, dS_x$$
$$-\varepsilon^2 k^2|\omega|u^0(x_0)\overline{v^0}(x_0) + \int_{\omega_\varepsilon(x_0)} \Big(\mathrm{div}\big(b_u^0 + u^0(x_0)\tfrac{k^2\rho\widehat{x}}{2}\big) - k^2 U_1^0\Big)\overline{v^0}\, dx$$

due to $\mathrm{div}(\nabla U_1^0) = -k^2 U_1^0$ and $\mathrm{div}(\frac{\rho\widehat{x}}{2}) = 1$, since $b_u^0 = -u^0(x_0)\frac{k^2\rho\widehat{x}}{2} + \mathrm{O}(\rho^2)$ according to (18) and (25).

Gathering like terms in view of the following asymptotic relations:

$$U_0^0 = V_0^0 = b_u^0 = \mathrm{O}(\varepsilon),\quad U_1^0 = b_u^0 + u^0(x_0)\frac{k^2\rho\widehat{x}}{2} = \mathrm{O}(\varepsilon^2)\quad \text{in } B_\varepsilon(x_0),$$
$$\|w^\varepsilon\|_{H^1(\omega_\varepsilon(x_0);\mathbb{C})} = \mathrm{O}(|1-\mathrm{Re}(\alpha)|),\quad \|q^\varepsilon\|_{H^1(\omega_\varepsilon(x_0);\mathbb{C})} = \mathrm{O}(|1-\alpha|\varepsilon^2),$$

we get the two-parameter representation of the first integral

$$\mathscr{I}(q^\varepsilon, v^0) = I_1^q + I_2^q + I_3^q, \tag{90}$$

with the following asymptotic terms:

$$I_1^q := (1-\alpha)\varepsilon^2 k^2 |\omega| u^0(x_0)\overline{v^0}(x_0),$$
$$I_2^q := -\int_{\omega_\varepsilon(x_0)} \big((1-\mathrm{Re}(\alpha))\nabla q^\varepsilon \cdot \nabla\overline{v^0} - (1-\alpha)k^2 q^\varepsilon \overline{v^0}\big)\,dx$$
$$-(1-\mathrm{Re}(\alpha))\int_{\partial\omega_\varepsilon(x_0)} n\cdot(b_u^0 + \nabla U_1^0)\overline{V_0^0}\,dS_x - \mathrm{i}\mathrm{Im}(\alpha)\int_{\omega_\varepsilon(x_0)} k^2(U_0^0\overline{v^0}(x_0)$$
$$+u^0(x_0)\overline{V_0^0})\,dx - \varepsilon\int_{B_\varepsilon(x_0)} \chi^\alpha_{\omega_\varepsilon(x_0)} k^2(\nabla u^0(x_0)\cdot w^\varepsilon)\overline{v^0}\,dx$$
$$= \mathrm{O}\big\{\big(|1-\mathrm{Re}(\alpha)|(|1-\alpha| + \mathrm{Re}(\alpha))\big)\varepsilon^3\big\},$$
$$I_3^q := -(1-\mathrm{Re}(\alpha))\int_{\omega_\varepsilon(x_0)} \Big(\mathrm{div}\big(b_u^0 + u^0(x_0)\tfrac{k^2\rho\widehat{x}}{2}\big) - k^2 U_1^0\Big)\overline{v^0}\,dx$$
$$-\mathrm{i}\mathrm{Im}(\alpha)\int_{\omega_\varepsilon(x_0)} k^2 U_0^0\overline{V_0^0}\,dx = \mathrm{O}\big(|1-\alpha|\varepsilon^4\big).$$

Second, inserting in $\mathscr{I}$ the asymptotic representations (41) for w^ε, (59) for q^ε, (76) and (79) for v^0 and ∇v^0, we have

$$\mathscr{I}(u^\varepsilon - u^0 - q^\varepsilon, v^0) = \int_{\partial B_\varepsilon(x_0)} \big(\tfrac{\partial(u^\varepsilon - u^0 - q^\varepsilon)}{\partial\rho}\overline{v^0} - (u^\varepsilon - u^0 - q^\varepsilon)\tfrac{\partial\overline{v^0}}{\partial\rho}\big)\,dS_x$$
$$= \varepsilon^2\int_{-\pi}^{\pi} \nabla u^0(x_0)\cdot\big\{\big(-\tfrac{A(\omega,\mathrm{Re}(\alpha))\widehat{x}}{2\pi\varepsilon} + \tfrac{\partial W^\varepsilon}{\partial\rho} + \varepsilon\sqrt{|\ln\varepsilon|}\tfrac{\partial u^1}{\partial\rho}\big)\big(\overline{v^0}(x_0)(1+a_0) + (\varepsilon + \tfrac{a_1}{k})$$
$$\times\nabla\overline{v^0}(x_0)\cdot\widehat{x} + \overline{V_1^0}\big) - \big(\tfrac{A(\omega,\mathrm{Re}(\alpha))\widehat{x}}{2\pi} + W^\varepsilon + \varepsilon\sqrt{|\ln\varepsilon|}u^1\big)\big(\nabla\overline{v^0}(x_0) + \overline{b_v^0} + \nabla\overline{V_1^0}\big)\cdot\widehat{x}\big\}\,d\theta.$$

To calculate the terms here we use the orthogonality (42) for W^ε, (77) for V_1^0, and the following asymptotic relations due to (18)–(20), (43), (80), which hold at $\partial B_\varepsilon(x_0)$:

$$\overline{b_v^0}\cdot\widehat{x} = \overline{v^0}(x_0)ka_0' + a_1'\nabla\overline{v^0}(x_0)\cdot\widehat{x},\quad u^1 = W^\varepsilon = \mathrm{O}(|1-\mathrm{Re}(\alpha)|),$$
$$a_0 = \mathrm{O}(\varepsilon^2),\quad a_0' = \mathrm{O}(\varepsilon),\quad a_1 = \mathrm{O}(\varepsilon^3),\quad a_1' = \mathrm{O}(\varepsilon^2).$$

Collecting like terms, as the result we get

$$\mathscr{I}(u^\varepsilon - u^0 - q^\varepsilon, v^0) = -I_1^w + I_2^w + I_3^w + I_4^w + I_5^w, \tag{91}$$

where the integral terms are

$$
\begin{aligned}
I_1^w &:= \varepsilon^2 \int_{-\pi}^{\pi} \nabla u^0(x_0)^\top \tfrac{A_{(\omega,\mathrm{Re}(\alpha))}\widehat{x}}{\pi} (\nabla \overline{v^0}(x_0) \cdot \widehat{x})\, d\theta = \varepsilon^2 \nabla u^0(x_0)^\top A_{(\omega,\mathrm{Re}(\alpha))} \nabla \overline{v^0}(x_0),\\
I_2^w &:= \varepsilon^3 \sqrt{|\ln \varepsilon|} \int_{-\pi}^{\pi} \nabla u^0(x_0) \cdot \big(\tfrac{\partial u^1}{\partial \rho} \overline{v^0}(x_0) - u^1 (\nabla \overline{v^0}(x_0) \cdot \widehat{x})\big)\, d\theta\\
&= \mathrm{O}\big(|1-\mathrm{Re}(\alpha)|\varepsilon^3 \sqrt{|\ln \varepsilon|}\big),\\
I_3^w &:= \varepsilon^2 \int_{-\pi}^{\pi} \nabla u^0(x_0) \cdot \big(\tfrac{\partial W^\varepsilon}{\partial \rho} \overline{V_1^0} - W^\varepsilon \tfrac{\partial \overline{V_1^0}}{\partial \rho}\big) d\theta = \mathrm{O}(|1-\mathrm{Re}(\alpha)|\varepsilon^3),\\
I_4^w &:= \varepsilon^3 \sqrt{|\ln \varepsilon|} \int_{-\pi}^{\pi} \nabla u^0(x_0) \cdot \big\{\tfrac{\partial u^1}{\partial \rho} \varepsilon \nabla \overline{v^0}(x_0) \cdot \widehat{x} - u^1 \big(\overline{v^0}(x_0) k a_0' + \tfrac{\partial \overline{V_1^0}}{\partial \rho}\big)\big\} d\theta\\
&= \mathrm{O}\big(|1-\mathrm{Re}(\alpha)|\varepsilon^4 \sqrt{|\ln \varepsilon|}\big),\\
I_5^w &:= \varepsilon^3 \sqrt{|\ln \varepsilon|} \int_{-\pi}^{\pi} \nabla u^0(x_0) \cdot \big\{\tfrac{\partial u^1}{\partial \rho} \big(\overline{v^0}(x_0) a_0 + \tfrac{a_1}{k\varepsilon} \nabla \overline{v^0}(x_0) \cdot \widehat{x} + \overline{V_1^0}\big)\\
&\quad - u^1 (a_1' \nabla \overline{v^0}(x_0) \cdot \widehat{x})\big\} d\theta + \tfrac{1}{2}\big(\tfrac{a_1}{\varepsilon k} + a_1'\big) I_1^w = \mathrm{O}(|1-\mathrm{Re}(\alpha)|\varepsilon^4).
\end{aligned}
$$

Finally, summing up $I_1^q - I_1^w = \varepsilon^2 J_1$, $I_2^w = J_2^{(\varepsilon,\alpha)}$, $I_3^q + I_3^w = J_3^{(\varepsilon,\alpha)}$, and $I_4^w = J_4^{(\varepsilon,\alpha)}$ in (89)–(91) gathers the asymptotic terms (84)–(88) and finishes the proof. □

We note that, for fixed α, from Theorem 7 it follows the topological derivative of the objective J following the terminology of [34]:

$$\mathrm{Re}\{J_1(\omega, x_0, \alpha)\} = \lim_{\varepsilon \searrow +0} \tfrac{1}{\varepsilon^2} \big(J(\omega, \varepsilon, x_0, \alpha) - J_0\big), \tag{92}$$

where $J_1(\omega, x_0, \alpha)$ is the first-order asymptotic term in (84). It is given explicitly by formula (85). For real $\alpha \in \mathbf{R}_+$, the topological derivative was found in [3, 5].

However, for fixed ε and complex $\alpha \in \mathbf{C}_+$, the leading term in (84) may change when varying the refractive index $|\alpha| \nearrow \infty$.

In fact, taking the limit as $\mathrm{Im}(\alpha) \nearrow \infty$, which corresponds to the sound-soft obstacle $\omega_\varepsilon(x_0)$ described by problem (10), in the next section we derive a derivative-free necessary optimality condition of the topology optimization (68).

3.2 Zero-Order Necessary Optimality Condition

Due to Theorem 5, the optimal objective $J(\omega^\star, \varepsilon^\star, x^\star, \alpha^\star) = 0$ in (68) for all feasible test parameters $(\omega^\star, \varepsilon^\star, x^\star, \alpha^\star) \in G \times \mathbf{C}_+$. We rewrite formula (84) of J due to (85):

$$J(\omega^\star,\varepsilon^\star,x^\star,\alpha^\star) = J_0 + (\varepsilon^\star)^2\Big(\mathrm{Re}\big\{-\nabla u^0(x^\star)^\top A_{(\omega^\star,\mathrm{Re}(\alpha^\star))}\nabla\overline{v^0}(x^\star)$$
$$+(1-\mathrm{Re}(\alpha^\star))k^2|\omega^\star|u^0(x^\star)\overline{v^0}(x^\star)\big\} + \mathrm{Im}(\alpha^\star)k^2|\omega^\star|\mathrm{Im}\{u^0(x^\star)\overline{v^0}(x^\star)\}\Big)$$
$$+\mathrm{O}\big(|1-\mathrm{Re}(\alpha^\star)|(\varepsilon^\star)^3\sqrt{|\ln\varepsilon^\star|}\big). \quad (93)$$

However, if $\mathrm{Im}(\alpha^\star) \nearrow \infty$, then zero optima of $J(\omega^\star,\varepsilon^\star,x^\star,\alpha^\star)$ can be preserved only when the complement $\mathrm{Im}\{u^0(x^\star)\overline{v^0}(x^\star)\}$ to $\mathrm{Im}(\alpha^\star)$ in (93) is zero.

This argument holds true also for finite optima $0 \neq J(\omega^\star,\varepsilon^\star,x^\star,\alpha^\star) < \infty$ when the test parameters are infeasible. Thus, we have proved the following result.

Theorem 8 *For test geometries* $(\omega^\star,\varepsilon^\star,x^\star)$ *and refractive index* $\mathrm{Im}(\alpha^\star) \nearrow \infty$, *the zero-order necessary optimality condition of (68) implies*

$$\mathrm{Im}\{u^0(x^\star)\overline{v^0}(x^\star)\} = 0 \quad (94)$$

expressed by the solution u^0 *to the primal background problem (14) and the solution* v^0 *to the dual background problem (72).*

Based on Theorem 8, in [24] the imaging function is introduced

$$f : \{u^\star \in G_u\} \mapsto C(\Omega;\mathbf{R}), \quad f_{u^\star}(x) := \mathrm{Im}\{u^0(x)\overline{v^0}(x)\} \quad (95)$$

over the set of feasible boundary measurements $u^\star \in G_u \subset L^2(\Gamma_N;\mathbf{C})$. Based on (95), the zero-level set $L_{=0}$ of $f_{u^\star}$ contains the test center $x^\star$:

$$x^\star \in L_{=0}(f_{u^\star}) := \{x \in \Omega : \quad f_{u^\star}(x) = 0\}. \quad (96)$$

The level-set description (96) based on the zero-order optimality condition (95) was implemented numerically and reported in [24]. We remark the following:

- Implementation of the imaging function $f_{u^\star}$ in (95) has low computational costs, since it needs to solve the background Helmholtz problems (14) and (72) in homogeneous medium.
- The center $x^\star$ can be detected as the intersection point of the zero-level sets

$$x^\star = \bigcap_{i=1}^d L_{=0}(f_{u_i^\star}) \quad (97)$$

  from two or three different measurements $u_i^\star$ in 2-D or 3-D, respectively.
- For low wave numbers k, formula (97) holds for arbitrary geometries $(\omega^\star,\varepsilon^\star,x^\star)$.
- The numerical result by (97) is stable to discretization and noise errors.
- Although Theorem 8 is stated for $\mathrm{Im}(\alpha^\star) \nearrow \infty$, formulas (95)–(97) are applicable numerically as well to finite refractive indexes $\alpha^\star \in \mathbf{C}_+$.

Finally, we illustrate formulas (95) and (97) analytically in 1-D and numerically in 2-D.

3.3 The Analytic Solution in One Dimension

Let $\Omega = (-R, R) \subset \mathbf{R}^1$ with $R > 0$, the size $\varepsilon > 0$ and the center x_0 be such that $|x_0| < R - \varepsilon$ implying the inhomogeneity $\omega_\varepsilon(x_0) = (x_0 - \varepsilon, x_0 + \varepsilon) \subset (-R, R)$ with the given refractive index $\alpha \in \mathbf{C}_+$.

Relying on the plane wave $u^0(x) = \exp(\mathrm{i}kx)$, we have the Neumann data $g(\pm R) = \pm \mathrm{i}k \exp(\pm \mathrm{i}kR)$. The Helmholtz problem (1)–(4) in this inhomogeneous medium turns into

$$
\begin{aligned}
-\big[\tfrac{d^2}{dx^2} + k^2\big]u^{(\varepsilon,\alpha)}(x) = 0 \quad &\text{for } x \in (-R, x_0 - \varepsilon) \cup (x_0 + \varepsilon, R), \\
\tfrac{du^{(\varepsilon,\alpha)}}{d(-x)}(-R) = g(-R), \quad &\tfrac{du^{(\varepsilon,\alpha)}}{dx}(R) = g(R), \qquad (98) \\
-\big[\mathrm{Re}(\alpha)\tfrac{d^2}{dx^2} + \alpha k^2\big]u^{(\varepsilon,\alpha)}(x) = 0 \quad &\text{for } |x - x_0| < \varepsilon, \qquad (99) \\
u^{(\varepsilon,\alpha)}((x_0 - \varepsilon)^-) - u^{(\varepsilon,\alpha)}((x_0 - \varepsilon)^+) &= 0, \\
u^{(\varepsilon,\alpha)}((x_0 + \varepsilon)^+) - u^{(\varepsilon,\alpha)}((x_0 + \varepsilon)^-) &= 0, \\
\tfrac{du^{(\varepsilon,\alpha)}}{d(-x)}((x_0 - \varepsilon)^-) - \mathrm{Re}(\alpha)\tfrac{du^{(\varepsilon,\alpha)}}{d(-x)}((x_0 - \varepsilon)^+) &= 0, \\
\tfrac{du^{(\varepsilon,\alpha)}}{dx}((x_0 + \varepsilon)^+) - \mathrm{Re}(\alpha)\tfrac{du^{(\varepsilon,\alpha)}}{dx}((x_0 + \varepsilon)^-) &= 0. \qquad (100)
\end{aligned}
$$

The general solution to (98) and (99) can be expressed as

$$
u^{(\varepsilon,\alpha)}(x) = \begin{cases} u^0(x) + c_1 \cosh(\mathrm{i}k(x + R)) & \text{for } x \in (-R, x_0 - \varepsilon), \\ u^0(x) + c_2 \cosh(\mathrm{i}k(x - R)) & \text{for } x \in (x_0 + \varepsilon, R), \\ c_3 \cosh(k\beta(x - x_0)) + c_4 \sinh(k\beta(x - x_0)) & \text{for } |x - x_0| < \varepsilon, \end{cases} \qquad (101)
$$

where $\beta \in \mathbf{C}$ is the complex root of the equation $\mathrm{Re}(\alpha)\beta^2 + \alpha = 0$ given by

$$
\mathrm{Re}(\beta) = \sqrt{\tfrac{1}{2}\Big(\sqrt{1 + \big(\tfrac{\mathrm{Im}(\alpha)}{\mathrm{Re}(\alpha)}\big)^2} - 1\Big)}, \quad \mathrm{Im}(\beta) = -\tfrac{2\mathrm{Im}(\alpha)}{\mathrm{Re}(\alpha)\mathrm{Re}(\beta)},
$$

and the coefficient vector $c \in \mathbf{C}^4$ is determined from the four jump conditions (100) as the solution of the following 4-by-4 matrix equations:

$$
\begin{pmatrix} \cosh(\mathrm{i}k(x_0 - \varepsilon + R)) & 0 & -\cosh(-k\beta\varepsilon) & -\sinh(-k\beta\varepsilon) \\ -\mathrm{i}\sinh(\mathrm{i}k(x_0 - \varepsilon + R)) & 0 & \mathrm{Re}(\alpha)\beta\sinh(-k\beta\varepsilon) & \mathrm{Re}(\alpha)\beta\cosh(-k\beta\varepsilon) \end{pmatrix} c = \begin{pmatrix} b_1 \\ b_2 \end{pmatrix},
$$

$$
\begin{pmatrix} 0 & \cosh(\mathrm{i}k(x_0 + \varepsilon - R)) & -\cosh(k\beta\varepsilon) & -\sinh(k\beta\varepsilon) \\ 0 & \mathrm{i}\sinh(\mathrm{i}k(x_0 + \varepsilon - R)) & -\mathrm{Re}(\alpha)\beta\sinh(k\beta\varepsilon) & -\mathrm{Re}(\alpha)\beta\cosh(k\beta\varepsilon) \end{pmatrix} c = \begin{pmatrix} b_3 \\ b_4 \end{pmatrix},
$$

$$
b = \big(-\exp(\mathrm{i}k(x_0 - \varepsilon), \mathrm{i}\exp(\mathrm{i}k(x_0 - \varepsilon), -\exp(\mathrm{i}k(x_0 + \varepsilon), -\mathrm{i}\exp(\mathrm{i}k(x_0 + \varepsilon))\big)^{\top}.
$$

For the boundary measurement determined according to (101):

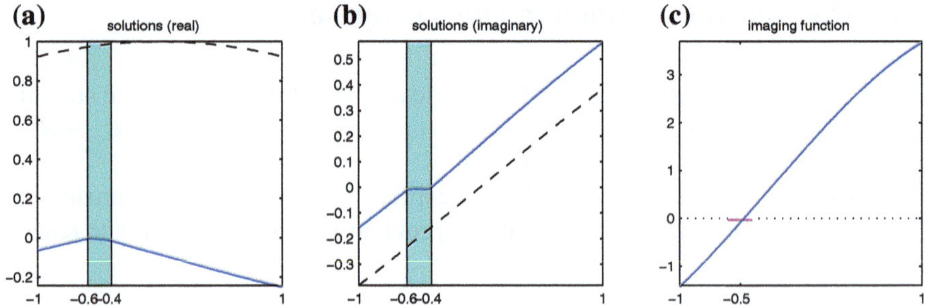

Fig. 1 The real part (**a**) and the imaginary part (**b**) of solutions of the Helmholtz problem in the homogeneous medium $u^0(x)$ (*dashed line*) and in the inhomogeneous medium $u^{(\varepsilon,\alpha)}(x)$ (*solid line*), and the imaging function $f_{u^\star}(x)$ (**c**) for the inhomogeneity $\omega_\varepsilon(x_0) = \{-0.6 < x < -0.4\}$ in $(-1, 1)$

$$u^\star(-R) = u^0(-R) + c_1, \quad u^\star(R) = u^0(R) + c_2,$$

the corresponding dual problem (72) possesses the solution

$$v^0(x) = \tfrac{1}{ik \sinh(i2kR)} \big(c_1 \cosh(ik(x - R)) + c_2 \cosh(ik(x + R))\big). \tag{102}$$

Using (102) the imaging function $f_{u^\star}(x) := \mathrm{Im}\{u^0(x)\overline{v^0}(x)\}$ can be calculated.

Based on these analytic formulas, in Fig. 1 we plot a typical result of numerical calculation of the problem for $R = 1, k = \frac{\pi}{2}, x_0 = -0.5, \varepsilon = 0.1$, and $\alpha = 2 + \mathrm{i}10^4$. The real and the imaginary parts of the solutions u^0 and $u^{(\varepsilon,\alpha)}$ are depicted in the plots (a) and (b), respectively, while in plot (c) we observe the imaging function $f_{u^\star}$ crossing at $x^\star \approx -0.48$ the inhomogeneity drawn along the x-axis.

3.4 *The Numerical Solution in Two Dimensions*

In Fig. 2, we show the numerical computation of the imaging function from (95) carried out in 2-D on the uniform grid with the mesh size $h = 2^{-6}$ over the unit square $\Omega = (0, 1)^2$. In this example, the shown inhomogeneity ω of size $\varepsilon = \sqrt{5}h$ centered at $x_0 = (3/8, 1/4)$ is sound-soft according to the particular case (10) of the forward Helmholtz problem.

Two boundary measurements $u_1^\star$ and $u_2^\star$ are synthesized when the inhomogeneity is illuminated by the plane wave $u^0(x) = e^{ik((x_1-1/2,x_2-1/2)\cdot(\cos\theta,\sin\theta))}$ in directions of $\theta_1 = \frac{\pi}{5}$ and $\theta_2 = \pi$, respectively. We note that the wave number should be chosen sufficiently small, in this example $k \le \frac{\pi}{2}$. In plots (a) and (b), there are depicted the corresponding imaging functions $f_{u_1^\star}$ and $f_{u_2^\star}$ interpolated linearly over the grid, together with its discrete zero-level sets $L_{=0}(f_{u_1^\star})$ and $L_{=0}(f_{u_2^\star})$ found by a narrow band technique, see e.g. [26]. The zero-level sets form straight lines crossing $\omega_\varepsilon(x_0)$.

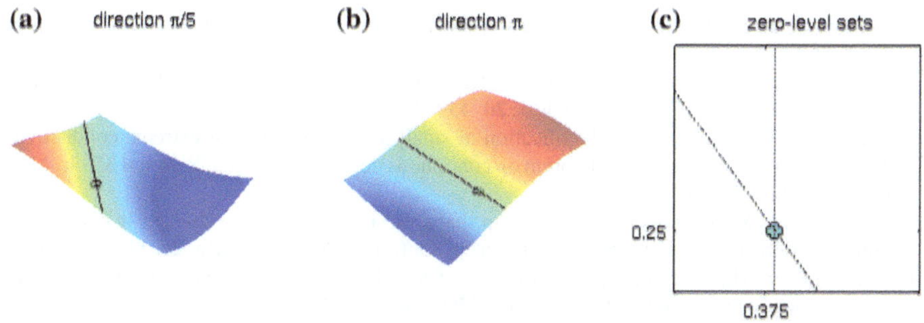

Fig. 2 Identification of the center x_0 of inhomogeneity from two boundary measurements in 2-D

They are depicted also in the test domain Ω in plot (c), where the intersection point $x^{\star}$ determined by (97) visually coincides with the test center x_0.

Further development of the optimality based imaging theory and numerical methods with respect to identification of the center of inhomogeneities in 2-D and 3-D from far-field measurements is the subject of a forthcoming paper.

Acknowledgments The results were obtained with the support of the Austrian Science Fund (FWF) project P26147-N26: "Object identification problems: numerical analysis" (PION), partial support of NAWI Graz, the Austrian Academy of Sciences (OeAW), and OeAD Scientific & Technological Cooperation (WTZ CZ 01/2016). The author thanks for invitation Organizers of the International Conference CoMFoS15: Mathematical Analysis of Continuum Mechanics and Industrial Applications, 16-18.11.2015, Kyushu University, Fukuoka, Japan.

References

1. Allaire, G., de Gournay, F., Jouve, F., Toader, A.-M.: Structural optimization using topological and shape sensitivity via a level set method. Control Cybern. **34**, 59–80 (2005)
2. Ammari, H., Griesmaier, R., Hanke, M.: Identification of small inhomogeneities: asymptotic factorization. Math. Comp. **76**, 1425–1448 (2007)
3. Ammari, H., Garnier, J., Jugnon, V., Kang, H.: Stability and resolution analysis for a topological derivative based imaging functional. SIAM J. Control Optim. **50**, 48–76 (2012)
4. Amrouche, C., Girault, V., Giroire, J.: Dirichlet and Neumann exterior problems for the n-dimensional Laplace operator: an approach in weighted Sobolev spaces. J. Math. Pures Appl. **76**, 55–81 (1997)
5. Amstutz, S.: Sensitivity analysis with respect to a local perturbation of the material property. Asymptotic Anal. **49**, 87–108 (2006)
6. Bellis, C., Bonnet, M., Cakoni, F.: Acoustic inverse scattering using topological derivative of far-field measurements-based L^2 cost functionals. Inverse Prob. **29**, 075012 (2013)
7. Brühl, M., Hanke, M., Vogelius, M.: A direct impedance tomography algorithm for locating small inhomogeneities. Numer. Math. **93**, 631–654 (2003)
8. Cakoni, F., Colton, D.: A Qualitative Approach to Inverse Scattering Theory, vol. 188. AMS, Springer (2014)
9. Carpio, A., Rapun, M.-L.: Solving inverse inhomogeneous problems by topological derivative methods. Inverse Prob. **24**, 045014 (2008)

10. Colton, D., Kress, R.: Inverse Acoustic and Electromagnetic Scattering Theory. Springer, New York (2013)
11. Duduchava, R., Tsaava, M.: Mixed boundary value problems for the Helmholtz equation in arbitrary 2D-sectors. Georgian Math. J. **20**, 439–467 (2013)
12. Friedman, A., Vogelius, M.: Identification of small inhomogeneities of extreme conductivity by boundary measurements: a theorem on continuous dependence. Arch. Rational Mech. Anal. **105**, 299–326 (1989)
13. Hintermüller, M., Kovtunenko, V.A.: From shape variation to topology changes in constrained minimization: a velocity method-based concept. Optimization Meth. Softw. **26**, 513–532 (2011)
14. Hintermüller, M., Laurain, A., Novotny, A.A.: Second-order topological expansion for electrical impedance tomography. Adv. Comput. Math. **36**, 235–265 (2012)
15. Ikehata, M., Itou, H.: An inverse problem for a linear crack in an anisotropic elastic body and the enclosure method. Inverse Prob. **24**, 025005 (2008)
16. Ikehata, M., Itou, H.: Extracting the support function of a cavity in an isotropic elastic body from a single set of boundary data. Inverse Prob. **25**, 10500 (2009)
17. Itou, H., Kovtunenko, V.A., Tani, A.: The interface crack with Coulomb friction between two bonded dissimilar elastic media. Appl. Math. **56**, 69–97 (2011)
18. Kaltenbacher, B., Neubauer, A., Scherzer, O.: Iterative Regularization Methods for Nonlinear Problems. de Gruyter, Berlin (2008)
19. Khludnev, A.M., Kovtunenko, V.A.: Analysis of Cracks in Solids. WIT-Press, Southampton (2000)
20. Khludnev, A.M., Kozlov, V.A.: Asymptotics of solutions near crack tips for Poisson equation with inequality type boundary conditions. Z. Angew. Math. Phys. **59**, 264–280 (2008)
21. Khludnev, A.M., Kovtunenko, V.A., Tani, A.: Evolution of a crack with kink and non-penetration. J. Math. Soc. Japan **60**, 1219–1253 (2008)
22. Kirsch, A.: An Introduction to the Mathematical Theory of Inverse Problems. Springer, New York (2011)
23. Kovtunenko, V.A.: High-order topological expansions for Helmholtz problems in 2d. In: Bergounioux, M. et al. (eds.) Topological Optimization. Radon Series on Computational and Applied Mathematics, vol. 17. de Gruyter, Berlin (2016), to appear
24. Kovtunenko, V.A., Kunisch, K.: High precision identification of an object: optimality conditions based concept of imaging. SIAM J. Control Optim. **52**, 773–796 (2014)
25. Kovtunenko, V.A., Leugering, G.: A shape-topological control problem for nonlinear crack—defect interaction: the anti-plane variational model. SIAM J. Control. Optim. **54**, 1329–1351 (2016)
26. Kovtunenko, V.A., Kunisch, K., Ring, W.: Propagation and bifurcation of cracks based on implicit surfaces and discontinuous velocities. Comput. Visual Sci. **12**, 397–408 (2009)
27. Lavrentiev, M.M., Avdeev, A.V., Lavrentiev Jr, M.M., Priimenko, V.I.: Inverse Problems of Mathematical Physics. VSP Publication, Utrecht (2003)
28. Maharani, A.U., Kimura, M., Azegami, H., Ohtsuka, K., Armanda, I.: Shape optimization approach to a free boundary problem. Recent Dev. Comput. Sci. **6**, 42–55 (2015) (Kanazawa e-Publishing)
29. Maz'ya, V.G., Nazarov, S.A., Plamenevski, B.A.: Asymptotic Theory of Elliptic Boundary Value Problems in Singularly Perturbed Domains. Birkhäuser, Basel (2000)
30. Nakamura, G., Potthast, R., Sini, M.: Unification of the probe and singular sources methods for the inverse boundary value problem by the no-response test. Commun. Partial. Differ. Equ. **31**, 1505–1528 (2006)
31. Neittaanmäki, P., Roach, G.F.: Weighted Sobolev spaces and exterior problems for the Helmholtz equation. Proc. Roy. Soc. Lond. Ser. A **410**, 373–383 (1987)
32. Park, K.-W., Lesselier, D.: MUSIC-type imaging of a thin penetrable inclusion from its multistatic response matrix. Inverse Prob. **25**, 075002 (2009)
33. Pauly, D., Repin, S.: Functional a posteriori error estimates for elliptic problems in exterior domains. J. Math. Sci. (N.Y.) **162**, 393–406 (2009)

34. Sokolowski, J., Zochowski, A.: On the topological derivative in shape optimization. SIAM J. Control Optim. **37**, 1251–1272 (1999)
35. Vekua, I.N.: On Metaharmonic Functions. Lecture Notes of TICMI, vol. 14. Tbilisi University Press (2013)

Solution of Shape Optimization Problem and Its Application to Product Design

Hideyuki Azegami

Abstract In this paper, we define shape optimization problems as problems of finding the shapes of domains in which boundary value problems of partial differential equations are defined. A domain mapping from an initial domain to a new domain is chosen as the design variable. Functionals of the design variable and the solution to the boundary value problem are used as cost functions. In this paper, the formulation of the shape optimization problem and a numerical method of solving the problem are presented. In addition, our subsequent works applying this method to product design are introduced: (1) shape optimization of a link mechanism, (2) shape optimization for suppressing brake squeal, (3) a method of designing beads in a shell structure, (4) shape optimization of a flow field to improve hydrodynamic stability, and (5) shape optimization of an electrostatic capacitive sensor.

1 Introduction

Problems of finding the shapes of domains in which boundary value problems of partial differential equations are defined are called shape optimization problems. For example, when we choose a linear elastic problem as the boundary value problem, a function representing the stiffness as the objective cost function, and a volume as the constraint cost function, the shape optimization problem becomes the problem of finding the shape of a linear elastic body that maximizes the stiffness under the volume constraint.

In engineering, shape optimization problems are categorized as parametric or non-parametric. In parametric shape optimization problems, a finite-dimensional vector is chosen as the design variable. For example, the vector may consist of parameters representing the shape of a linear elastic body such as the diameter, length, or thickness. Sometimes the nodal coordinates of a finite element model, positions of

H. Azegami (✉)
Graduate School of Information Science, Nagoya University,
A4-2(780) Furo-cho, Chikusa-ku, Nagoya 464-8601, Japan
e-mail: azegami@is.nagoya-u.ac.jp

H. Itou et al. (eds.), *Mathematical Analysis of Continuum Mechanics and Industrial Applications*, Mathematics for Industry 26,
DOI 10.1007/978-981-10-2633-1_6

control points for nonuniform rational B-spline curves or surfaces, or undetermined multipliers in a linear combination of basis vectors to determine the boundary are used as the parametric design variables. In this formulation, increasing the number of design parameters corresponds to increasing the number of dimensions of the vector space. Consequently, the optimization problem becomes more difficult to solve with increasing number of design parameters.

On the other hand, in nonparametric shape optimization problems, a displacement denoting the domain variation from an initial domain to a new domain is selected as the design variable. Because the displacement is a function, the nonparametric shape optimization problem becomes a functional optimization problem. In this problem, the number of design parameters is infinite. Thus, we can expect a finer result than that for parametric optimization problems.

The basic concept of the Gâteaux derivative of a cost function with respect to domain variation was presented early in the twentieth century [10, 17]. The derivative was called the shape derivative of the cost function. Subsequently, many researchers have studied the shape derivative (see the citations in [2]). However, the nonparametric shape optimization problem is found to be irregular because the regularities of the shape derivatives of the cost functions are less than the regularity required to maintain the original regularity of the initial domain. Methods of compensating for the lack of regularity of the shape derivative have been presented. Mohammadi and Pironneau proposed a method of remaking a smooth boundary by using the Laplace operator on the boundary [12]. Another method that introduced the Tikhonov regularization term into the cost function was presented by Yamada et al. in the level set approach [18].

To ensure regularity, the authors have developed another method that uses the gradient method in the Hilbert space $H^1\left(\Omega; \mathbb{R}^d\right)$ for $d \in \{2, 3\}$-dimensional domains Ω in which boundary value problems of partial differential equations are defined [1, 5, 6]. In this method, domain variation is obtained as a solution to a boundary value problem of an elliptic partial differential equation, such as a linear elastic problem defined in the current domain using the Neumann condition with the negative value of the shape derivative on the boundary. Because the Neumann condition can be considered as a fictitious traction, we called this method the traction method. This method is similar to methods producing domain variations by fictitious forces [8, 14, 19]. However, those methods are formulated using parametric design variables of nodal displacements, whereas the traction method is formulated using nonparametric design variables of domain variation. Moreover, the traction method is essentially different from the method using the fictitious linear elastic solution, in which the shape derivative is used as the Dirichlet condition [9].

Applications of the traction method to various shape optimization problems in engineering have been reported (see the citations in [2]). Moreover, we previously presented a similar method, referred to as the H^1 gradient method, for a topology optimization problem of density type [4]. Then, in the context of the H^1 gradient method for the topology optimization problem, we refer to the traction method as the H^1 gradient method for the shape optimization problem [2]. The definition of the H^1 gradient method and the solution obtained using this method are described herein.

On the basis of this background, this paper introduces an outline of the formulation and solution of a shape optimization problem, and recent results for product design obtained using this solution.

In this paper, we use the notation $W^{s,p}\left(\Omega_0; \mathbb{R}^d\right)$ to represent the Sobolev space for the set of functions defined in Ω_0, which has a value of $\mathbb{R}^d$ and is $s \in [0, \infty]$ times differentiable and $p \in [1, \infty]$-th order Lebesgue integrable. Moreover, $L^p\left(\Omega_0; \mathbb{R}^d\right)$ and $H^s\left(\Omega_0; \mathbb{R}^d\right)$ are denoted by $W^{0,p}$ and $W^{s,2}\left(\Omega_0; \mathbb{R}^d\right)$, respectively.

2 Outline of Shape Optimization Problem and Its Solution

Let us assume that a boundary value problem of partial differential equations is given as a linear elastic problem in a $d \in \{2, 3\}$-dimensional bounded domain $\Omega_0 \subset \mathbb{R}^d$, the boundary $\partial\Omega_0$ of which is required to be at least a Lipschitz boundary, i.e., locally the graph of a Lipschitz continuous function. We assume that the domain varies by deformation $\boldsymbol{\phi} \in \mathscr{D} \subset X$ as shown in Fig. 1 and becomes $\Omega(\boldsymbol{\phi}) = (\mathbf{i} + \boldsymbol{\phi})(\Omega_0)$, where $\mathbf{i}$ denotes the identity mapping. Here, the linear space X and the admissible set $\mathscr{D}$ of $\boldsymbol{\phi}$ are given by

$$X = \left\{\boldsymbol{\phi} \in H^1\left(\mathbb{R}^d; \mathbb{R}^d\right) \mid \boldsymbol{\phi} = \mathbf{0}_{\mathbb{R}^d} \text{ on } \bar{\Omega}_{\mathrm{C0}}\right\}, \tag{1}$$

$$\mathscr{D} = \left\{\boldsymbol{\phi} \in X \cap W^{1,\infty}\left(\mathbb{R}^d; \mathbb{R}^d\right) \mid \text{conditions for bijection}\right\}, \tag{2}$$

where $\bar{\Omega}_{\mathrm{C0}}$ is a subset of $\bar{\Omega}_0$ ($\bar{\ }$ denotes closure) previously derived from the design demands ($\bar{\Omega}_{\mathrm{C0}} = \emptyset$ in the case of Fig. 1). In the definition of X, the definition of $\boldsymbol{\phi} \in X$ is extended from Ω_0 to $\mathbb{R}^d$ on the basis of Calderón's extension theorem in order to fix the definition of the function in the process of domain variation. $H^1\left(\mathbb{R}^d; \mathbb{R}^d\right)$ is selected because a Hilbert space is required in order to define a gradient method later. In the following, we use the notation $(\,\cdot\,)(\boldsymbol{\phi})$ to represent $\{(\mathbf{i} + \boldsymbol{\phi})(\mathbf{x}) \mid \mathbf{x} \in (\,\cdot\,)_0\}$.

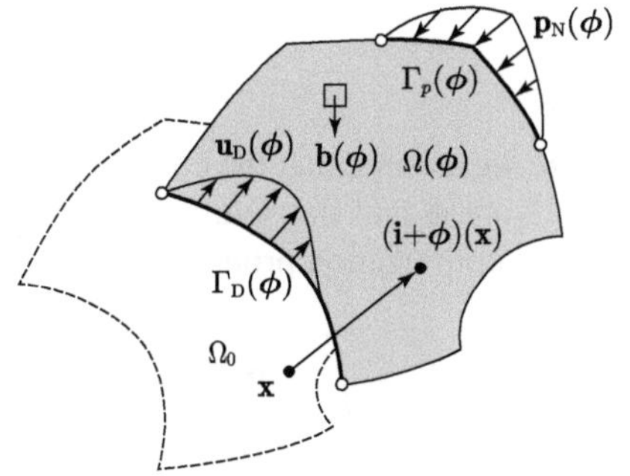

Fig. 1 Initial domain $\Omega_0 \subset \mathbb{R}^d$ and domain variation $\boldsymbol{\phi} : \mathbb{R}^d \to \mathbb{R}^d$

For $\boldsymbol{\phi} \in \mathscr{D}$, we define a linear elastic problem in $\Omega(\boldsymbol{\phi})$. Let $\mathbf{u} : \Omega(\boldsymbol{\phi}) \to \mathbb{R}^d$ be an elastic displacement, $\mathbf{E}(\mathbf{u}) = \left\{\nabla\mathbf{u}^{\mathrm{T}} + \left(\nabla\mathbf{u}^{\mathrm{T}}\right)^{\mathrm{T}}\right\}/2$ be a linear strain, $\mathbf{S}(\mathbf{u}) = \mathbf{C}(\boldsymbol{\phi})\mathbf{E}(\mathbf{u})$ be a stress, and $\mathbf{C}(\boldsymbol{\phi}) : \Omega(\boldsymbol{\phi}) \to \mathbb{R}^{d\times d\times d\times d}$ denote a stiffness. We denote the outer unit normal by $\boldsymbol{\nu}$.

Problem 2.1 (*Linear elastic problem*) For $\boldsymbol{\phi} \in \mathscr{D}$ and given functions $\mathbf{b}(\boldsymbol{\phi})$, $\mathbf{p}_{\mathrm{N}}(\boldsymbol{\phi})$, and $\mathbf{u}_{\mathrm{D}}(\boldsymbol{\phi})$, as shown in Fig. 1, and $\mathbf{C}(\boldsymbol{\phi})$, find $\mathbf{u}$ such that

$$-\nabla^{\mathrm{T}}\mathbf{S}(\boldsymbol{\phi}, \mathbf{u}) = \mathbf{b}^{\mathrm{T}}(\boldsymbol{\phi}) \quad \text{in } \Omega(\boldsymbol{\phi}), \tag{3}$$

$$\mathbf{S}(\boldsymbol{\phi}, \mathbf{u})\boldsymbol{\nu} = \mathbf{p}_{\mathrm{N}}(\boldsymbol{\phi}) \quad \text{on } \Gamma_p(\boldsymbol{\phi}), \tag{4}$$

$$\mathbf{S}(\boldsymbol{\phi}, \mathbf{u})\boldsymbol{\nu} = \mathbf{0}_{\mathbb{R}^d} \quad \text{on } \Gamma_{\mathrm{N}}(\boldsymbol{\phi}) \setminus \bar{\Gamma}_p(\boldsymbol{\phi}), \tag{5}$$

$$\mathbf{u} = \mathbf{u}_{\mathrm{D}}(\boldsymbol{\phi}) \quad \text{on } \Gamma_{\mathrm{D}}(\boldsymbol{\phi}). \tag{6}$$

When the given functions are well defined, $\mathbf{u} - \mathbf{u}_{\mathrm{D}}$ is found uniquely in

$$U = \left\{\mathbf{u} \in H^1\left(\Omega(\boldsymbol{\phi}); \mathbb{R}^d\right) \mid \mathbf{u} = \mathbf{0}_{\mathbb{R}^d} \text{ on } \Gamma_{\mathrm{D}}(\boldsymbol{\phi})\right\}. \tag{7}$$

For later use, we define the Lagrange function of Problem 2.1 as

$$\begin{aligned}\mathscr{L}_{\mathrm{M}}(\boldsymbol{\phi}, \mathbf{u}, \mathbf{v}) = &\int_{\Omega(\boldsymbol{\phi})} (-\mathbf{S}(\mathbf{u}) \cdot \mathbf{E}(\mathbf{v}) + \mathbf{b} \cdot \mathbf{v})\,\mathrm{d}x + \int_{\Gamma_p(\boldsymbol{\phi})} \mathbf{p}_{\mathrm{N}} \cdot \mathbf{v}\,\mathrm{d}\gamma \\ &+ \int_{\Gamma_{\mathrm{D}}(\boldsymbol{\phi})} \{(\mathbf{u} - \mathbf{u}_{\mathrm{D}}) \cdot (\mathbf{S}(\mathbf{v})\boldsymbol{\nu}) + \mathbf{v} \cdot (\mathbf{S}(\mathbf{u})\boldsymbol{\nu})\}\,\mathrm{d}\gamma,\end{aligned} \tag{8}$$

where $\mathbf{v} \in U$ is introduced as the Lagrange multiplier. If $\mathbf{u}$ is the solution to Problem 2.1, the weak form of Problem 2.1 is written as

$$\mathscr{L}_{\mathrm{M}}(\boldsymbol{\phi}, \mathbf{u}, \mathbf{v}) = 0$$

for all $\mathbf{v} \in U$.

In a shape optimization problem, to find a solution $\boldsymbol{\phi}$ in $\mathscr{D}$, we need a condition that $\mathbf{u} - \mathbf{u}_{\mathrm{D}}$ belongs to

$$\mathscr{S} = U \cap W^{1,\infty}\left(\Omega(\boldsymbol{\phi}); \mathbb{R}^d\right). \tag{9}$$

Here we assume that the conditions for $\mathbf{u} - \mathbf{u}_{\mathrm{D}} \in \mathscr{S}$ are satisfied.

Using $\boldsymbol{\phi}$ and the solution $\mathbf{u}$ to Problem 2.1, we define the cost functions for a shape optimization problem as

$$f_0(\boldsymbol{\phi}, \mathbf{u}) = \int_{\Omega(\boldsymbol{\phi})} \mathbf{b} \cdot \mathbf{u}\,\mathrm{d}x + \int_{\Gamma_{\mathrm{N}}(\boldsymbol{\phi})} \mathbf{p}_{\mathrm{N}} \cdot \mathbf{u}\,\mathrm{d}\gamma - \int_{\Gamma_{\mathrm{D}}(\boldsymbol{\phi})} \mathbf{u}_{\mathrm{D}} \cdot (\mathbf{S}(\mathbf{u})\boldsymbol{\nu})\,\mathrm{d}\gamma, \tag{10}$$

$$f_1(\boldsymbol{\phi}) = \int_{\Omega(\boldsymbol{\phi})} \mathrm{d}x - c_1. \tag{11}$$

Here, f_0 is called the mean compliance representing the deformability of the linear elastic body. Further, f_1 is the cost function for the volume constraint, where c_1 is a constant for the volume constraint. The shape optimization problem of the linear elastic body is defined as follows.

Problem 2.2 (*Mean compliance minimization problem*) For f_0 and f_1, find $\Omega(\boldsymbol{\phi})$ such that

$$\min_{\boldsymbol{\phi}\in\mathscr{D}} \{ f_0(\boldsymbol{\phi}, \mathbf{u}) \mid f_1(\boldsymbol{\phi}) \le 0, \ \mathbf{u} - \mathbf{u}_{\mathrm{D}} \in \mathscr{S}, \ \text{Problem 2.1} \} .$$

The shape derivatives of the cost functions are evaluated as follows. Because f_0 is a functional of $\mathbf{u}$, which is the solution to Problem 2.1, we define the Lagrange function for f_0 as

$$\mathscr{L}_0(\boldsymbol{\phi}, \mathbf{u}, \mathbf{v}_0) = f_0(\boldsymbol{\phi}, \mathbf{u}) + \mathscr{L}_{\mathrm{M}}(\boldsymbol{\phi}, \mathbf{u}, \mathbf{v}_0) . \tag{12}$$

We denote $\boldsymbol{\varphi} \in X$ as the domain variation and take the derivative of $\mathscr{L}_0$ with respect to arbitrary variation $(\boldsymbol{\varphi}, \mathbf{u}', \mathbf{v}'_0) \in X \times U^2$ of $(\boldsymbol{\phi}, \mathbf{u}, \mathbf{v}_0) \in \mathscr{D} \times \mathscr{S}^2$. Then, we have the following form:

$$\begin{aligned} \mathscr{L}'_0(\boldsymbol{\phi}, \mathbf{u}, \mathbf{v}_0)\left[\boldsymbol{\varphi}, \mathbf{u}', \mathbf{v}'_0\right] &= \mathscr{L}_{0\boldsymbol{\phi}'}(\boldsymbol{\phi}, \mathbf{u}, \mathbf{v}_0)\left[\boldsymbol{\varphi}\right] \\ &+ \mathscr{L}_{0\mathbf{u}}(\boldsymbol{\phi}, \mathbf{u}, \mathbf{v}_0)\left[\mathbf{u}'\right] + \mathscr{L}_{0\mathbf{v}_0}(\boldsymbol{\phi}, \mathbf{u}, \mathbf{v}_0)\left[\mathbf{v}'_0\right] . \end{aligned} \tag{13}$$

For the third term of (13), we have

$$\mathscr{L}_{0\mathbf{v}_0}(\boldsymbol{\phi}, \mathbf{u}, \mathbf{v}_0)\left[\mathbf{v}'_0\right] = \mathscr{L}_{\mathrm{M}\mathbf{v}_0}(\boldsymbol{\phi}, \mathbf{u}, \mathbf{v}_0)\left[\mathbf{v}'_0\right] = \mathscr{L}_{\mathrm{M}}\left(\boldsymbol{\phi}, \mathbf{u}, \mathbf{v}'_0\right) . \tag{14}$$

Then, if $\mathbf{u}$ is the solution to Problem 2.1, the third term of (13) becomes 0. Moreover, replacing the second term of (13) with 0, we have a weak form of an adjoint problem to determine $\mathbf{v}_0$. In this case, we have

$$\mathscr{L}_{0\mathbf{u}}(\boldsymbol{\phi}, \mathbf{u}, \mathbf{v}_0)\left[\mathbf{u}'\right] = \mathscr{L}_{\mathrm{M}}\left(\boldsymbol{\phi}, \mathbf{v}_0, \mathbf{u}'\right) . \tag{15}$$

Then, if we assume the self-adjoint relation

$$\mathbf{v}_0 = \mathbf{u}, \tag{16}$$

the second term of (13) becomes 0.

Under the above assumptions for $\mathbf{u}$ and $\mathbf{v}_0$, we have the shape derivative of f_0 from the first term. Here, we assume that $\mathbf{b}$, $\mathbf{p}_{\mathrm{N}}$, $\mathbf{u}_{\mathrm{D}}$, and $\mathbf{C}$ are fixed in the initial domain, which means that they vary with domain variation. Then, applying Propositions A.1 and A.2, which are shown in the Appendix, and using the notation $\tilde{f}_0(\boldsymbol{\phi})$ to represent $f_0(\boldsymbol{\phi}, \mathbf{u}(\boldsymbol{\phi}))$, we have

$$\tilde{f}_0' (\boldsymbol{\phi}) [\boldsymbol{\varphi}] = \mathscr{L}_{0\boldsymbol{\phi}'} (\boldsymbol{\phi}, \mathbf{u}, \mathbf{v}_0) [\boldsymbol{\varphi}] = \langle \mathbf{g}_0, \boldsymbol{\varphi} \rangle$$
$$= \int_{\Omega(\boldsymbol{\phi})} \left(\mathbf{G}_{\Omega 0} \cdot \boldsymbol{\nabla} \boldsymbol{\varphi}^{\mathrm{T}} + g_{\Omega 0} \boldsymbol{\nabla} \cdot \boldsymbol{\varphi} \right) \mathrm{d}x + \int_{\Gamma_p(\boldsymbol{\phi})} \mathbf{g}_{p0} \cdot \boldsymbol{\varphi} \, \mathrm{d}\gamma + \int_{\partial \Gamma_p(\boldsymbol{\phi}) \cup \Theta(\boldsymbol{\phi})} \mathbf{g}_{\partial p 0} \cdot \boldsymbol{\varphi} \, \mathrm{d}\varsigma, \tag{17}$$

where

$$\mathbf{G}_{\Omega 0} = 2 \mathbf{S} (\mathbf{u}) \left(\boldsymbol{\nabla} \mathbf{u}^{\mathrm{T}} \right)^{\mathrm{T}}, \tag{18}$$
$$g_{\Omega 0} = -\mathbf{S} (\mathbf{u}) \cdot \mathbf{E} (\mathbf{u}) + 2 \mathbf{b} \cdot \mathbf{u}, \tag{19}$$
$$\mathbf{g}_{p0} = 2\kappa \left(\mathbf{p}_{\mathrm{N}} \cdot \mathbf{u} \right) \boldsymbol{\nu}, \tag{20}$$
$$\mathbf{g}_{\partial p 0} = 2 \left(\mathbf{p}_{\mathrm{N}} \cdot \mathbf{u} \right) \boldsymbol{\tau}. \tag{21}$$

Here, the Dirichlet condition in Problem 2.1 was used. $\Theta (\boldsymbol{\phi})$ is the set of corner points in $\Gamma (\boldsymbol{\phi})$ for $d = 2$ and the vertices and edges for $d = 3$, $\boldsymbol{\tau}$ denotes the tangent, and $\kappa = \boldsymbol{\nabla} \cdot \boldsymbol{\nu}$. We can confirm that $\mathbf{g}_0$ is included in X' (the dual space of X) when $\mathbf{u} \in \mathscr{S}$, and we call it the shape gradient of f_0 as the meaning of the Fréchet derivative.

On the other hand, the shape derivative $f_1 (\boldsymbol{\phi})$ is obtained using Proposition A.1 as

$$f_1' (\boldsymbol{\phi}) \left[\boldsymbol{\varphi} \right] = \langle \mathbf{g}_1, \boldsymbol{\varphi} \rangle = \int_{\Omega(\boldsymbol{\phi})} \boldsymbol{\nabla} \cdot \boldsymbol{\varphi} \, \mathrm{d}x. \tag{22}$$

When we assume that $\mathbf{b}$, $\mathbf{p}_{\mathrm{N}}$, $\mathbf{u}_{\mathrm{D}}$, and $\mathbf{C}$ are fixed in $\mathbb{R}^d$, that is, that they are independent of the domain variation, we have other results for $\mathbf{g}_0$ and $\mathbf{g}_1$ given by boundary integrals [3].

Using $\mathbf{g}_0$ and $\mathbf{g}_1$, we can apply an iterative algorithm based on a gradient method in X, which we call the H^1 gradient method, to solve Problem 2.2. The H^1 gradient method for $\mathbf{g}_i$ ($i \in \{0, 1\}$) is to find the domain variation $\boldsymbol{\varphi}_{gi} \in X$ as the solution of the following problem. Hereafter, we denote $\tilde{f}_0 (\boldsymbol{\phi}) = f_0 (\boldsymbol{\phi}, \mathbf{u} (\boldsymbol{\phi}))$ as $f_0 (\boldsymbol{\phi})$ and consider a problem minimizing $f_0 (\boldsymbol{\phi})$ under constraints $f_1 (\boldsymbol{\phi}) \leq 0, \ldots, f_m (\boldsymbol{\phi}) \leq 0$.

Problem 2.3 (H^1 *gradient method of domain variation type*) Let $a_X : X \times X \to \mathbb{R}$ be a bounded coercive bilinear form such that there exist $\alpha_X > 0$ and $\beta_X > 0$ that satisfy

$$a_X (\boldsymbol{\varphi}, \boldsymbol{\varphi}) \geq \alpha_X \|\boldsymbol{\varphi}\|_X^2, \quad |a_X (\boldsymbol{\varphi}, \boldsymbol{\psi})| \leq \beta_X \|\boldsymbol{\varphi}\|_X \|\boldsymbol{\psi}\|_X \tag{23}$$

for all $\boldsymbol{\varphi} \in X$ and $\boldsymbol{\psi} \in X$. For $\mathbf{g}_i \in X'$, find $\boldsymbol{\varphi}_{gi} \in X$ such that

$$a_X \left(\boldsymbol{\varphi}_{gi}, \boldsymbol{\psi} \right) = - \langle \mathbf{g}_i, \boldsymbol{\psi} \rangle \tag{24}$$

for all $\boldsymbol{\psi} \in X$.

The solution $\boldsymbol{\varphi}_{gi}$ decreases f_i because

$$f_i(\boldsymbol{\phi} + \boldsymbol{\varphi}_{gi}) - f_i(\boldsymbol{\phi}) = \langle \mathbf{g}_i, \boldsymbol{\varphi}_{gi} \rangle + o(\|\boldsymbol{\varphi}_{gi}\|_X) = -\langle \boldsymbol{\varphi}_{gi}, \boldsymbol{\varphi}_{gi} \rangle + o(\|\boldsymbol{\varphi}_{gi}\|_X)$$
$$\leq -\alpha \|\boldsymbol{\varphi}_{gi}\|_X^2 + o(\|\boldsymbol{\varphi}_{gi}\|_X)$$

holds for small $\boldsymbol{\varphi}_{gi}$. A simple choice for a_X is

$$a_X(\boldsymbol{\varphi}, \boldsymbol{\psi}) = \int_{\Omega(\boldsymbol{\phi})} \left\{ (\boldsymbol{\nabla} \boldsymbol{\varphi}^{\mathrm{T}}) \cdot (\boldsymbol{\nabla} \boldsymbol{\psi}^{\mathrm{T}}) + c_\Omega \boldsymbol{\varphi} \cdot \boldsymbol{\psi} \right\} \mathrm{d}x, \tag{25}$$

where c_Ω is a positive constant. For the regularity of $\boldsymbol{\varphi}_{gi}$, we have $\boldsymbol{\varphi}_{gi} \in \mathscr{D}$ under appropriate conditions (Theorem 9.1 in [2]).

To solve Problem 2.2, we use an iterative method of varying the domain using the solution $\boldsymbol{\varphi}_g$ to a sequential quadratic approximation problem, which does not mean to use second derivatives, defined as follows.

Problem 2.4 (*SQ approximation*) For $\boldsymbol{\phi}_k \in \mathscr{D}$, let $\mathbf{g}_i$ be given for $i \in I_{\mathrm{A}}(\boldsymbol{\phi}_k) = \{i \in \{1, \ldots, m\} \mid f_i(\boldsymbol{\phi}_k) \geq 0\}$. Find $\boldsymbol{\varphi}_g$ such that

$$q(\boldsymbol{\varphi}_g) = \min_{\boldsymbol{\varphi} \in X} \left\{ q(\boldsymbol{\varphi}) = \frac{c_a}{2} a_X(\boldsymbol{\varphi}, \boldsymbol{\varphi}) + \langle \mathbf{g}_0, \boldsymbol{\varphi} \rangle \;\middle|\; f_i(\boldsymbol{\phi}_k) + \langle \mathbf{g}_i, \boldsymbol{\varphi} \rangle \leq 0,\ i \in I_{\mathrm{A}}(\boldsymbol{\phi}_k) \right\}.$$

We can find the solution $\boldsymbol{\varphi}_g$ as follows. We define a Lagrange function

$$\mathscr{L}_{\mathrm{S}}(\boldsymbol{\varphi}, \boldsymbol{\lambda}_{k+1}) = q(\boldsymbol{\varphi}) + \sum_{i \in I_{\mathrm{A}}(\boldsymbol{\phi}_k)} \lambda_{i\,k+1} (f_i(\boldsymbol{\phi}_k) + \langle \mathbf{g}_i, \boldsymbol{\varphi} \rangle), \tag{26}$$

where $\lambda_{i\,k+1}$ are the Lagrange multipliers. The Karush–Kuhn–Tucker conditions at the minimum point $\boldsymbol{\varphi}_g$ are given as

$$c_a a_X(\boldsymbol{\varphi}_g, \boldsymbol{\varphi}) + \langle \mathbf{g}_0, \boldsymbol{\varphi} \rangle + \sum_{i \in I_{\mathrm{A}}(\boldsymbol{\phi}_k)} \lambda_{i\,k+1} \langle \mathbf{g}_i, \boldsymbol{\varphi} \rangle = 0, \tag{27}$$

$$f_i(\boldsymbol{\phi}_k) + \langle \mathbf{g}_i, \boldsymbol{\varphi}_g \rangle \leq 0 \quad \text{for } i \in I_{\mathrm{A}}(\boldsymbol{\phi}_k), \tag{28}$$

$$\lambda_{i\,k+1} (f_i(\boldsymbol{\phi}_k) + \langle \mathbf{g}_i, \boldsymbol{\varphi}_g \rangle) = 0 \quad \text{for } i \in I_{\mathrm{A}}(\boldsymbol{\phi}_k), \tag{29}$$

$$\lambda_{i\,k+1} \geq 0 \quad \text{for } i \in I_{\mathrm{A}}(\boldsymbol{\phi}_k) \tag{30}$$

for all $\boldsymbol{\varphi} \in X$. Here, we assume

$$\boldsymbol{\varphi}_g = \boldsymbol{\varphi}_g(\boldsymbol{\lambda}_{k+1}) = \boldsymbol{\varphi}_{g0} + \sum_{i \in I_{\mathrm{A}}(\boldsymbol{\phi}_k)} \lambda_{i\,k+1} \boldsymbol{\varphi}_{gi}, \tag{31}$$

where $\boldsymbol{\varphi}_{g0}, \boldsymbol{\varphi}_{gi_1}, \ldots, and\ \boldsymbol{\varphi}_{gi_{|I_{\mathrm{A}}|}}$ are the solutions of the H^1 gradient method for each $\mathbf{g}_0, \mathbf{g}_{i_1}, \ldots, and\ \mathbf{g}_{i_{|I_{\mathrm{A}}|}}$ as

$$c_a a_X \left(\boldsymbol{\varphi}_{gi}, \boldsymbol{\psi}\right) = -\left\langle \mathbf{g}_i, \boldsymbol{\psi} \right\rangle \tag{32}$$

for all $\boldsymbol{\psi} \in X$, and $\boldsymbol{\lambda}_{k+1} \in \mathbb{R}^{|I_A|}$ are unknown parameters. Then, we can find that (27) holds for $\boldsymbol{\varphi}_g$ in (31), and that (28) becomes a linear system to determine $\boldsymbol{\lambda}_{k+1}$, when "$\le$" is replaced by "$=$," as follows:

$$\left(\left\langle \mathbf{g}_i, \boldsymbol{\varphi}_{gj} \right\rangle\right)_{(i,j)\in I_A^2(\boldsymbol{\phi}_k)} \left(\lambda_{j\,k+1}\right)_{j\in I_A(\boldsymbol{\phi}_k)} = -\left(f_i\left(\boldsymbol{\phi}_k\right) + \left\langle \mathbf{g}_i, \boldsymbol{\varphi}_{g0} \right\rangle\right)_{i\in I_A(\boldsymbol{\phi}_k)}. \tag{33}$$

In this case, we assume an algorithm to set $\lambda_{i\,k+1} = 0$ for

$$i \in I_I\left(\boldsymbol{\phi}_k\right) = \left\{i \in I_A\left(\boldsymbol{\phi}_k\right) \mid \lambda_{i\,k+1} < 0\right\},$$

replace $I_A\left(\boldsymbol{\phi}_k\right)$ with $I_A\left(\boldsymbol{\phi}_k\right) \setminus I_I\left(\boldsymbol{\phi}_k\right)$, and resolve the linear system of (28). When $I_I\left(\boldsymbol{\phi}_k\right)$ becomes $\emptyset$, the Karush–Kuhn–Tucker conditions are satisfied for small $\boldsymbol{\varphi}_g$. Moreover, when $f_i\left(\boldsymbol{\phi}_k\right) = 0$ for all $i \in I_A\left(\boldsymbol{\phi}_k\right)$, $\boldsymbol{\lambda}_{k+1}$ are determined independently of the magnitude of $\boldsymbol{\varphi}_g$.

A simple algorithm for solving a shape optimization problem is shown below.

1. Set Ω_0 and $\boldsymbol{\phi}_0 = \mathbf{i}$ as $f_1\left(\boldsymbol{\phi}_0\right) \le 0, \ldots, f_m\left(\boldsymbol{\phi}_0\right) \le 0$. Set $c_a, \varepsilon_0, \varepsilon_1, \ldots, \varepsilon_m$ appropriately. Set $k = 0$.
2. Solve the state determination problem at $\boldsymbol{\phi}_k$ (Fig. 2a), and compute $f_0\left(\boldsymbol{\phi}_k\right)$, $f_1\left(\boldsymbol{\phi}_k\right), \ldots, f_m\left(\boldsymbol{\phi}_k\right)$. Set $I_A\left(\boldsymbol{\phi}_k\right) = \left\{i \in \{1, \ldots, m\} \mid f_i\left(\boldsymbol{\phi}_k\right) \ge -\varepsilon_i\right\}$.
3. Solve adjoint problems at $\boldsymbol{\phi}_k$ (Fig. 2a), and compute $\mathbf{g}_0, \mathbf{g}_{i_1}, \ldots, \mathbf{g}_{i_{|I_A|}}$.
4. Solve $\boldsymbol{\varphi}_{g0}, \boldsymbol{\varphi}_{gi_1}, \ldots, \boldsymbol{\varphi}_{gi_{|I_A|}}$ using (32) (Fig. 2b).
5. Solve $\boldsymbol{\lambda}_{k+1}$ using (33). When $I_I\left(\boldsymbol{\phi}_k\right) \ne \emptyset$, replace $I_A\left(\boldsymbol{\phi}_k\right) \setminus I_I\left(\boldsymbol{\phi}_k\right)$ with $I_A\left(\boldsymbol{\phi}_k\right)$, and resolve (33) until $I_I\left(\boldsymbol{\phi}_k\right) = \emptyset$.
6. Compute $\boldsymbol{\varphi}_g$ using (31), set $\boldsymbol{\phi}_{k+1} = \boldsymbol{\phi}_k + \boldsymbol{\varphi}_g$ (Fig. 2c), and compute $f_0\left(\boldsymbol{\phi}_{k+1}\right)$, $f_1\left(\boldsymbol{\phi}_{k+1}\right), \ldots, f_m\left(\boldsymbol{\phi}_{k+1}\right)$. Set $I_A\left(\boldsymbol{\phi}_{k+1}\right) = \left\{i \in \{1, \ldots, m\} \mid f_i\left(\boldsymbol{\phi}_{k+1}\right) \ge -\varepsilon_i\right\}$.
7. Assess $\left|f_0\left(\boldsymbol{\phi}_{k+1}\right) - f_0\left(\boldsymbol{\phi}_k\right)\right| \le \varepsilon_0$.
   - If "Yes," proceed to (8).
   - If "No," replace $k+1$ with k and return to (3).
8. Stop the algorithm.

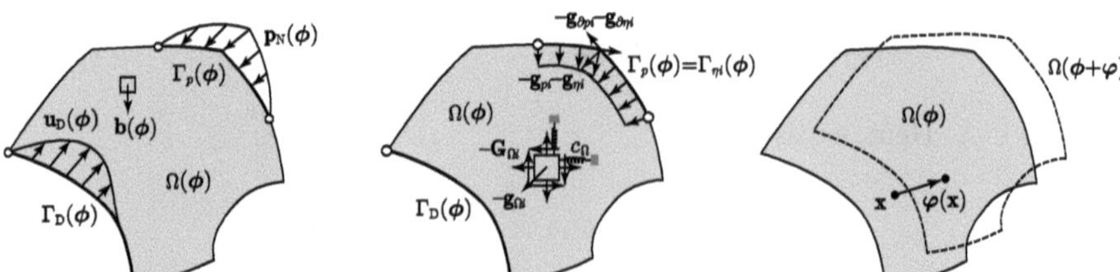

(a) State determination and adjoint problems (PDE) (b) H^1 gradient method (PDE) (c) Reshaping

Fig. 2 H^1 gradient method for shape optimization. *PDE* partial differential equation

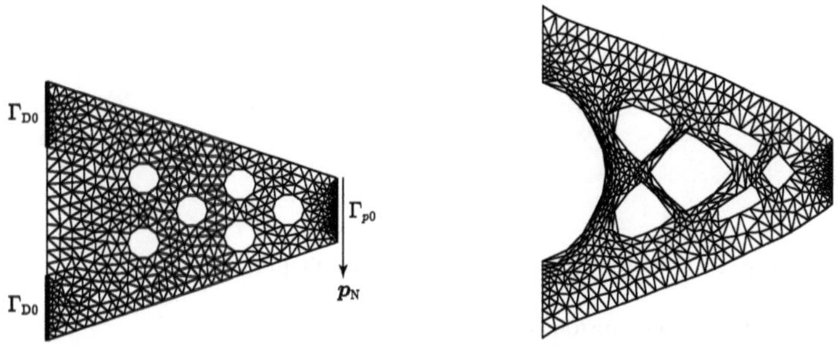

(a) Initial shape and boundary condition (b) Optimized shape

Fig. 3 Numerical result for mean compliance minimization problem

Figure 3 shows a numerical result for Problem 2.2 computed by a program developed using FreeFEM++ [11].

3 Shape Optimization for a Link Mechanism

Next, we introduce recent results obtained using this approach relating to product design. In this section, we show a result for a link mechanism [7] (Fig. 4). We define cost functions with negative-signed external work and a function for a volume constraint. Figure 5 shows a numerical result for a piston crank mechanism. The external force $\mathbf{p}_3 = (-0.2, 0)^{\mathrm{T}}$ N/mm, initial state $\theta_1(0) = 60°, \theta_2(0) = 150°$, initial angular velocity 0.2618 rad/s, and terminal time t_{T} are determined such that $\theta_1(t_{\mathrm{T}}) = 73°$ for the initial shape.

(a) Boundary conditions (b) Rigid motion of Ω_{01}

Fig. 4 Link mechanism

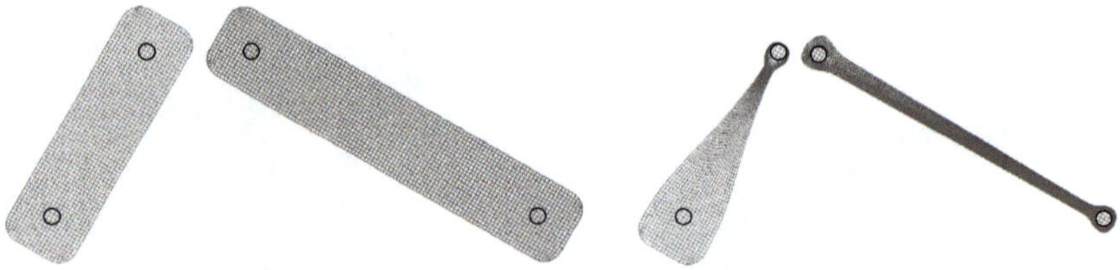

(a) Initial shapes (b) Optimum shapes

Fig. 5 Numerical result for a piston crank mechanism

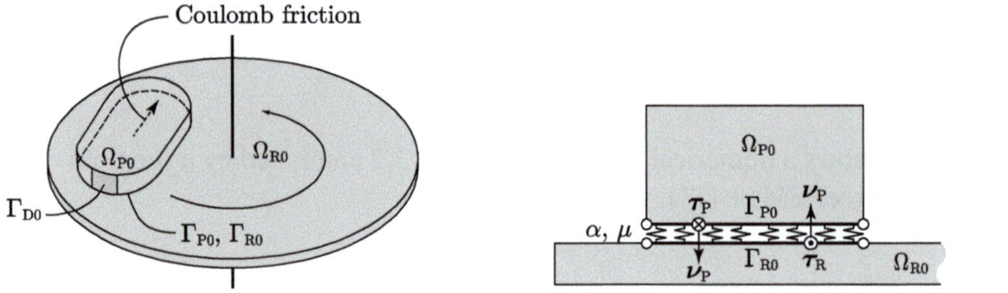

(a) Pad and rotor (b) Coulomb friction between Γ_{R0} and Γ_{P0}

Fig. 6 Brake model

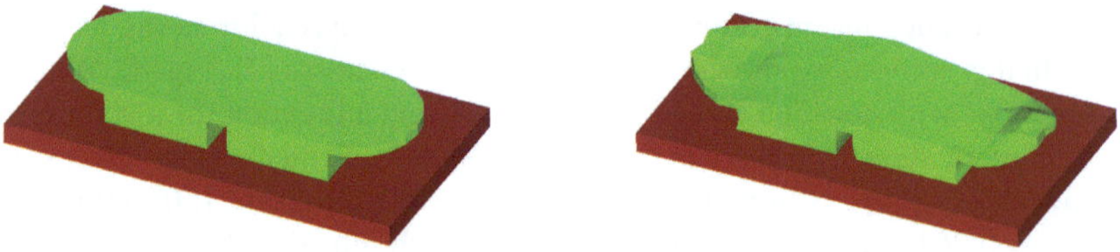

(a) Initial shape (b) Optimized shape

Fig. 7 Numerical result for a brake model

4 Shape Optimization for Suppressing Brake Squeal

The second example is an application to a brake squeal problem [16]. Brake squeal is considered to be a self-excited vibration phenomenon caused by friction between the rotor and the pad, which are indicated by $\boldsymbol{\tau}_P$ and $\boldsymbol{\tau}_R$ in Fig. 6b. The cause has been considered to be natural vibration in which the real part of a complex eigenvalue is positive. Then, we formulate a natural vibration problem of a pad and rotor that are linear elastic bodies between which Coulomb friction acts as the state determination problem. The maximum real part of the eigenvalues is used as the objective function. For a constraint function, we use the function for the volume constraint. Table 1 compares the eigenvalues of the initial and optimized shapes. In this table, the real part of the third eigenvalue is reduced to a negative value. Figure 7 compares the shapes.

Table 1 Comparison of complex eigenvalues of initial and optimized brake models

k	Re	Im
	Initial shape	
1	−1.692E+01	8.022947E+03
2	−1.444E+01	9.438261E+03
3	8.613E+00	1.249724E+04
4	−2.944E+01	1.437360E+04
5	−5.783E+01	1.629984E+04
6	−5.356E+01	2.168113E+04
7	−5.195E+01	2.394771E+04
8	−6.593E+01	2.573753E+04
9	−6.325E+01	2.711726E+04
10	−6.896E+01	2.893466E+04
	Optimized shape	
1	−1.647E+01	7.745197E+03
2	−1.765E+01	1.027973E+04
3	−1.163E+01	1.110440E+04
4	−3.048E+01	1.503565E+04
5	−4.185E+01	2.092213E+04
6	−5.070E+01	2.186379E+04
7	−6.588E+01	2.671747E+04
8	−7.522E+01	2.756015E+04
9	−7.540E+01	3.137934E+04
10	−7.658E+01	3.320161E+04

5 Method of Designing Beads in Shell Structure

Convexo-concave shapes with a small height in a shell structure are called beads. We formulate a shape optimization problem of beads in a shell structure [15] (Fig. 8). To create beads, the variation of the shell structure in the out-of-plane direction is restricted using the sigmoid function. The mean compliance is used as the objective function. Figure 9 shows the result for a shell cantilever with two twisting loads.

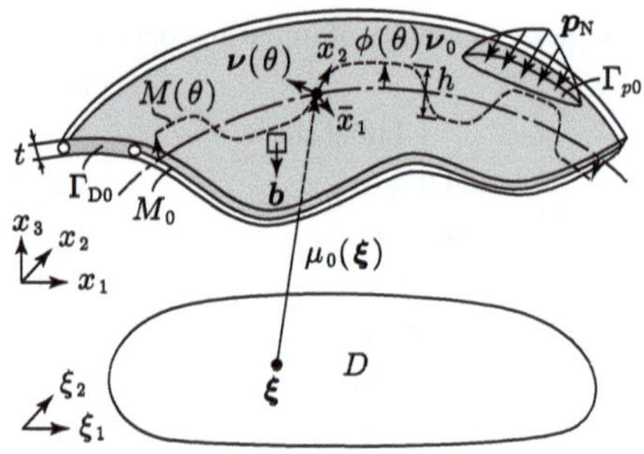

Fig. 8 Initial shell model and boundary conditions

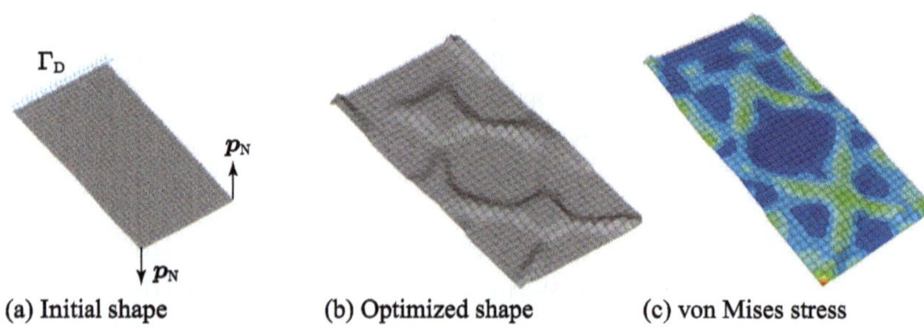

Fig. 9 Numerical result for a shell cantilever with two twisting loads

6 Shape Optimization of Flow Field to Improve Hydrodynamic Stability

The critical Reynolds number $\mathrm{Re_c}$ for the transition from steady flow to unsteady flow has been investigated for Poiseuille flow with sudden expansion and is reportedly around $\mathrm{Re_c} = 40$. In this study, we formulate a shape optimization problem of a flow field for delaying the transition from laminar flow to turbulent flow [13]. The maximum real part of every eigenvalue of the linear disturbance problem is defined as the objective function. Figure 10 shows the resulting shape for a two-dimensional Poiseuille flow with sudden expansion. Figure 11 shows that the critical Reynolds number increased from 40.24 in the initial domain to 45.49 in the optimized domain. The results reveal that the critical Reynolds number increases with iteration of the reshaping.

7 Shape Optimization of an Electrostatic Capacitive Sensor

An electrostatic capacitive sensor is used to detect fingers. To enhance the detection performance, we need to increase the difference in the capacitance between the cases with and without fingers. Thus, we formulated a shape optimization problem

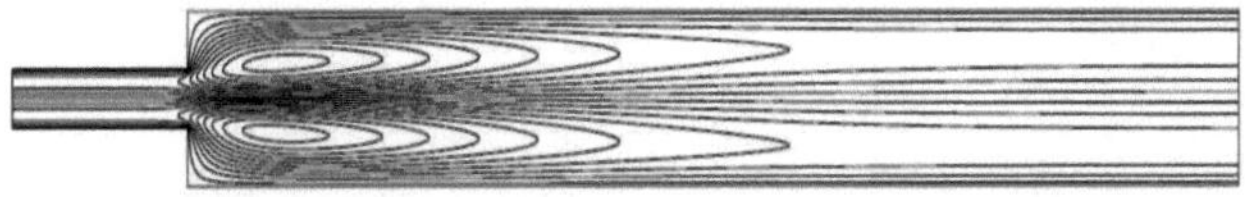

(a) Initial shape

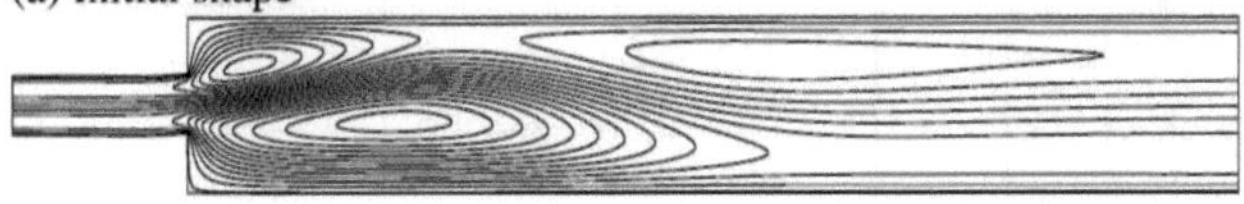

(b) Optimized shape

Fig. 10 Numerical result for Poiseuille flow with sudden expansion at Re = 45

Fig. 11 Real part of λ_1 with respect to the Reynolds number

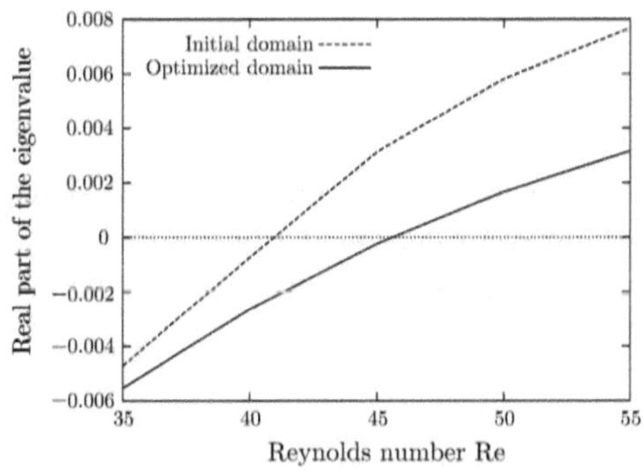

Fig. 12 Initial domain $\Omega_0 = D_0 \setminus (\bar{E}_0 \cup \bar{G}_0)$ for the electrostatic field problem

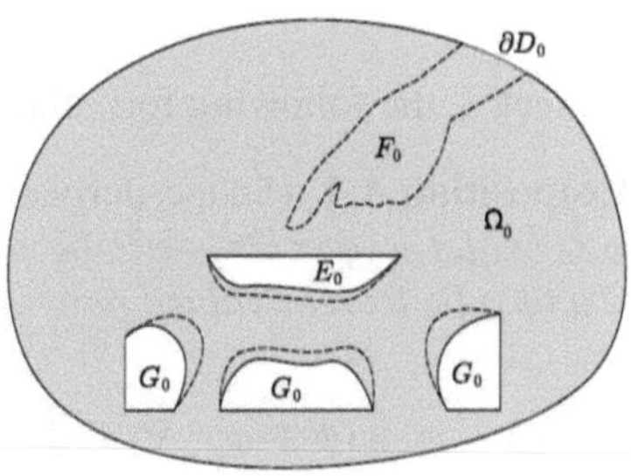

by using a cost function of the negative-signed squared H^1 norm of the difference between the solutions of two electrostatic field problems. The first problem is a basic electrostatic field problem consisting of a sensing electrode, an earth electrode, and air. The second problem adds fingers to the first problem, as shown in Fig. 12. The volume of the sensing electrode is used as a constraint cost function. Figure 13 shows the result for a numerical example.

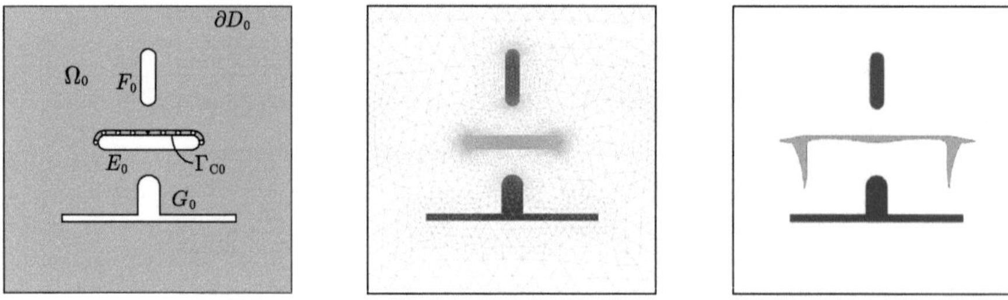

(a) Problem setting (b) Initial model (c) Optimized model

Fig. 13 Two-dimensional electrostatic field with protruding earth electrode

8 Summary

In this paper, the formulation of the shape optimization problem and a numerical solution were presented using the mean compliance minimization problem of a linear elastic body. Then our recent works applying this approach to product design were introduced.

Appendix

In Sect. 2, the following formulae are used (Propositions 4.4 and 4.7 in [2]).

Proposition A.1 (Shape derivative of domain integral) *For a domain variation* $\boldsymbol{\phi} \in \mathscr{D}$, *let* $u(\boldsymbol{\phi}) : \mathbb{R}^d \to \mathbb{R}$ *be a function varying with the domain and* $h(u(\boldsymbol{\phi}), \nabla u(\boldsymbol{\phi}))$ *be its real-valued function. Writing* $\mathbf{z} = \mathbf{x} + \boldsymbol{\varphi}(\mathbf{x})$, *let*

$$f(\boldsymbol{\phi} + \boldsymbol{\varphi}, u(\boldsymbol{\phi} + \boldsymbol{\varphi}), \nabla_z u(\boldsymbol{\phi} + \boldsymbol{\varphi})) = \int_{\Omega(\boldsymbol{\phi}+\boldsymbol{\varphi})} h(u(\boldsymbol{\phi} + \boldsymbol{\varphi}), \nabla_z u(\boldsymbol{\phi} + \boldsymbol{\varphi}))\, \mathrm{d}z$$

for arbitrary domain variation $\boldsymbol{\varphi} \in X$ *from* $\boldsymbol{\phi}$. *Then, the shape derivative (Fréchet derivative with respect to domain variation) of* f *is given by*

$$f'(\boldsymbol{\phi}, u, \nabla u)[\boldsymbol{\varphi}] = \int_{\Omega(\boldsymbol{\phi})} \{h_u(u, \nabla u)[u'] + h_{\nabla u}(u, \nabla u)[\nabla u' - \nabla \boldsymbol{\varphi}^{\mathrm{T}} \nabla u] + h(u, \nabla u) \nabla \cdot \boldsymbol{\varphi}\} \mathrm{d}x, \tag{34}$$

where $u(\boldsymbol{\phi})$ *and* $u'(\boldsymbol{\phi})[\boldsymbol{\varphi}]$ *are written as* u *and* u', *respectively, and* $u'(\boldsymbol{\phi})[\boldsymbol{\varphi}]$ *is the shape derivative of* u *with respect to the domain variation* $\boldsymbol{\varphi} \in X$.

Proposition A.2 (Shape derivative of boundary integral) *For a domain variation* $\boldsymbol{\phi} \in \mathscr{D}$, *let* $u(\boldsymbol{\phi}) : \mathbb{R}^d \to \mathbb{R}$ *be a function varying with the domain and*

$h(u(\boldsymbol{\phi}), \partial_\nu u(\boldsymbol{\phi}))$ be its real-valued function on $\Gamma(\boldsymbol{\phi}) \subset \partial\Omega(\boldsymbol{\phi})$. Writing $\partial_\mu = \boldsymbol{\mu} \cdot \boldsymbol{\nabla}_z$, where $\boldsymbol{\mu}$ denotes the normal to $\partial\Omega(\boldsymbol{\phi} + \boldsymbol{\varphi})$, let

$$f\left(\boldsymbol{\phi} + \boldsymbol{\varphi}, u(\boldsymbol{\phi} + \boldsymbol{\varphi}), \partial_\mu u(\boldsymbol{\phi} + \boldsymbol{\varphi})\right) = \int_{\Gamma(\boldsymbol{\phi}+\boldsymbol{\varphi})} h\left(u(\boldsymbol{\phi} + \boldsymbol{\varphi}), \partial_\mu u(\boldsymbol{\phi} + \boldsymbol{\varphi})\right) \mathrm{d}\zeta$$

for an arbitrary domain variation $\boldsymbol{\varphi} \in H^2\left(\mathbb{R}^d; \mathbb{R}^d\right)$ from $\boldsymbol{\phi}$. Then, the shape derivative of f is given by

$$f'(\boldsymbol{\phi}, u, \partial_\nu u)\left[\boldsymbol{\varphi}\right] = \int_{\Gamma(\boldsymbol{\phi})} \Big(h_u(u, \partial_\nu u)\left[u'\right] + h_{\partial_\nu u}(u, \partial_\nu u)\left[\partial_\nu u' + w(\boldsymbol{\varphi}, u)\right]$$
$$+ h(u, \partial_\nu u)(\boldsymbol{\nabla} \cdot \boldsymbol{\varphi})_\tau\Big)\mathrm{d}\gamma,$$

where $u(\boldsymbol{\phi})$ and $u'(\boldsymbol{\phi})\left[\boldsymbol{\varphi}\right]$ are written as u and u', respectively, and

$$w(\boldsymbol{\varphi}, u) = \left[\left\{\boldsymbol{\nu} \cdot \left(\boldsymbol{\nabla}\boldsymbol{\varphi}^{\mathrm{T}}\boldsymbol{\nu}\right)\right\}\boldsymbol{\nu} - \left\{\left(\boldsymbol{\nabla}\boldsymbol{\varphi}^{\mathrm{T}} + \left(\boldsymbol{\nabla}\boldsymbol{\varphi}^{\mathrm{T}}\right)^{\mathrm{T}}\right\}\boldsymbol{\nu}\right] \cdot \boldsymbol{\nabla} u,$$
$$(\boldsymbol{\nabla} \cdot \boldsymbol{\varphi})_\tau = \boldsymbol{\nabla} \cdot \boldsymbol{\varphi} - \boldsymbol{\nu} \cdot \left(\boldsymbol{\nabla}\boldsymbol{\varphi}^{\mathrm{T}}\boldsymbol{\nu}\right).$$

Moreover, if $\Gamma(\boldsymbol{\phi})$ is of piecewise C^2 class, the shape derivative of f is given by

$$f'(\boldsymbol{\phi}, u, \partial_\nu u)\left[\boldsymbol{\varphi}\right] = \int_{\Gamma(\boldsymbol{\phi})} \Big(h_u(u, \partial_\nu u)\left[u'\right] + h_{\partial_\nu u}(u, \partial_\nu u)\left[\partial_\nu u' + w(\boldsymbol{\varphi}, u)\right]$$
$$+ \kappa h(u, \partial_\nu u)\boldsymbol{\nu} \cdot \boldsymbol{\varphi} - \boldsymbol{\nabla}_\tau h(u, \partial_\nu u) \cdot \boldsymbol{\varphi}_\tau\Big)\mathrm{d}\gamma$$
$$+ \int_{\partial\Gamma(\boldsymbol{\phi}) \cup \Theta(\boldsymbol{\phi})} h(u, \partial_\nu u)\boldsymbol{\tau} \cdot \boldsymbol{\varphi}\,\mathrm{d}\varsigma, \tag{35}$$

where $\Theta(\boldsymbol{\phi})$ is the set of corner points in $\Gamma(\boldsymbol{\phi})$ for $d = 2$ and the vertices and edges for $d = 3$, $\boldsymbol{\tau}$ denotes the tangent, and $\kappa = \boldsymbol{\nabla} \cdot \boldsymbol{\nu}$.

References

1. Azegami, H.: A solution to domain optimization problems (in Japanese). Trans. Jpn. Soc. Mech. Eng., Ser. A **60**(574), 1479–1486 (1994)
2. Azegami, H.: Regularized solution to shape optimization problem (in Japanese). Trans. Jpn. Soc. Ind. Appl. Math. **23**(2), 83–138 (2014)
3. Azegami, H., Fukumoto, S., Aoyama, T.: Shape optimization of continua using NURBS as basis functions. Struct. Multidiscip. Optim. **47**(2), 247–258 (2013)
4. Azegami, H., Kaizu, S., Takeuchi, K.: Regular solution to topology optimization problems of continua. JSIAM Lett. **3**, 1–4 (2011)
5. Azegami, H., Takeuchi, K.: A smoothing method for shape optimization: traction method using the Robin condition. Int. J. Comput. Methods **3**(1), 21–33 (2006)

6. Azegami, H., Wu, Z.Q.: Domain optimization analysis in linear elastic problems: approach using traction method. JSME Int. J. Ser. A **39**(2), 272–278 (1996)
7. Azegami, H., Zhou, L., Umemura, K., Kondo, N.: Shape optimization for a link mechanism. Struct. Multidiscip. Optim. **48**(1), 115–125 (2013)
8. Belegundu, A.D., Rajan, S.D.: A shape optimization approach based on natural design varaibles and shape functions. Comput. Methods Appl. Mech. Eng. **66**, 87–106 (1988)
9. Choi, K.K., Kim, N.H.: Structural Sensitivity Analysis and Optimization, vol. 1 and 2. Springer, New York (2005)
10. Hadamard, J.: Mémoire des savants etragers. Oeuvres de J. Hadamard, chap. Mémoire sur le probléme d'analyse relatif á l'équilibre des plaques élastiques encastrées, Mémoire des savants etragers, Oeuvres de J. Hadamard, pp. 515–629. CNRS, Paris (1968)
11. Hecht, F.: New development in freefem++. J. Numer. Math. **20**(3–4), 251–265 (2012)
12. Mohammadi, B., Pironneau, O.: Applied Shape Optimization for Fluids. Oxford University Press, Oxford, New York (2001)
13. Nakazawa, T., Azegami, H.: Shape optimization of flow field improving hydrodynamic stability. Jpn. J. Ind. Appl. Math. (2015)
14. Scherer, M., Denzer, R., Steinmann, P.: A fictitious energy approach for shape optimization. Int. J. Numer. Meth. Eng. **82**, 269–302 (2010)
15. Shintani, K., Azegami, H.: A design method of beads in shell structure using non-parametric shape optimization method. In: Proceedings of the the ASME 2014 International Design Engineering Technical Conferences & Computers and Information in Engineering Conference (IDETCCIE 2014) (eBook), pp. 1–8 (2014)
16. Shintani, K., Azegami, H.: Shape optimization for suppressing brake squeal. Struct. Multidiscip. Optim. **50**(6), 1127–1135 (2014)
17. Sokolowski, J., Zolésio, J.P.: Introduction to Shape Optimization: Shape Sensitivity Analysis. Springer, New York (1992)
18. Yamada, T., Izui, K., Nishiwaki, S., Takezawa, A.: A topology optimization method based on the level set method incorporating a fictitious interface energy. Comput. Methods Appl. Mech. Eng. **199**, 2876–2891 (2010)
19. Yao, T.M., Choi, K.K.: 3-D shape optimal design and automatic finite element regridding. Int. J. Numer. Meth. Eng. **28**, 369–384 (1989)

Shape Optimization by GJ-Integral: Localization Method for Composite Material

Kohji Ohtsuka

Abstract GJ-integral $J_\omega(u, \mu) = P_\omega(u, \mu) + R_\omega(u, \mu)$ is the tool for shape sensitivity analysis of singular points in boundary value problem for partial differential equations, that is, GJ-integral takes value 0 if the solution u is regular inside ω for any vector field μ. The variation of energies with respect to the movement of singular points are expressed by $R_\omega(u, \mu)$ having finite value even if u has not regularity inside ω. We can solve shape optimization problems with respect to the set of singular points using GJ-integral and H^1 gradient method (Azegami's method). Here the singular points are the points on the boundary and on the interface of different materials. This paper provides a brief introduction to the history and basic theorems on GJ-integral. We also give extended results for composite material and its application to the shape optimization problem with some numerical examples by finite element analysis.

1 Introduction

GJ-integral (generalized J-integral) was proposed first in the paper [6] about 3D fracture problem [8], and is derived from J-integral in 2D fracture problem, called also *path-independent integral* [3, 11]

$$J = \int_C \left\{ \widehat{W}(\varepsilon) dx_2 - \sigma_{ij} n_j (\partial u_i / \partial x_1) dl \right\}$$

that represents the variation of the potential energies with respect to crack extension,

$$-\lim_{\ell \to 0} \frac{\Delta \mathscr{E}}{\Delta \ell} = J \tag{1}$$

K. Ohtsuka
Hiroshima Kokusai Gakuin University, 6-20-1 Nakano Aki-ku,
Hiroshima 739-0321, Japan
e-mail: ohtsuka@otlab.jp

H. Itou et al. (eds.), *Mathematical Analysis of Continuum Mechanics and Industrial Applications*, Mathematics for Industry 26,
DOI 10.1007/978-981-10-2633-1_7

if the crack is lying on the x_1-axis and extends straightforward in the case that body force is zero near the crack tip, where $\Delta\mathscr{E}$ stands for the potential energy difference by the crack extension, $\Delta\ell$ the length of newly created crack, C a closed curve surrounding the crack tip described in contra-clockwise, dl the line element of C, $\varepsilon(u) = \left(\varepsilon_{ij}(u)\right)$, $\varepsilon_{ij}(u) = \frac{1}{2}\left(\partial u_i/\partial x_j + \partial u_j/\partial x_i\right)$ the strain tensor, $u = (u_1, u_2)$ the displacement vector, the stress–strain relation is linear and homogeneous, that is, $\sigma_{ij}(u) = C_{ijkl}\varepsilon_{kl}(u)$ with the Hooke's tensor C_{ijkl} is constant, and $\widehat{W}(\varepsilon(u)) = \frac{1}{2}\sigma_{ij}(u)\varepsilon_{ij}(u)$. We can rewrite (1) as follows: By setting $C = \partial\omega$ with a domain ω containing the crack tip

$$J_\omega(u, \mathbf{e}_1) = \int_{\partial\omega} \left\{\widehat{W}(\varepsilon(u))(\mathbf{e}_1 \cdot \mathbf{n}) - \sigma_{ij}n_j(\mathbf{e}_1 \cdot \nabla u)\right\} dl \tag{2}$$

where $\mathbf{n}$ is the unit outward normal of C and $\mathbf{e}_1 = (1, 0)$ the direction of the crack extension. In 3D case and general situation, the crack surface Σ will be 2D-manifold (surface) with the edge $\partial\Sigma$ (crack tip in 2D), ω a tubular neighborhood of $\partial\Sigma$, $\mathbf{e}_1$ will be replaced with the vector field μ obtained from the crack extension, the body force f nonzero near $\partial\Sigma$ and the stress–strain relation is not homogeneous, that is, $\sigma_{ij}(x, u) = C_{ijkl}(x)\varepsilon_{kl}(u)$, $C_{ijkl} \in C^2(\mathbf{R}^3)$, and $\widehat{W}(x, \varepsilon(u)) = \frac{1}{2}\sigma_{ij}(x, u)\,\varepsilon_{ij}(u)$. In such situation, GJ-integral was proposed as

$$J_\omega(u, \mu) = P_\omega(u, \mu) + R_\omega(u, \mu), \tag{3}$$

$$\begin{aligned}
P_\omega(u, \mu) &= \int_{\partial(\omega\cap\Omega_\Sigma)} \left\{\widehat{W}(x, \varepsilon)(\mu \cdot \mathbf{n}) - T(x, u) \cdot (\mu \cdot \nabla u)\right\} ds, \\
R_\omega(u, \mu) &= -\int_{\omega\cap\Omega_\Sigma} \left\{(\mu \cdot \nabla_x)\widehat{W}(x, \varepsilon) + f \cdot (\mu \cdot \nabla u)\right\} \\
&\quad + \int_{\omega\cap\Omega_\Sigma} \left\{\sigma_{ij}(u)\,(\partial_j\mu_k)(\partial_k u_i) - \widehat{W}(x, \varepsilon)\,\mathrm{div}\mu\right\} dx, \\
T_i(x, u) &= \sigma_{ij}(x, u)\, n_j, \quad T(x, u) = (T_i(x, u))_{i=1,2,3}.
\end{aligned}$$

where Ω is a domain in $\mathbf{R}^3$ and $\Omega_\Sigma = \Omega \setminus \Sigma$ and $\mathbf{n} = (n_1, n_2, n_3)$ the outward unit normal of $\partial\omega$. In GJ-integral, the domain integral $R_\omega(u, \mu)$ appears because of the extension $P_\omega(u, \mu)$ depending on ω and it should be noticed that $R_\omega(u, \mu)$ is well-defined for the weak solution $u \in W^{1,2}(\Omega_\Sigma; \mathbf{R}^3)$.

More generalization of GJ-integral and some theorems on GJ-integral are given in Sect. 2. Section 3 extends Theorem 2 to composite material. Section 4 provides the method to solve shape optimization problem and its finite element analysis of shape optimization with examples is given in Sect. 5.

By $W^{s,p}(\Omega; \mathbf{R}^d)$, $0 \le s$, $1 \le p \le \infty$, $d = 1, 2, \ldots$, we denote Sobolev space of vector-valued function defined on Ω equipped with a norm $\|\cdot\|_{s,p,\Omega}$ that is a combination of L^p-norms of the function itself and its (fractional) derivatives up to s and $W^{s,p}(\Omega) = W^{s,p}(\Omega; \mathbf{R}^1)$. We use the summation convention.

2 GJ-Integral

In 1985 [7], GJ-integral has been extended as the tool for shape sensitivity analysis of sets of singular points in boundary value problems for partial differential equations. Let us consider the linear elastic system as partial differential equation defined in a domain $\Omega \subset \mathbf{R}^3$, and the elasticity (ε, σ) with body force f which is fixed on a part Γ_D of the boundary $\partial\Omega$ and surface force g is given on $\Gamma_N = \partial\Omega \setminus \overline{\Gamma_D}$, whose Hooke's tensor $C_{ijkl} \in C^2(\mathbf{R}^3)$ has the property

$$C_{ijkl}\xi_{ij}\xi_{kl} \geq \alpha\xi_{ij}\xi_{ij} \quad \text{for all } \xi_{ij} \in \mathbf{R} \quad \text{with } \alpha > 0$$

Consider the perturbation $\Omega(t) = \varphi_t(\Omega)$ of Ω by $[t \mapsto \varphi_t] \in C^2([0, \varepsilon); W^{2,\infty}(\mathbf{R}^3; \mathbf{R}^3))$ with the parameter t and a positive number $\varepsilon > 0$ such that $\varphi_0(x) = x$.

For given force $f \in W^{1,2}(\mathbf{R}^3; \mathbf{R}^3)$ and surface force $g \in W^{1,2}(\mathbf{R}^3; \mathbf{R}^3)$, find the displacement vector $u(t) \in V_t(\Omega(t))$, $\Omega(t) = \varphi_t(\Omega)$ which minimizes the functional

$$\mathscr{E}(v; f, g, \Omega(t)) = \int_{\Omega(t)} \left\{\widehat{W}(x, \varepsilon(v)) - f \cdot v\right\} dx - \int_{\Gamma_N(t)} g \cdot v \, ds \tag{4}$$

over the space

$$V_t(\Omega(t)) = \left\{v \in W^{1,2}(\mathbf{R}^3; \mathbf{R}^3) : v = 0 \quad \text{on } \Gamma_D(t)\right\}, \quad \Gamma_D(t) = \varphi_t(\Gamma_D) \tag{5}$$

where f, g are the restrictions to $\Omega(t)$ and $\Gamma_N(t) = \varphi_t(\Gamma_N)$, respectively.

GJ-integral is defined as the same formula in (3), except Ω_Σ is replaced by Ω, where ω is an arbitrary open set in $\mathbf{R}^3$ and an arbitrary vector field $\mu \in W^{1,\infty}(\mathbf{R}^3; \mathbf{R}^3)$. The important property of GJ-integral is to take the value 0 if the displacement u is regular inside $\omega \cap \Omega$.

Theorem 1 [7, Theorem 3.4] *Let u be the solution of Problem (4) at $t = 0$. If $u|_{\omega\cap\Omega}$ belongs to $W^{2,2}(\omega \cap \Omega; \mathbf{R}^3)$, then it holds that*

$$J_\omega(u, \mu) = 0 \quad \text{for all } \mu \in W^{1,\infty}(\mathbf{R}^3; \mathbf{R}^3) \tag{6}$$

We can get the following theorem corresponding to (1).

Theorem 2 [9, 10] *Let $V_t(\Omega(t))$ be the admissible space given in (5) or generally the subspace in $W^{1,2}(\Omega; \mathbf{R}^3)$ such that*

(H1) the pushforward $\varphi_{t*}v(x) = v(\varphi_t^{-1}(x))$, $x \in \Omega(t)$ become one-to-one mapping $[v \mapsto \varphi_{t*}v]$ from $V(\Omega)$ onto $V_t(\Omega(t))$.

Then, for the solution u of Problem (4) we have

$$\frac{d}{dt}\mathscr{E}(u(t); f, \Omega(t))\Big|_{t=0} = -R_\Omega(u, \mu_\varphi) - \int_{\partial\Omega} f \cdot u(\mu_\varphi \cdot \mathbf{n})\, ds \tag{7}$$

$$\mu_\varphi(x) = \frac{d}{dt}\varphi_t(x)\Big|_{t=0} \tag{8}$$

Combining Theorems 1 and 2 we have the following if $u \in W^{2,2}(\Omega; \mathbf{R}^3)$,

$$\frac{d}{dt}\mathscr{E}(u(t); f, \Omega(t))\Big|_{t=0} = P_\Omega(u, \mu_\varphi) - \int_{\partial\Omega} f \cdot u(\mu \cdot \mathbf{n})\, ds \tag{9}$$

This means that the shape sensitivity of the energy depends only on $\partial\Omega$. But, for some f and g, there will be a singularity at $\overline{\Gamma_D} \cap \overline{\Gamma_N}$. In such case, taking the open neighborhood B_ε of $\overline{\Gamma_D} \cap \overline{\Gamma_N}$, we can decompose such as

$$\begin{aligned}
-R_\Omega(u, \mu_\varphi) &= P_{\Omega\setminus\overline{B_\varepsilon}}(u, \mu_\varphi) - R_{B_\varepsilon}(u, \mu_\varphi)\\
&= \int_{\partial\Omega\setminus B_\varepsilon} \left\{\widehat{W}(x, \varepsilon)(\mu_\varphi \cdot \mathbf{n}) - T(x, u) \cdot (\mu_\varphi \cdot \nabla u)\right\} ds - \tilde{J}_{B_\varepsilon}(u, \mu_\varphi)\\
\tilde{J}_{B_\varepsilon}(u, \mu_\varphi) &= \int_{\partial\Omega\cap B_\varepsilon} \left\{\widehat{W}(x, \varepsilon)(\mu_\varphi \cdot \mathbf{n}) - T(x, u) \cdot (\mu_\varphi \cdot \nabla u)\right\} ds\\
&\quad + R_{B_\varepsilon}(u, \mu_\varphi)
\end{aligned} \tag{10}$$

Taking B_ε so that the volume $|B_\varepsilon|$ tend to 0 as $\varepsilon \to 0$, we have that the limit $\tilde{J}(u, \mu_\varphi) = \lim_{\varepsilon\to 0} \tilde{J}_{B_\varepsilon}(u, \mu_\varphi)$ expresses the shape sensitivity of the singular points $\overline{\Gamma_D} \cap \overline{\Gamma_N}$, if the limit exists. The part $-R_\Omega(u, \mu_\varphi) + \tilde{J}(u, \mu_\varphi)$ is the shape sensitivity of $\partial\Omega \setminus (\overline{\Gamma_D} \cap \overline{\Gamma_N})$.

3 Shape Sensitivity of Composite Materials

Assume that $\Omega \subset \mathbf{R}^3$ is decomposed into a finite number of pairwise disjoint subdomains $\Omega_\kappa \subset \Omega, \kappa = 1, \ldots, K$ with local Lipschitz property, such that $\overline{\Omega} = \cup_{\kappa=1}^K \overline{\Omega_\kappa}$. Let Γ_D be a part of $\partial\Omega$ and $\Gamma_N = \partial\Omega \setminus \overline{\Gamma_D}$. For $1 \le \kappa \le K$, let $u|_{\Omega_\kappa}$ be denoted by u^κ, $\sigma^\kappa(x, \varepsilon) = (C^\kappa_{ijkl}|_{\Omega_\kappa})(x)\varepsilon_{kl}$, $C^\kappa_{ijkl} \in C^2(\mathbf{R}^3)$, and $\widehat{W}_\kappa(x, \varepsilon(u^\kappa)) = \frac{1}{2}\sigma^\kappa_{ij}(x, \varepsilon(u^\kappa))\varepsilon_{ij}(u^\kappa)$ in Ω_κ. Here the Hooke's tensor $C_{ijkl}(x)$, $x \in \Omega$ defined by $C_{ijkl}(x) = (C^\kappa_{ijkl}|_{\Omega_\kappa})(x)$, $x \in \Omega_\kappa$ permit discontinuity. We consider the mathematical model of composite material (transmission problem): For given functions f, g, find u, $u^\kappa = u|_{\Omega_\kappa}$ such that

$$\begin{cases} -\mathrm{div}(\sigma_{ij}^{\kappa}(x,\varepsilon(u^{\kappa}))) = f(x) \quad x \in \Omega_{\kappa}, 1 \le i \le M \\ u^{\kappa} = u^{\ell} \quad \text{on } \Gamma_{\kappa\ell} = \partial\Omega_{\kappa} \cap \partial\Omega_{\ell} \\ \sigma^{\kappa}(x,\varepsilon(u^{\kappa}))\mathbf{n}^{\kappa} = -\sigma^{\ell}(x,\varepsilon(u^{\ell}))\mathbf{n}^{\ell} \quad \text{on } \Gamma_{\kappa\ell} \\ u = 0 \quad \text{on } \Gamma_D \\ \sigma^{\kappa}(x,\varepsilon(u^{\kappa}))\mathbf{n} = g \quad \text{on } \Gamma_N \cap \partial\Omega_{\kappa} \end{cases} \tag{11}$$

where $\mathbf{n}^{\kappa}$ denote the outward unit normal of Ω_{κ}. $\mathbf{n} = (n_j)$ the outward unit normal of $\Omega, \sigma\mathbf{n} = (\sigma_{ij}n_j)$ for the tensor $\sigma = (\sigma_{ij})$. The displacement vector u is the minimizer of the functional

$$\mathscr{E}(v; f, g, \cup\Omega_{\kappa}) = \sum_{\kappa=1}^{K} \int_{\Omega_{\kappa}} \left\{\widehat{W}_{\kappa}(x, \varepsilon(v^{\kappa})) - f \cdot v^{\kappa}\right\} dx - \int_{\Gamma_N} g \cdot v \, ds$$

over the subspace $V(\Omega)$ of $W^{1,2}(\Omega; \mathbf{R}^3)$.

Let us consider the mapping $\varphi_t(x) = x + t\mu$ for $\mu \in W^{1,\infty}(\mathbf{R}^3; \mathbf{R}^3)$. Here we assumed that $\mu(x) = 0$ on $\mathrm{supp}\, g$,

$$\mathrm{supp}\, g = \overline{\{x \in \Gamma_N;\ g(x) \neq 0\}}$$

We consider the variational problem with the parameter t: For the given functions $f, g \in W^{1,2}(\mathbf{R}^3; \mathbf{R}^3)$, find $u(t)$ minimizing the functional

$$\mathscr{E}(v; f, g, \cup\Omega_{\kappa}(t)) = \sum_{\kappa=1}^{K} \int_{\Omega_{\kappa}(t)} \left\{\widehat{W}_{\kappa}(x, \varepsilon(v^{\kappa})) - f \cdot v^{\kappa}\right\} dx - \int_{\Gamma_N(t)} g \cdot v \, ds$$

over the subspace $v \in V_t(\Omega(t))$ of $W^{1,2}(\Omega(t); \mathbf{R}^3)$, where $\Omega_{\kappa}(t) = \varphi_t(\Omega_{\kappa})$.

Theorem 3 *If $V(\Omega)$ and $V_t(\Omega(t))$ satisfy the condition (H1), then*

$$\frac{d}{dt}\mathscr{E}(u(t); f, g, \cup\Omega_{\kappa}(t))\Big|_{t=0} = -\sum_{\kappa=1}^{K} \left\{ R_{\Omega_{\kappa}}(u^{\kappa}, \mu_{\phi}) + \int_{\partial\Omega\cap\partial\Omega_{\kappa}} f \cdot u^{\kappa}(\mu_{\phi} \cdot \mathbf{n})\, ds \right\} \tag{12}$$

Proof By the assumption (H1), $u(t) = \varphi_{t*}u_0$ where u_0 is the minimizer of the functional

$$\begin{aligned} \widetilde{\mathscr{E}}(v; f, g, \varphi_t) &= \left\{ \sum_{\kappa=1}^{K} \int_{\Omega_{\kappa}(t)} \left\{\widehat{W}_{\kappa}(x, \varepsilon(\varphi_{t*}v^{\kappa})) - f \cdot \varphi_{t*}v^{\kappa}\right\} dx - \int_{\Gamma_N(t)} g \cdot \varphi_{t*}v^{\kappa}\, ds_t \right\} \\ &= \sum_{\kappa=1}^{K} \int_{\Omega_{\kappa}} \left\{\widehat{W}_{\kappa}(\varphi_t(x), \widetilde{\varepsilon}(v^{\kappa})) - \varphi_t^* f \cdot v^{\kappa}\right\} D(\varphi_t) dx \\ &\quad - \sum_{\kappa=1}^{K} \int_{\Gamma_N} \varphi_t^* g \cdot v^{\kappa} S(\varphi_t)\, ds \end{aligned} \tag{13}$$

where $\widetilde{\varepsilon}_{ij}(v^\kappa) = \varepsilon_{ij}(\varphi_{t*}v^\kappa) \circ \varphi_t$, $D(\varphi_t) = |\det \nabla\varphi_t^T|$, and $S(\varphi_t) = |M(\varphi_t)\mathbf{n}|$ with the cofactor matrix $M(\varphi_t)$ of the Jacobian matrix $\nabla\varphi_t^T$ and the pullback $\varphi_t^* f(x) = f(\varphi_t(x))$. By [10, Theorem 3], it was proved that

$$\begin{aligned}\frac{d}{dt}\mathscr{E}(u(t); f, g, \cup\Omega_\kappa(t))\Big|_{t=0} &= \frac{d}{dt}\widetilde{\mathscr{E}}(u; f, g, \varphi_t)\Big|_{t=0} \\ &= -\sum_{\kappa=1}^{K}\left\{R_{\Omega_\kappa}(u^\kappa, \mu_\phi) + \int_{\partial\Omega_\kappa} f \cdot u^\kappa(\mu_\phi \cdot \mathbf{n}^\kappa)\, ds\right\} \end{aligned} \tag{14}$$

Since $\cup_{\kappa=1}^K \partial\Omega_\kappa = (\cup_{\kappa<\ell\leq K}\Gamma_{\kappa\ell}) \cup (\cup_{\kappa=1}^K \partial\Omega \cap \partial\Omega_\kappa)$, here $\Gamma_{\kappa\ell} = \partial\Omega_\kappa \cap \partial\Omega_\ell \cap \Omega$, we have

$$\begin{aligned}\sum_{\kappa=1}^{K}\int_{\partial\Omega_\kappa} f \cdot u^\kappa(\mu \cdot \mathbf{n}^\kappa)\, ds &= \sum_{\kappa=1}^{K}\int_{\partial\Omega_\kappa\cap\partial\Omega} f \cdot u^\kappa(\mu_\phi \cdot \mathbf{n}^\kappa)\, ds \\ &\quad + \sum_{\kappa<\ell\leq K}\int_{\Gamma_{\kappa\ell}} \{f \cdot u^\kappa(\mu_\phi \cdot \mathbf{n}^\kappa + \mu_\phi \cdot \mathbf{n}^\ell)\}\, ds \end{aligned} \tag{15}$$

However $\mathbf{n}^\kappa = -\mathbf{n}^\ell$ on $\Gamma_{\kappa\ell}$, this means that (14) and (15) leads (12). □

4 Shape Optimization

Let $u(\cup\Omega_\kappa)$ be the solution of Problem (11), $\overline{\Omega} = \cup_{\kappa=1}^K \overline{\Omega_\kappa}$. The shape optimization problem is as follows: For a given cost functional $F(u(\cup\Omega_\kappa))$, find another domains Ω_κ^o such that

$$F(u(\cup\Omega_\kappa^o)) \leq F(u(\cup\Omega_\kappa)) \tag{16}$$

for the solution $u(\cup\Omega_\kappa^o)$ of (11) defined on Ω^o, $\overline{\Omega^o} = \cup_{\kappa=1}^K \overline{\Omega_\kappa^o}$.

Here we consider the two cost functional

1. Energy functional: $F(u(\cup\Omega_\kappa)) = \mathscr{E}(u(\cup\Omega_\kappa); f, g, \cup\Omega_\kappa)$
2. Mean compliance: $F(u(\cup\Omega_\kappa)) = \sum_{\kappa=1}^{K}\left\{\int_{\Omega_\kappa} f \cdot u^\kappa\, dx + \int_{\Gamma_N\cap\partial\Omega_\kappa} g \cdot u^\kappa\, ds\right\}$

$$\text{i.e. } F(u(\cup\Omega_\kappa)) = -2\mathscr{E}(u(\cup\Omega_\kappa); f, g, \cup\Omega_\kappa) \tag{17}$$

By the use of the adjoint method, we can apply GJ-integral for the type $F(u(\cup\Omega_\kappa)) = \int_\Omega G(u(\cup\Omega_\kappa))\, dx$ for a given function $G(\xi) \in C^2(\mathbf{R}^m)$, $m = 1, 2, \ldots$. In this article, we assume that F is given by the above energy or mean compliance.

Let $dF(u(\cup\Omega_\kappa))[\mu]$ be the shape derivative of F, that is, for a vector field $\mu \in \widetilde{W}(\Omega; \mathbf{R}^3) \subset W^{1,2}(\Omega; \mathbf{R}^3)$

$$dF(u(\cup\Omega_\kappa))[\mu] = \lim_{t\to 0} t^{-1}\{F(u(\cup\Omega_\kappa(t))) - F(u(\cup\Omega_\kappa))\} \tag{18}$$
$$\Omega_\kappa(t) = \{x + t\mu(x) : x \in \Omega_\kappa\}$$

$dF(u(\cup\Omega_\kappa))[\mu]$ is written by GJ-integral as follows:

$$\begin{cases} -\sum_{\kappa=1}^{K}\left\{R_{\Omega_\kappa}(u^\kappa,\mu) + \int_{\Gamma_N\cap\partial\Omega_\kappa} f\cdot u^\kappa(\mu\cdot\mathbf{n}^\kappa)\,ds\right\} & \text{(Energy)} \\ 2\sum_{\kappa=1}^{K}\left\{R_{\Omega_\kappa}(u^\kappa,\mu) + \int_{\Gamma_N\cap\partial\Omega_\kappa} f\cdot u^\kappa(\mu\cdot\mathbf{n}^\kappa)\,ds\right\} & \text{(Mean compliance)} \end{cases} \tag{19}$$

Consider the variational formula(H^1-gradient method [2]): Find the vector field $\mu^o = (\mu_1^o, \mu_2^o, \mu_3^o) \in \widetilde{W}(\Omega;\mathbf{R}^3)$ such that

$$b(\mu^o,\eta) = -dF(u(\cup\Omega_\kappa))[\eta] \qquad \text{for all } \eta = (\eta_1,\eta_2,\eta_3) \in \widetilde{W}(\Omega;\mathbf{R}^3) \tag{20}$$
$$b(\zeta,\eta) = \sum_{i=1}^{3}\int_\Omega \{\nabla\zeta_i\cdot\nabla\eta_i + \zeta_i\eta_i\}\,dx$$

The map $[\eta \mapsto dF(u(\cup\Omega_\kappa))[\eta]]$ is continuous on $W^{1,\infty}(\Omega;\mathbf{R}^3)$, but we need the regularity of u in order to make the continuous extension of $[\eta \mapsto dF(u(\cup\Omega_\kappa))[\eta]]$ on $W^{1,2}(\Omega;\mathbf{R}^3)$. If $\mu^o \in W^{1,\infty}(\Omega;\mathbf{R}^3)$, then putting $\Omega_\kappa^o(t) = \{x + t\mu^o(x) : x \in \Omega_\kappa\}$, we have

$$\begin{aligned} F(u(\cup\Omega_\kappa^o(t))) &= F(u(\cup\Omega_\kappa)) + t\,dF(u(\cup\Omega_\kappa))[\mu^o] + o(t) \\ &= F(u(\cup\Omega_\kappa)) - t\,b(\mu^o,\mu^o) + o(t) \end{aligned}$$

Since $b(\mu^o,\mu^o) = \|\mu^o\|_{1,2,\Omega}^2$, there is a constant $\varepsilon > 0$ such that

$$F(u(\cup\Omega_\kappa^o(t))) \le F(u(\cup\Omega_\kappa)) \qquad \text{for all } 0 \le t \le \varepsilon \tag{21}$$

So that we can take $\{x + \varepsilon\mu^o(x) : x \in \Omega_\kappa\}$ for Ω_κ^o in (16).

5 Finite Element Analysis

In this section, Ω is the polygonal or polyhedral domain and $V_h(\Omega)$ the finite element space contained in $W^{1,2}(\Omega;\mathbf{R}^d)$, $d = 2, 3$. The finite element approximation u_h of the solution u of Problem (11) has the estimation by Céa's lemma

$$\|u - u_h\|_{1,2,\Omega} \le C\|u - P_h u\|_{1,2,\Omega}, \quad \|u - P_h u\|_{1,2,\Omega} = \inf_{v_h\in V_h(\Omega)} \|u - v_h\|_{1,2,\Omega}$$

with a constant $C > 0$ independent of h, where P_h is the projection from $W^{1,2}(\Omega; \mathbf{R}^d)$ to $V_h(\Omega)$. For each $\kappa = 1, \ldots, K$, $u^\kappa \in W^{1+\alpha,2}(\Omega_\kappa, \mathbf{R}^d)$ with a number $0 < \alpha \le 1$ (see e.g., [5]). The projection P_h has the estimations

$$\begin{aligned} \|(1 - P_h)v\|_{1,2,\Omega_\kappa} &\le \|v\|_{1,2,\Omega_\kappa} \quad \text{for all } v \in W^{1,2}(\Omega_\kappa; \mathbf{R}^d) \\ \|(1 - P_h)w\|_{1,2,\Omega_\kappa} &\le Ch\|w\|_{1,2,\Omega_\kappa} \quad \text{for all } w \in W^{2,2}(\Omega_\kappa; \mathbf{R}^d) \end{aligned} \tag{22}$$

with a number $C > 0$ independent of h. Applying the interpolation of operators [1, 7.23] to $1 - P_h$, we then have

$$\|(1 - P_h)u^\kappa\|_{1,2,\Omega_\kappa} \le Ch^\alpha \|u^\kappa\|_{1+\alpha,2,\Omega_\kappa} \tag{23}$$

with a constant $C > 0$ independent of h. For $R_{\Omega_\kappa}(\cdot, \mu)$, $\mu \in W^{1,\infty}(\Omega, \mathbf{R}^d)$, by the deformation such as

$$\sigma^\kappa_{ij}(u^\kappa)\sigma_{ij}(u^\kappa) - \sigma^\kappa_{ij}(u^\kappa_h)\sigma_{ij}(u^\kappa_h) = \sigma^\kappa_{ij}(u^\kappa - u^\kappa_h)\sigma_{ij}(u^\kappa) + \sigma^\kappa_{ij}(u^\kappa_h)\sigma_{ij}(u^\kappa - u^\kappa_h)$$

and the use of Schwarz's inequality, we have the estimation

$$|R_{\Omega_\kappa}(u^\kappa, \mu) - R_{\Omega_\kappa}(u^\kappa_h, \mu)| \le C\|u^\kappa - u^\kappa_h\|_{1,2,\Omega_\kappa} \tag{24}$$

with a constant C depending on $\|\mu\|_{1,\infty,\Omega}$, $\|u\|_{1,2,\Omega}$. By (23) and (24) and with the help of the trace theorem [1, 5.36], we can show that the shape derivatives $dF(u_h(\cup\Omega_\kappa))[\mu]$ approximate $dF(u(\cup\Omega_\kappa))[\mu]$ as $h \to 0$ if $\mu \in W^{1,\infty}(\Omega; \mathbf{R}^d)$.

In shape optimization of interfaces, there are cases where we make $\partial\Omega_\kappa$ change under the condition fixed to $\partial\Omega$, when $\overline{\Omega_\kappa} \subset \Omega$. In such case, for an open neighborhood ω of Ω_κ, we use the cut-off function $\chi_\omega \in C^1(\mathbf{R}^d)$, $\chi_\omega \equiv 1$ on ω and $\operatorname{supp} \chi_\omega \subset \Omega$, and apply $R_{\Omega_\kappa}(u^\kappa, \chi_\omega\mu)$ instead of $R_{\Omega_\kappa}(u^\kappa, \mu)$ in (19), which is called *localization method*. Here we proved that $J_\omega(u, \mu) = R_\Omega(u, \chi_\omega\mu)$.

In shape optimization of interfaces $\Gamma_{\kappa\ell}$, we must avoid the influence from the boundary $\partial\Omega$. On that account, there are two methods. First is to fix the shape of $\partial\Omega$ (see Example 1). Second is to use localization method (see Example 2). In the calculation, we used FreeFem++ [4].

Example 1 First example is the bar illustrated as in Fig. 1, $\Omega = (0, 9) \times (-1, 1)$, $\Omega_1 = (0, 4.5) \times (-1, 1)$, $\Omega_2 = (4.5, 9) \times (-1, 1)$, and assumed that the material is hard in Ω_1 and soft in Ω_2. Here we find μ^o such that $\mu^o = 0$ on Γ_D and $\mu^o_2 = 0$ on Γ_N, the body force $f = (0, -1)$ in Ω and the surface force $g = (0, 0)$ on Γ_N. Initially, the energy is -7.10564 and the mean compliance is $\int_\Omega f \cdot u\, dx = 14.2113$. In energy minimization problem, Ω_1 will be going out, but in mean compliance problem Ω_2 will be going out.

1. Energy minimization: In Fig. 2, the vector fields μ^o after 5, 20, 55 times iteration and the meshes after 5, 20, 60 times iteration are shown. The "volume" shows the volume of Ω_2. In the final shape, the energy becomes -19.4386.

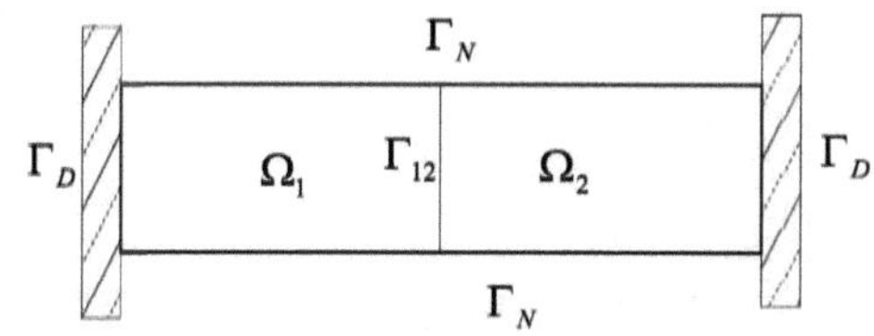

Fig. 1 The bar $(0, 9) \times (-1, 1)$ is fixed on two sides Γ_D and stress-free on the upper and lower side Γ_N. Young's modulus and Poisson's radio are $E = 200, 20$ and $\nu = 0.25, 0.3$ in Ω_1, Ω_2 respectively. Here $\Gamma_{12} = \{4.5\} \times (-1, 1)$

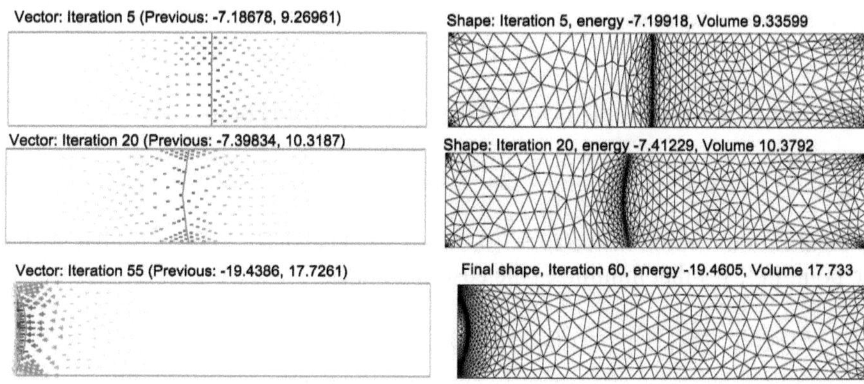

Fig. 2 Energy minimization

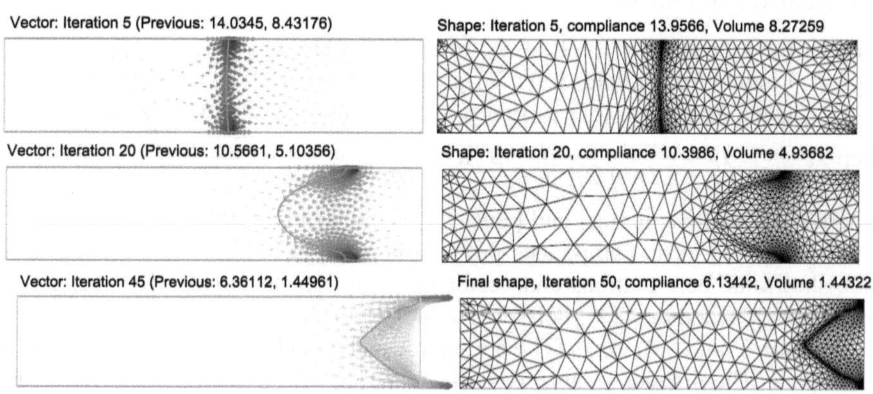

Fig. 3 Mean compliance minimization

2. Mean compliance minimization: In Fig. 3, the vector fields μ^o after 5, 20, 45 times iteration and the meshes after 5, 20, 50 times iteration are shown. In the final shape, the mean compliance becomes 6.13442.

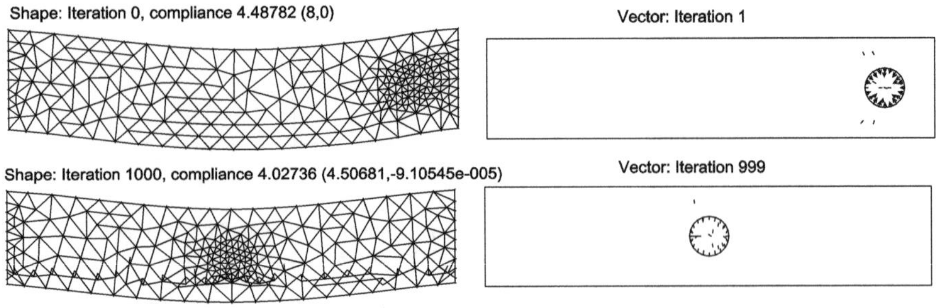

Fig. 4 Mean compliance minimization by localization method

Example 2 Second example is to find the optimal location of the disc

$$\Omega_2 = \{(x_1, x_2) : (x_1 - x_c)^2 + (x_2 - y_c)^2 < 0.4^2\}$$

by localization method inside $\Omega = (0, 9) \times (-1, 1)$ and $\Omega_1 = \Omega \setminus \overline{\Omega_2}$, that is, find the point (x_c^o, y_c^o) that minimizes the cost functional.

$$\begin{aligned} &F(u(\Omega_1^o \cup \Omega_2^o)) \leq F(u(\Omega_1 \cup \Omega_2)) \\ &\Omega_2^o = \{(x_1, x_2) : (x_1 - x_c^o)^2 + (x_2 - y_c^o)^2 < 0.4^2\} \quad \Omega_1^o = \Omega \setminus \overline{\Omega_2^o} \end{aligned} \tag{25}$$

The method is as follows:

Step 1: Find the solution $u(\Omega_1 \cup \Omega_2)$ of Problem (11).
Step 2: Find the vector field μ^o by H^1-gradient method (20) by localization method using $\chi_\omega \equiv 1$ on $\{(x_1, x_2) : (x_1 - x_c)^2 + (x_2 - y_c)^2 < 0.6^2\}$.
Step 3: $x_c^o = x_c + \dfrac{\varepsilon}{\mathrm{Len}(\partial\Omega_2)} \displaystyle\int_{\partial\Omega_2} \mu_1^o \, ds, \quad y_c^o = y_c + \dfrac{\varepsilon}{\mathrm{Len}(\partial\Omega_2)} \displaystyle\int_{\partial\Omega_2} \mu_2^o \, ds$

where ε is a positive number and $\mathrm{Len}(\partial\Omega_2)$ is the length of $\partial\Omega_2$.

The result in mean compliance problem is given in Fig. 4, where the left-hand side shows the deformations and vector fields in right-hand side. The material is hard (Young's modulus 200, Poisson's ratio 0.25) in Ω_1 and soft (Young's modulus 20, Poisson's ratio 0.3) in Ω_2. The body force and boundary condition is same as in Example 1. The first compliance is 4.48782 and $(x_c, y_c) = (8, 0)$. It was necessary to make ε small not to go out. After 1000 times of iteration, the compliance becomes 4.02736 and $(x_c^o, y_c^o) = (4.50681, -9.10545 \times 10^{-5})$ shows almost the center of the bar. In the case that the material is soft in Ω_1 and hard in Ω_2, Ω_2^o want to go out of Ω.

Acknowledgements I would like to express my gratitude to Prof. Kimura and to Prof. O. Pironneau and Prof. F. Hecht for FreeFem++. I am deeply grateful to Prof. Azegami who provided the knowledge on shape optimization. This work was supported by JSPS KAKENHI Grant Number 16K05285.

References

1. Adams, R.A.: Sobolev Spaces, 2nd edn. Academic Press (2003)
2. Azegami, H., Wu, Z.: Domain optimization analysis in linear elastic problems: Approach using traction method. JSME Int. J. Ser. A **39**(2), 272–278 (1996)
3. Cherepanov, G.P.: On crack propagation in continuous media. Prikl. Math. Mekh. **31**, 476–493 (1967)
4. Hecht, F.: New development in FreeFem++. J. Numer. Math. **20**(3-4), 251–265 (2012). 65Y15, http://www.freefem.org. Accessed 25 Feb 2016
5. Knees, D., Sändig, A.-M.: Regularity of elastic fields in composites. In: Lecture Notes in Applied and Computational Mechanics, vol. 28, pp. 331–360. Springer (2006)
6. Ohtsuka, K.: Generalized J-integral and three dimensional fracture mechanics I. Hiroshima Math. J. **11**, 21–52 (1981)
7. Ohtsuka, K.: Generalized J-integral and its applications. I.—Basic theory. Jpn. J. Appl. Math. **2**, 329–350 (1985)
8. Ohtsuka, K.: Mathematical theory on perturbation of singular points in continuum mechanics and its application to fracture and shape optimization, Mathematics for Industry Research, No. 2, http://www.imi.kyushu-u.ac.jp/files/imipublishattachment/file/math_550a1d94872c6.pdf
9. Ohtsuka, K., Khludnev, A.: Generalized J-integral method for sensitivity analysis of static shape design. Control Cybern. **29**, 513–533 (2000)
10. Ohtsuka, K., Kimura, M.: Differentiability of potential energies with a parameter and shape sensitivity analysis for nonlinear case: the p-Poisson problem. Jpn. J. Ind. Appl. Math. **29** (2012)
11. Rice, J.R.: A path-independent integral and the approximate analysis of strain concentration by notches and cracks. J. Appl. Mech. **35**, 379–386 (1968)

Shape Optimization Approach by Traction Method to Inverse Free Boundary Problems

Shogen Shioda, Ahsani Ummi Maharani, Masato Kimura, Hideyuki Azegami and Kohji Ohtsuka

1 Introduction

The importance of the optimal shape design has been increasing in the present industrial design due to the request to make their production more efficient. Actually, a number of techniques and algorithms for shape optimization have been proposed in engineering and industry [1, 11, 12].

Mathematical analysis is playing an essential role there, mainly in the following three processes: (1) choice of a cost functional, (2) derivation of the shape derivative of the cost functional, (3) shape deformation to reduce the cost functional. Concerning the choice of the cost functional, mathematical study for the existence and uniqueness of the minimizer is important. For the second process which is often called the shape sensitivity analysis, there have been many mathematical studies. However, not so many mathematical studies have dealt with the shape deformation process. In this paper, we focus on the traction method which was proposed by one of the authors

S. Shioda · A.U. Maharani · M. Kimura (✉)
Kanazawa University, Kanazawa, Japan
e-mail: mkimura@se.kanazawa-u.ac.jp

S. Shioda
e-mail: sho02345@stu.kanazawa-u.ac.jp

A.U. Maharani
Bandung Institute of Technology, Bandung, Indonesia
e-mail: maharaniahsani@gmail.com

H. Azegami
Nagoya University, Nagoya, Japan
e-mail: azegami@is.nagoya-u.ac.jp

K. Ohtsuka
Hiroshima Kokusai Gakuin University, Hiroshima, Japan
e-mail: ohtsuka@edu.hkg.ac.jp

H. Itou et al. (eds.), *Mathematical Analysis of Continuum Mechanics and Industrial Applications*, Mathematics for Industry 26,
DOI 10.1007/978-981-10-2633-1_8

for stable shape deformation (see detail in [2]). The traction method and its variants are widely used in the optimal shape design in engineering and industry.

We consider an optimal shape design approach to a free boundary problem (1) arising from inverse potential problems, where the cost functional is defined by a boundary integral. We derive its shape derivative and give a weak formulation of the traction method. The geometrical difficulty caused by the presence of a curvature term can be dissolved by introducing the weak form.

For a special case of the free boundary problem (1) in plane which will be considered in Sect. 5, all the solutions are classified and described by using the conformal mapping method [4, 8]. We give several numerical examples and compare them with the exact solutions. Often in industrial examples, because of many constraints and complicated geometry, precise property or even existence and uniqueness of the exact solution is not clear in most cases. In this sense, our target problem (1) is appropriate for the mathematical study of the optimal shape design problem.

The outline of this paper is as follows. In Sect. 2, the free boundary problem and our optimal shape design approach to it are described. We derive the shape derivative of the cost functional in Sect. 3, and give a weak formulation of the traction method in Sect. 4. In Sect. 5, we introduce some numerical examples and show efficiency of the method.

2 Free Boundary Problem and Shape Optimization

Let $d \geq 2$ be an integer. We suppose that $\mu \in L^2(\mathbb{R}^d)$ be a given function with a compact support, and p and q are given smooth functions defined on $\mathbb{R}^d$. We define

$$G := \{\Omega \mid \Omega \text{ is a bounded Lipschitz domain in } \mathbb{R}^d, \operatorname{supp}(\mu) \subset \Omega\}.$$

We consider the following free boundary problem in $\mathbb{R}^d$.

Problem 2.1 Find (u, Ω) such that

$$\begin{cases} -\Delta u = \mu \text{ in } \Omega \\ u = p \text{ on } \Gamma \\ \dfrac{\partial u}{\partial n} = q \text{ on } \Gamma, \end{cases} \tag{1}$$

where $\Omega \in G$ and n is the outward unit normal vector on the boundary $\Gamma := \partial\Omega$.

Problem 2.1 and similar problems appear in mathematical theory of quadrature surface [4, 10] and in engineering problems such as electromagnetic casting [5, 9] and equilibrium shape of plasma [6].

We remark that Problem 2.1 is ill-posed in general. For example, if $\mu(x) \geq 0$ and $q(x) > 0$ then there is no solution since Γ has to satisfy

$$\int_{\mathbb{R}^d} \mu \, dx = -\int_\Omega \Delta u \, dx = -\int_\Gamma q \, ds.$$

We apply the following shape optimization approach to this problem. We fix $\beta > 0$ and define $\tilde{q}(x) := \beta p(x) + q(x)$. Then for given $\Omega \in G$ with $\Gamma = \partial\Omega$, we denote by $u_\Omega \in H^1(\Omega)$ a unique weak solution to the following boundary value problem:

$$u_\Omega : \begin{cases} -\Delta u = \mu & \text{in } \Omega \\ \beta u + \dfrac{\partial u}{\partial n} = \tilde{q} & \text{on } \Gamma. \end{cases} \tag{2}$$

The weak form is given as

$$u_\Omega \in H^1(\Omega) : \ a_\Omega(u, v) = l_\Omega(v) \quad (\forall v \in H^1(\Omega)), \tag{3}$$

where

$$a_\Omega(u, v) := \int_\Omega \nabla u \cdot \nabla v \, dx + \beta \int_\Gamma uv \, ds, \quad l_\Omega(v) := \int_\Omega \mu v \, dx + \int_\Gamma \tilde{q} v \, ds$$

If $u_\Omega = p$ on Γ then (u_Ω, Ω) is a solution for Problem 2.1. We define a cost functional as follows:

$$J(\Omega) := \frac{1}{2} \int_\Gamma |u_\Omega - p|^2 ds \quad (\Omega \in G). \tag{4}$$

We consider the optimal shape design problem to minimize the cost functional $J(\Omega)$ among $\Omega \in G$. We remark that (u_Ω, Ω) is a solution to Problem 2.1 if and only if $J(\Omega) = 0$ and $u = u_\Omega$.

3 Variation Formula of the Cost Functional

In this section, we consider the variation of $J(\Omega)$ with respect to Ω, in other words, the shape derivative (shape gradient) of $J(\Omega)$. This step is usually called the shape sensitivity analysis. We use the Lagrange multiplier method to derive the variation formula.

We suppose $\Omega \in G$. For a vector field $V \in W^{1,\infty}(\mathbb{R}^d, \mathbb{R}^d)$ and for $t > 0$, we define

$$\Omega(t) := \{x + tV(x) \mid x \in \Omega\} \qquad (0 \le t < t_0). \tag{5}$$

For sufficiently small $t > 0$, $\Omega(t)$ belongs to G. We put $\Gamma := \partial\Omega$ and $\Gamma(t) := \partial\Omega(t)$. We have the following variational formulas for integrals on $\Omega(t)$ and $\Gamma(t)$.

Proposition 3.1 *Let f, g be functions defined in $\mathbb{R}^d$. We suppose that Γ, V, f, g are sufficiently smooth and denote the outward unit normal vector on Γ by n. Then we have*

$$\frac{d}{dt}\int_{\Omega(t)} f(x)\,dx\Big|_{t=0} = \int_{\Gamma}(V\cdot n) f\,ds, \tag{6}$$

$$\frac{d}{dt}\int_{\Gamma(t)} g(x)\,ds\Big|_{t=0} = \int_{\Gamma}(\nabla g\cdot V + g\,\mathrm{div}_{\Gamma} V)\,ds, \tag{7}$$

$$= \int_{\Gamma}(V\cdot n)\left(g\kappa + \frac{\partial g}{\partial n}\right)ds, \tag{8}$$

where $\kappa = \kappa_1 + \cdots + \kappa_{d-1}$ is the sum of principal curvatures of Γ.

The surface divergence div_{Γ} [7, 12] is defined as

$$\mathrm{div}_{\Gamma} V := \mathrm{div} V - \frac{\partial V}{\partial n}\cdot n \quad \text{on } \Gamma,$$

for V of C^1-class. Proofs of these formulas are found in (2.168), (2.172) and (2.174) of [12], for example. We also refer to [7] for (6) and (8).

Using this proposition, we derive the shape derivative of the cost functional J. For $\Omega \in G$ and $u, v \in H^1(\Omega)$, we define functionals J_0, L_M and L as follows.

$$\begin{aligned} J_0(\Omega, u) &:= \frac{1}{2}\int_{\Gamma}|u-p|^2\,ds,\\ L_M(\Omega, u, v) &:= a_{\Omega}(u, v) - l_{\Omega}(v),\\ L(\Omega, u, v) &:= J_0(\Omega, u) + L_M(\Omega, u, v). \end{aligned} \tag{9}$$

Since u_{Ω} is the weak solution defined by (3), for all $v \in H^1(\Omega)$, we have

$$L_M(\Omega, u_{\Omega}, v) = 0, \quad J(\Omega) = J_0(\Omega, u_{\Omega}) = L(\Omega, u_{\Omega}, v).$$

We fix sufficiently large $R > 0$ such that $\overline{\Omega(t)} \subset B_R := \{x \in \mathbb{R}^d;\ |x| < R\}$. We take $u(t) = u(\cdot, t) \in H^1(B_R)$ which satisfies $u(t)|_{\Omega(t)} = u_{\Omega(t)}$ and also put $u_0 := u(0)$. Then, for arbitrary $v_0 \in H^1(B_R)$, from $J(\Omega(t)) = L(\Omega(t), u(t), v_0)$, we obtain

$$\frac{d}{dt}J(\Omega(t))\Big|_{t=0} = A_1 + A_2,$$

$$A_1 := \frac{d}{dt}L(\Omega(t), u_0, v_0)\Big|_{t=0}, \quad A_2 := \frac{d}{dt}L(\Omega, u(t), v_0)\Big|_{t=0}.$$

In the following, we assume that the right-hand sides of A_1 and A_2 are sufficiently smooth with respect t. We put $\varphi_0(x) := \frac{\partial}{\partial t}u(x,t)|_{t=0}$ and suppose that $\varphi_0 \in H^1(B_R)$.

For the term A_2, we have

$$\begin{aligned} A_2 &= \frac{d}{dt}\left\{\frac{1}{2}\int_\Gamma |u(t)-p|^2\,ds + a_\Omega(u(t),v_0) - l_\Omega(v_0)\right\}\Big|_{t=0} \\ &= \int_\Gamma \varphi_0(u_0-p)\,ds + a_\Omega(\varphi_0, v_0) \\ &= a_\Omega(v_0,\varphi_0) - \tilde{l}_\Omega(\varphi_0) \end{aligned}$$

where $\tilde{l}_\Omega(\varphi) := \int_\Gamma \varphi(p-u_\Omega)ds$ for $\varphi \in H^1(\Omega)$. We choose v_0 such that

$$a_\Omega(v_0,\varphi) = \tilde{l}_\Omega(\varphi) \qquad (\forall\varphi \in H^1(\Omega)).$$

Then $A_2 = 0$ follows. In other words, $v_0|_\Omega = v_\Omega$ holds, where v_Ω is a unique weak solution of the adjoint problem:

$$v_\Omega : \begin{cases} \Delta v = 0 & \text{in } \Omega \\ \dfrac{\partial v}{\partial n} + \beta v = p - u_\Omega & \text{on } \Gamma. \end{cases} \tag{10}$$

We call problems (2) and (10) the primal problem and its adjoint problem, respectively.

Hence, we obtain $\frac{d}{dt}J(\Omega(t))|_{t=0} = A_1 = \frac{d}{dt}L(\Omega(t),u_0,v_0)|_{t=0}$. For simplicity, we define

$$f := \nabla u_0 \cdot \nabla v_0, \qquad g := \frac{1}{2}|u_0-p|^2 + \beta u_0 v_0 - \tilde{q} v_0.$$

Then $L(\Omega(t),u_0,v_0)$ is written in the form:

$$L(\Omega(t),u_0,v_0) = \int_{\Omega(t)} (f-\mu v_0)\,dx + \int_{\Gamma(t)} g\,ds. \tag{11}$$

If f and g are enough smooth, the variation of $J(\Omega)$ is derived from (11) using Proposition 3.1 as follows.

$$\begin{aligned} \frac{d}{dt}J(\Omega(t))\Big|_{t=0} &= \frac{d}{dt}L(\Omega(t),u_0,v_0)\Big|_{t=0} \\ &= \int_\Gamma \{(V\cdot n)f + V\cdot\nabla g + (\mathrm{div}_\Gamma V)g\}\,ds \qquad (12) \\ &= \int_\Gamma V\cdot n\left(f + g\kappa + \frac{\partial g}{\partial n}\right) ds. \qquad (13) \end{aligned}$$

4 Traction Method

In this section, we give a brief introduction of the traction method proposed by [2]. The traction method is also called H^1 gradient method. Under the setting of (5), the variation of a cost functional $J(\Omega)$ has the following form in general:

$$\frac{d}{dt}J(\Omega(t))\Big|_{t=0} = \int_\Gamma V \cdot n\, B_n\, ds = \int_\Gamma V \cdot B\, ds,$$

where $B_n : \Gamma \to \mathbb{R}$ and $B : \Gamma \to \mathbb{R}^d$ are independent of the choice of V. We call B_n a scalar variation of $J(\Omega)$ and $B(= B_n n)$ a vector-valued variation. For $J(\Omega)$ defined by (4), from (12) and (13), we have

$$\begin{aligned} B_n &= f + g\kappa + \frac{\partial g}{\partial n} \quad \text{on } \Gamma, \\ \int_\Gamma B \cdot V\, ds &= \int_\Gamma \{(V \cdot n) f + V \cdot \nabla g + (\mathrm{div}_\Gamma V)\, g\}\, ds, \end{aligned} \tag{14}$$

for a smooth vector field V.

In the conventional method of shape optimization, the boundary $\Gamma = \partial\Omega$ is modified in the normal direction by using the scalar variation B_n on Γ. Namely, for sufficiently small $\eta > 0$, we deform the boundary as

$$\Gamma_{\text{new}} := \{x - \eta B_n(x) n(x);\ x \in \Gamma\}.$$

Let V be a smooth extension of the vector field $-B_n n$. Then, formally, we have

$$\begin{aligned} J(\Omega_{\text{new}}) &= J(\Omega(\eta)) \\ &= J(\Omega) + \eta \frac{d}{dt} J(\Omega(t))|_{t=0} + O(\eta^2) \\ &= J(\Omega) + \eta \int_\Gamma (V \cdot n) B_n\, ds + O(\eta^2) \\ &= J(\Omega) - \eta \int_\Gamma |B_n|^2\, ds + O(\eta^2) < J(\Omega), \end{aligned} \tag{15}$$

for small η, if $B_n \not\equiv 0$. However, this method often fails numerically due to strong instability and some additional regularization technique is required. In particular, if the cost functional is defined by a boundary integral like (4), the scalar variation B_n of $J(\Omega)$ contains the curvature term κ of Γ due to the formula (8). Besides, we cannot avoid regeneration or adaptation of finite element mesh of Ω_{new}.

On the other hand, the main idea of the traction method is that we treat domain Ω as a virtual elastic body and repeat small deformation using traction force on the boundary given by the vector-valued variation B of $J(\Omega)$.

Let D be a Lipschitz subdomain with $\overline{D} \subset \Omega$. We solve the following artificial elasticity problem.

$$\begin{cases} -\mathrm{div}\, \sigma[w] = 0 & \text{in } \Omega \setminus \overline{D} \\ \sigma[w]n = -B & \text{on } \Gamma \\ w = 0 & \text{on } \partial D, \end{cases} \tag{16}$$

where $w(x) \in R^d$ is a displacement $\overline{\Omega} \backslash D$. The strain tensor is defined by $e[w] = (\nabla^{\mathrm{T}} w + \nabla w^{\mathrm{T}})/2 \in \mathbb{R}^{d\times d}_{\mathrm{sym}}$ and the stress tensor $\sigma[w]$ is defined by $\sigma[w] = \lambda(\mathrm{div}\, w)I + 2\mu e[w]$ with suitable positive Lamé constants λ and μ.

The weak form of this problem is

$$w \in H^1_D(\Omega;\, \mathbb{R}^d): \int_\Omega \sigma[w] : e[\varphi]\, dx = -\int_\Gamma B \cdot \varphi\, ds \quad (\forall \varphi \in H^1_D(\Omega;\, \mathbb{R}^d)),$$

where $H^1_D(\Omega;\, \mathbb{R}^d) := \{w \in H^1(\Omega;\, \mathbb{R}^d);\ w|_D = 0\}$.

We solve the above weak form numerically by using the finite element method. The surface traction term $\int_\Gamma B \cdot \varphi\, ds$ can be evaluated by using the formula (14). This enables us not to treat the curvature term numerically.

We deform not only the boundary but also the whole domain by w as follows:

$$\Omega_{\mathrm{new}} = \{x + \eta w(x) |\, x \in \Omega\},$$

with small $\eta > 0$. A similar estimate to (15) also formally holds:

$$\begin{aligned} J(\Omega_{\mathrm{new}}) &= J(\Omega) + \eta \int_\Gamma B \cdot w\, ds + O(\eta^2) \\ &= J(\Omega) - \eta \int_\Omega \sigma[w] : e[w]\, dx + O(\eta^2) < J(\Omega), \end{aligned} \tag{17}$$

for small $\eta > 0$, if $|B_n| = |B| \not\equiv 0$.

Summarizing the above, the algorithm for Problem 2.1 by the traction method becomes:

1. Choose an initial domain and generate its finite element mesh.
2. Solve u_Ω and v_Ω using finite element method.
3. Solve the artificial elasticity problem (16) using finite element method.
4. Deform the domain by $\Omega_{\mathrm{new}} = \{x + \eta w(x) |\, x \in \Omega\}$ together with the nodal points of the mesh.
5. Repeat 2–4 until Ω converges.

Some numerical examples of the traction method are shown in the next section. Several benefits of this method will be summarized in the conclusion.

5 Numerical Examples

We consider the following free boundary problem which arises in the theory of quadrature surface [4, 10]. We choose N points $\xi_j \in \mathbb{R}^2$ $(j = 1, \ldots, N)$ in $\mathbb{R}^2$, and denote the Dirac measure at ξ_j by δ_{ξ_j}. We define $\mu := \sum_{j=1}^{N} \alpha\delta_{\xi_j}$, where $\alpha > 0$. We consider Problem 2.1 with $d = 2$, $p = 0$, $q = -1$ and the above μ in this section. This μ does not belong to $L^2(\mathbb{R}^2)$ but a notion of weak solutions can be defined [4, 8]. We also remark that we can always replace μ by $\mu^\varepsilon := \sum_{j=1}^{N} \alpha\delta^\varepsilon_{\xi_j}$, where

$$\delta^\varepsilon_\xi(x) := \begin{cases} \dfrac{1}{\pi\varepsilon^2} & |x-\xi| < \varepsilon \\ 0 & |x-\xi| \geq \varepsilon. \end{cases}$$

for sufficiently small $\varepsilon > 0$. It is shown from the property of harmonic functions that the solution (u, Ω) is affected by the replacement of μ by μ^ε only on the support of μ^ε.

For the cases $N = 1$ and $N = 2$, all the weak solutions to Problem 2.1 are classified by using the conformal mapping method by Henrot [4] as follows. Also in [8], we can find the definition of the weak solution, detail proofs of the theorems, some figures of the solutions.

Suppose $N = 1$. Then (u, Ω) is a weak solution of Problem 2.1, if and only if

$$\begin{cases} \Omega = B\left(\xi_1, \frac{\alpha}{2\pi}\right) \\ u = \dfrac{\alpha}{2\pi}\log\dfrac{\alpha}{2\pi|x-\xi_1|}, \end{cases}$$

where $B(\xi, r) := \{x \in \mathbb{R}^2;\ |x-\xi| < r\}$ for $\xi \in \mathbb{R}^2$.

For the case $N = 2$, we denote the unit disc in $\mathbb{C}$ by $D_0 := B(0, 1) \subset \mathbb{R}^2 \cong \mathbb{C}$ and define a conformal mapping on D_0 by

$$\Phi_a(z) := \frac{\alpha(1-a^4)}{4\pi a^2}\left[\frac{-2z}{z^2 - 1/a^2} + a\log\frac{1/a + z}{1/a - z}\right] \tag{18}$$

for $0 < a < 1$.

Table 1 Number of the solutions

$\alpha/(2c)$	0	...	2.300...	...	2.827...	...	π	...
# connected convex solution	0	0	0	0	1	1	1	1
# connected nonconvex solution	0	0	1	2	1	1	0	0
# connected solution	0	0	1	2	2	2	1	1
# disconnected solution	0	1	1	1	1	1	0	0

Theorem 5.1 (Henrot [4]) *Let $N = 2$. We suppose that $c > 0$ and $\xi_1, \xi_2 \in \mathbb{C} \cong \mathbb{R}^2$ are given as $\xi_1 = c$ and $\xi_2 = -c$.*

1. *(u, Ω) is a weak solution to* (1) *and Ω is connected, if and only if there exists $a \in (0, 1)$ such that $c = \Phi_a(a)$, $\Omega = \Phi_a(D_0)$, and $u(\xi) = w(\Phi_a^{-1}(\xi))$, $\xi \in \Omega$, where*

$$w(z) := \frac{\alpha}{2\pi} \log \left| \frac{1 - a^2 z^2}{z^2 - a^2} \right| \qquad (z \in D_0). \tag{19}$$

2. *(u, Ω) is a weak solution to* (1) *and Ω is disconnected if and only if $\frac{\alpha}{2\pi} < c$ and Ω consists from two discs $\Omega = B(c, \frac{\alpha}{2\pi}) \cup B(-c, \frac{\alpha}{2\pi})$.*

Due to Theorem 5.1, we can classify the number of solutions depending on the value α/c as shown in Table 1.

In the following three numerical examples, we choose $N = 2$ and $\alpha = 3$. We also take numerical parameters as $\varepsilon = 0.1$ and $\eta = 0.1$. We use finite element method software FreeFem++ [3] with P2 element together with the algorithm of traction method explained in Sect. 4. The step 4 to compute Ω_{new} in the traction method is realized using the function movemesh() of FreeFem++. We also redivide the domain by using the function adaptmash(Th, 1, hmax = 0.02) in following numerical calculation at every M iterations.

Figures 1 and 2 are the numerical examples both for the case $\alpha = 3$, $c = 0.625$. In this case, there are just two exact solutions, which are both nonconvex. The exact solutions are drawn in the figures for comparison. The initial shape of Ω is a circle in Fig. 1 and is a dumbbell domain in Fig. 2. Since there are two solutions, the limit shape can depend on the choice of the initial shape. We can see the numerical solutions converge to different exact solutions. The history of the cost functionals are also shown in the graphs. In the iteration 2000 of Fig. 1 and in the iteration 6000 of Fig. 2, although the value of cost seem to be enough small, the shape of the solutions has not converged completely. This shows that it is not easy to judge the convergence of the shape only from the value of the cost functional. We have chosen $M = 500$ and $M = 4000$ in the simulation of Figs. 1 and 2, respectively.

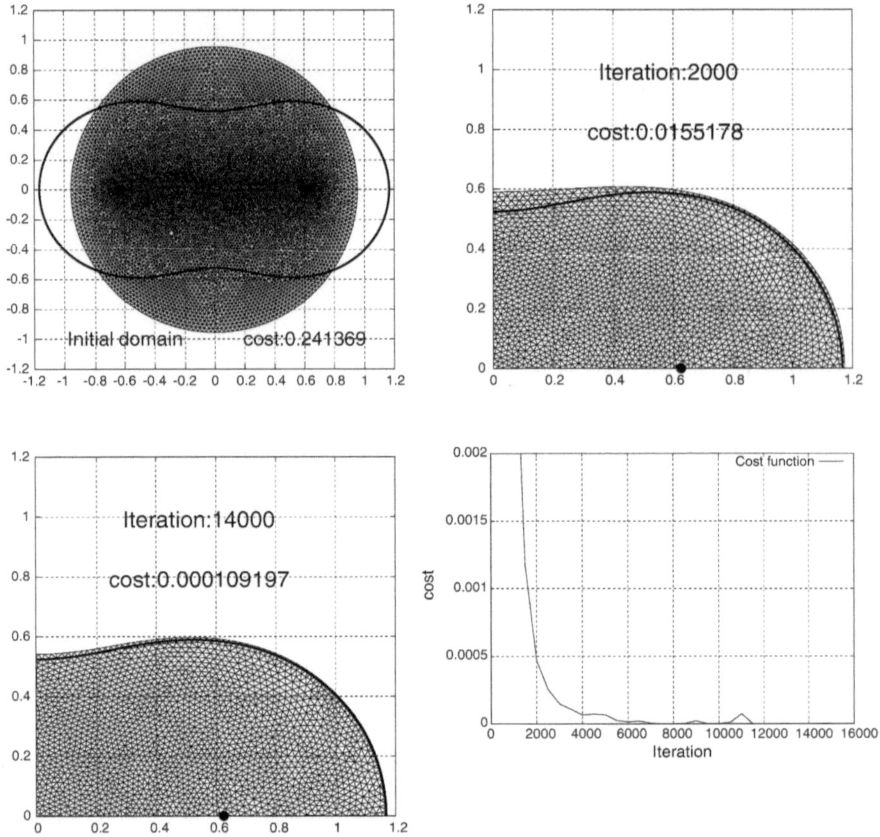

Fig. 1 $\alpha = 3$, $c = 0.625$, the shape of Ω and the finite element mesh at iteration 0, 2000, 140,000 and the history of the cost functional

In Fig. 3, we computed the case $\alpha = 3$, $c = 0.5306$. In this case, there are also two exact solutions [8], one is convex and the other is an hourglass form with a very thin neck. To catch the hourglass solution, we had to start from a very close domain to the exact one as shown in Fig. 3. We could capture the thin neck shape numerically as shown in the enlarged view at the lower left in Fig. 3 with the help of adaptmesh() with $M = 2000$ in this case.

6 Conclusions

We investigated a shape optimization approach to free boundary problems governed by the Poisson equation using the traction method. Although the scalar variation of the cost functional includes the curvature term, its geometric difficulty is dissolved

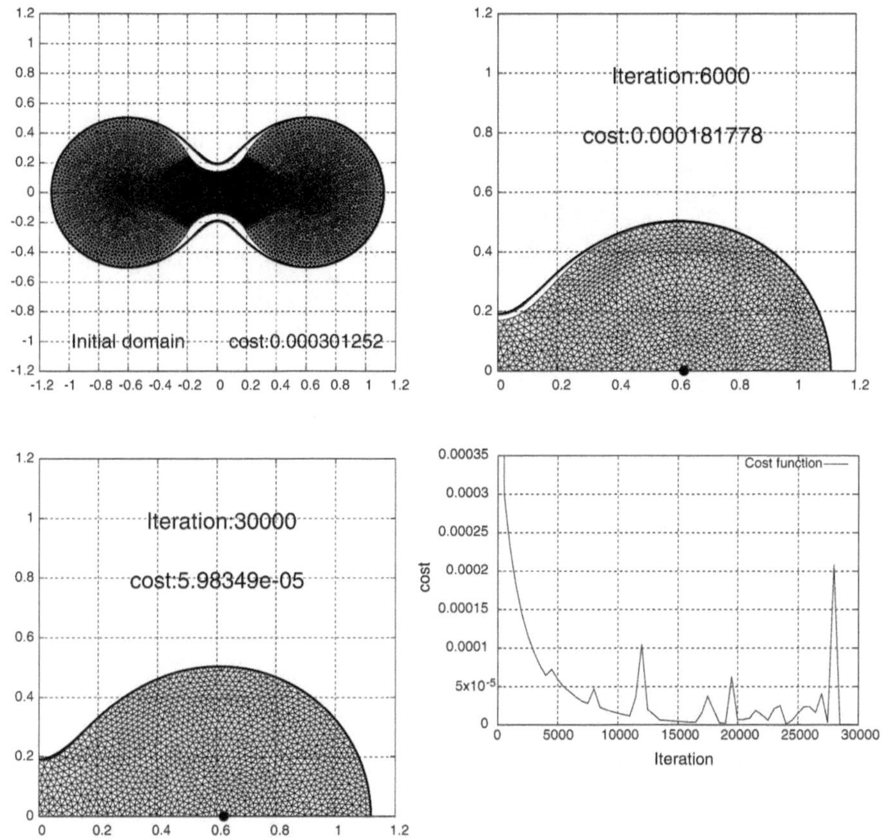

Fig. 2 $\alpha = 3, c = 0.625$, the shape of Ω and the finite element mesh at iteration 0, 6000, 30,000 and the history of the cost functional

by using the weak form of the vector-valued variation combined with the traction method. Besides, based on the positivity of the Neumann-to-Dirichlet map of elasticity problem (i.e., traction-to-displacement map), the traction method realizes both regularizing effect and decreasing property of the cost functional $J(\Omega)$ as shown in (17). Moreover, the frequency of mesh generation can be reduced since the finite element mesh on the new domain is automatically generated in the traction method. We also remark that the use of the elasticity problem has an effect to prevent the triangle element from degenerating.

Through the test problems for which the exact solutions are known, we could check the efficiency of the traction method numerically in Sect. 5. We can observe that the obtained domain approaches to one of the exact solutions in the iterations and the limit shape depends on the choice of the initial domain.

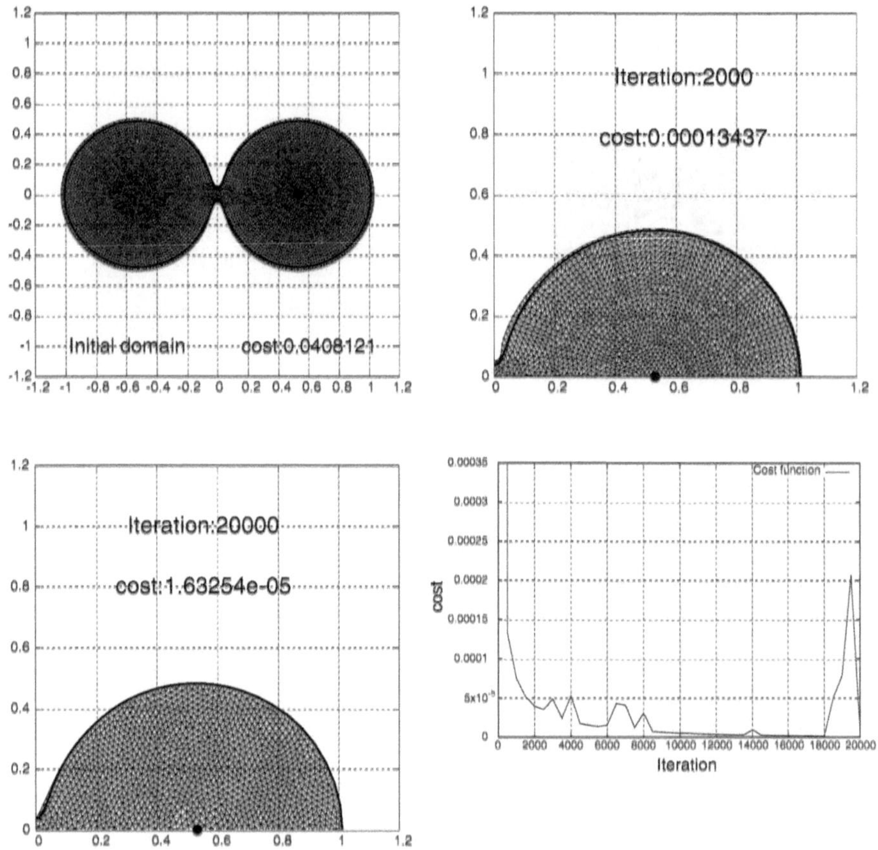

Fig. 3 $\alpha = 3$, $c = 0.5306$, the shape of Ω and the finite element mesh at iteration 0, 2000, 20,000 and the history of the cost functional

Mathematical study for the traction method such as its convergence and stability analysis is desirable and important even from the viewpoint of engineering. We think that the framework given in this work provides a good mathematical problem for such further analysis, although it seems to be more challenging at present.

References

1. Allaire, G.: Shape optimization by the homogenization method. Applied Mathematical Sciences, vol. 146, Springer (2002)
2. Azegami, H., Takeuchi, K.: A smoothing method for shape optimization: traction method using the Robin condition. Int. J. Comput. Methods **3**(1), 21–33 (2006)
3. Hecht, F.: New development in FreeFem++. J. Numer. Math. **20**(3-4), 251–265 (2012)

4. Henrot, A.: Subsolutions and supersolutions in a free boundary problem. Ark. Mat. **32**, 78–98 (1994)
5. Henrot, A., Pierre, M.: About existence of equilibria in electromagnetic casting. Q. Appl. Math. **49**(3), 563–575 (1991)
6. Kawarada, H., Sawaguri, T., Imai, H.: An approximate resolution of a free boundary problem appearing in the equilibrium plasma by means of conformal mapping. Jpn. J. Appl. Math. **6**, 331–340 (1989)
7. Kimura, M.: Geometry of hypersurfaces and moving hypersurfaces in R^m—for the study of moving boundary problems—. Jindřich Nečas Center for Mathematical Modeling Lecture notes Volume IV, Topics in Mathematical Modeling, pp. 39–93 (2008)
8. Maharani, A.U., Kimura, M., Azegami, H., Ohtsuka, K., Armanda, I.: Shape optimization approach to a free boundary problem. Recent Dev. Comput. Sci. **6**, 42–55 (2015)
9. Morisue, T., Yajima, T., Kume, T., Fujimori, S.: Analysis of electromagnetic force for shaping the free surface of a molten metal in a cold crucible. IEEE Trans. Magn. **29**(2), 1562–1565 (1993)
10. Onodera, M.: Geometric flows for quadrature identities. Math. Ann. **361**, 77–106 (2015)
11. Pironneau, O.: Optimal Shape Design for Elliptic Systems. Springer (1984)
12. Sokolowski, J., Zolésio, J.-P.: Introduction to shape optimization: shape sensitivity analysis. Springer (1992)

Part III
Earthquakes and Inverse Problems

Synthesis of Seismic Wave Envelopes Based on the Markov Approximation

Kentaro Emoto

Abstract In this study, we use stochastic methods to analyze complex short-period seismograms reflecting small-scale heterogeneities in the Earth. We consider an ensemble of random velocity-fluctuated media and the statistical characteristics of wave propagation. The Markov approximation is a stochastic method and a multiple forward-scattering approximation. In the Markov approximation, we neglect wide-angle and conversion scatterings and directly calculate the statistical average wave envelopes. Even though we cannot model the latter part of the envelope because we neglect wide-angle scattering, we can adequately describe the initial part of the envelope, i.e., from the onset to near the peak arrival time. We can estimate the statistical properties of the small-scale heterogeneities in the Earth by analyzing the envelope broadening effect. The Markov approximation was developed in optics and was introduced to seismology in the late 1980s. Here, on the basis of the Markov approximation, we summarize the development of envelope modeling and describe a method to calculate envelopes on a layered random heterogeneous media.

1 Introduction

Short-period (<1 s) seismograms show complex waveforms with long-lasting tails following the direct wave. This tail in such seismograms is called a coda wave. Coda waves have random phases and comprise scattered waves. In this study, we consider seismic wave scattering resulting from continuous velocity fluctuations, such as the diffraction effect, rather than those resulting from distributed discrete scatterers. We focus on the amplitude or energy of the seismograms, i.e., the envelopes, by neglecting the phase information. An example of short-period seismogram envelopes is shown in Fig. 1. At short propagation distances, we can clearly see the sharp onsets of the P- and S-waves. The duration of the envelope and the time interval

K. Emoto (✉)
Graduate School of Science, Tohoku University,
Aramaki-Aza Aoba 6-3, Aoba-ku, Sendai 980-8578, Japan
e-mail: emoto@zisin.gp.tohoku.ac.jp

H. Itou et al. (eds.), *Mathematical Analysis of Continuum Mechanics and Industrial Applications*, Mathematics for Industry 26,
DOI 10.1007/978-981-10-2633-1_9

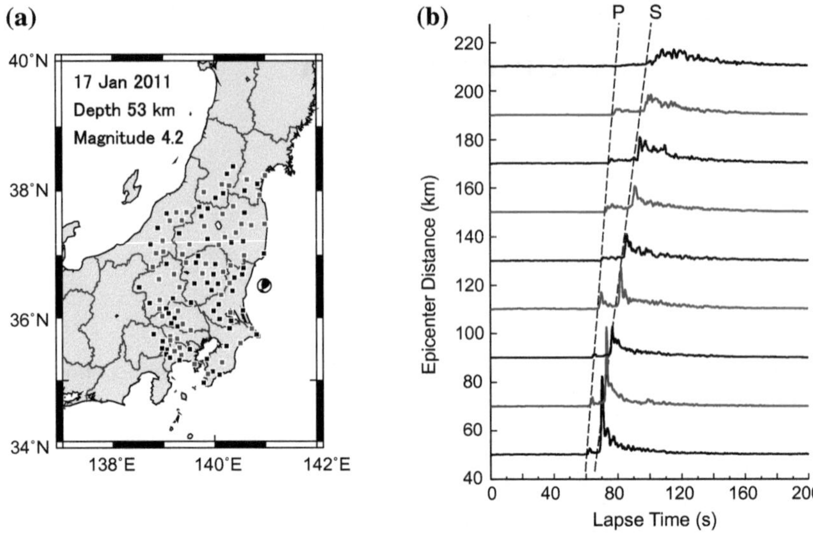

Fig. 1 **a** Map of the Hi-net stations (*squares*) and the location of the epicenter. **b** Root mean square envelopes for different epicentral distances. The *dashed lines* indicate the P- and S-wave onsets. We applied a bandpass filter with a central frequency of 2 Hz to the transverse component of the seismograms. Then, we stacked the MS envelope over stations that had similar epicentral distances with an interval of 20 km

between the onset and the peak amplitude arrival increase with epicentral distance. This phenomenon is known as the envelope broadening effect, which results from the existence of small-scale heterogeneities in the lithosphere.

It is impossible to deterministically estimate the small-scale velocity fluctuations that affect short-period seismograms because we do not have a sufficient distribution of earthquakes and stations. Therefore, we assume random velocity fluctuations. We consider an ensemble of random heterogeneous media and discuss the statistical properties of wave propagation. In seismology, several statistical methods that calculate the mean square (MS) envelopes in random media have been used to estimate the random heterogeneities in the Earth [28]. The Markov approximation is a statistical method to calculate MS envelopes in random heterogeneous media and a multiple forward-scattering approximation. Because we are neglecting the back- and wide-angle scatterings, we cannot model the latter coda part of the envelope; however, we can construct the envelope near its peak amplitude. In this study, we review the development of the Markov approximation and its applications in seismology. For simplicity, we restrict the derivation of the Markov approximation to the case of two-dimensional (2D) random heterogeneous media.

2 Random Media

We define a random heterogeneous medium as having the following form:

$$\alpha(\mathbf{x}) = \alpha_0 \left[1 + \xi(\mathbf{x})\right], \tag{1}$$

where α is the P-wave velocity, α_0 is the average P-wave velocity, ξ is a random variable, and $\mathbf{x} = (x, z)$. The mean propagation direction is z and the transverse direction is x. We consider an ensemble of random fluctuations $\{\xi\}$, where the braces indicate a set of random variables. The ensemble average of the fluctuation is zero: $\langle\xi\rangle = 0$, where $<>$ denotes the ensemble average. The S-wave velocity and the mass density are also fluctuated in the same manner. We assume a weak fluctuation, this means that the variance of ξ is sufficiently small. We characterize the statistical property of the fluctuation using the autocorrelation function (ACF) of ξ:

$$R(\mathbf{x}) = \langle\xi(\mathbf{x}')\xi(\mathbf{x}+\mathbf{x}')\rangle = \begin{cases} \varepsilon^2 e^{-r^2/a^2} & \text{Gaussian type} \\ \varepsilon^2 e^{-r/a} & \text{Exponential type} \\ \dfrac{\varepsilon^2 2^{1-\kappa}}{\Gamma(\kappa)} \left(\dfrac{r}{a}\right)^{\kappa} K_\kappa\left(\dfrac{r}{a}\right) & \text{von Karman type} \end{cases} \tag{2}$$

where a is the correlation distance, ε is the root mean square fractional fluctuation, $r = |\mathbf{x}|$, $\Gamma(\kappa)$ is a gamma function, and K_κ is a modified Bessel function of the second kind of order κ. A von Karman-type fluctuation with $\kappa = 0.5$ corresponds to the exponential type fluctuation. The power spectrum density function (PSDF) of the von Karman (exponential) type shows power-law decay for large wavenumbers. A von Karman-type random medium has much smaller scale heterogeneities compared with the Gaussian type. According to an analysis of the well-logging data, the PSDF of the velocity fluctuations in the shallow crust follows a power-law decay function with respect to the wavenumber [30].

3 The Born Approximation

First, we check the scattering pattern based on the Born approximation [1, 28]. We consider a localized velocity fluctuation that generates a scattered wave for an incident plane wave (Fig. 2a). We can write the total wave field as the sum of the incident and scattered waves. When the velocity fluctuation is weak, we can assume that the amplitude of the scattered wave is much smaller than that of the incident wave. In this case, we can neglect the multiple scatterings at the localized fluctuation. Therefore, there is no interaction between the scattered wave and the inhomogeneity. The scattered wavefield satisfies the wave equation, where the source term is the product of the fluctuation and the incident wavefield. This single-scattering approximation

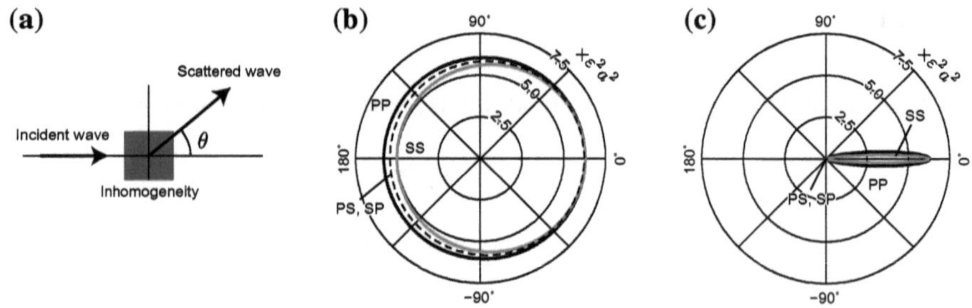

Fig. 2 **a** A schematic illustration of the Born approximation. The scattering angle is θ. **b** The contribution of the PSDF term to the scattering coefficients for different types of scatterings: P–P (*solid black line*), P–S and S–P (*black dashed line*), and S–S (*solid gray line*). The angle denotes the scattering angle θ and the radius corresponds to the amplitude of the PSDF term scaled by $\varepsilon^2 a^2$. The wavelength is larger than the correlation distance, $ak_s = 0.2$. **c** Same as (**b**) but for the case of $ak_s = 10$

is called the Born approximation. Based on the Born approximation, we can calculate the scattering coefficient, which represents the scattering power and has the dimension of the reciprocal length. According to [14], the scattering coefficient can be written as

$$
\begin{aligned}
g_{PP}(\theta) &= \frac{\gamma\, {k_s}^3}{8\pi} P\left(\frac{2k_s}{\gamma}\sin(\theta/2)\right)\left|X_r^{PP}\right|^2 \\
g_{PS}(\theta) &= \frac{{k_s}^3}{8\pi} P\left(\frac{k_s}{\gamma}\sqrt{1+\gamma^2-2\gamma\cos\theta}\right)\left|X_\theta^{PS}\right|^2 \\
g_{SP}(\theta) &= \frac{\gamma\, {k_s}^3}{8\pi} P\left(\frac{k_s}{\gamma}\sqrt{1+\gamma^2-2\gamma\cos\theta}\right)\left|X_r^{SP}\right|^2 \\
g_{SS}(\theta) &= \frac{{k_s}^3}{8\pi} P(2k_s\sin(\theta/2))\left|X_\theta^{SS}\right|^2 ,
\end{aligned}
\tag{3}
$$

where g_{PP}, g_{PS}, g_{SP}, and g_{SS} indicate the P–P, P–S, S–P, and S–S scattering coefficients, respectively. The ratio of the P- to S-wave velocity is $\gamma = \alpha_0/\beta_0$, where β_0 is the average S-wave velocity, $k_s = 2\pi/\lambda$ is the S-wave wavenumber, where λ is the wavelength, and P is the PSDF. X is a function of the scattering angle θ and represents the scattering pattern, e.g., X_r^{PP} indicates the r component of the P–P scattering pattern. The contribution of the wavenumber to the scattering coefficient is controlled by the PSDF term because X is independent of the wavenumber. We show the contribution of the PSDF term in Fig. 2. When the wavelength is larger than the correlation distance (Fig. 2b), the contributions of the PSDF term are almost isotropic for all scattering types. Conversely, when the wavelength is shorter than the correlation distance (Fig. 2c), the contribution of the small scattering angle is dominant (narrow-angle scattering) for P–P and S–S scatterings. The amplitudes of the P–S and S–P conversion scatterings are negligible compared to the narrow-angle scattering of the P–P and S–S scatterings. In the derivation of the Markov approximation, we assume $ak_s \gg 1$. Therefore, we can consider the P- and S-waves separately because the conversion scattering can be negligible and focus in particular on the narrow-angle scattered waves.

4 The Markov Approximation

The Markov approximation was originally developed in optics in the 1970s. Sato was the first to apply the Markov approximation to observed seismograms in seismology [21]. We summarize the history of the development of the Markov approximation in Table 1.

In this section, we briefly describe the derivation of MS envelopes based on the Markov approximation for the case of an incidence of a plane P-wavelet in random 2D media, as in [8]. For a detailed description of the derivation, see [28]. We assume that the incident plane P-wavelet vertically propagating in the positive z-direction originates in the homogeneous medium ($z < 0$). For the P-wave, the scalar potential ϕ satisfies the wave equation as $\nabla^2\phi - (1/\alpha^2)\ddot{\phi} = 0$. In the Markov approximation, we neglect the derivative of the medium parameters, which appears in the case of the elastic wave propagation in heterogeneous media. When the wavelength is shorter than the correlation distance, forward scattering is dominant; therefore, we can expand the scalar potential using plane waves as follows:

$$\phi(x, z, t) = \frac{1}{2\pi} \int_{-\infty}^{\infty} d\omega e^{ik_0 z - i\omega t} \frac{1}{ik_0} U(x, z, \omega), \tag{4}$$

where the wavenumber $k_0 = \omega/\alpha_0$. We can neglect the second derivative of U with respect to z because the variation of U is slow in the z-direction. Thus, U satisfies the parabolic equation as follows:

$$2ik_0\partial_z U + \partial_x{}^2 U - 2k_0{}^2\xi U = 0, \tag{5}$$

where $\partial_z = \partial/\partial z$, $\partial_x = \partial/\partial x$ and we neglect the higher order terms of ξ because $|\xi|$ is sufficiently small in most cases. To calculate the MS envelope, we introduce the two-frequency mutual coherence function (TFMCF) defined in [32] as

$$\Gamma_2(x', x'', z, \omega', \omega'') \equiv \langle U(x', z, \omega')U^*(x'', z, \omega'')\rangle, \tag{6}$$

where U^* is the complex conjugate of U. From Eqs. (5) and (6), we obtain

$$2i\partial_z\Gamma_2 + \left(\frac{\partial_{x'}{}^2}{k_0'} - \frac{\partial_{x''}{}^2}{k_0''}\right)\Gamma_2 - 2\langle(k_0'\xi(x', z) - k_0''(x'', z))U(x', z, \omega')U^*(x'', z, \omega'')\rangle = 0, \tag{7}$$

Table 1 Development of the Markov approximation

Medium	Dimension, ACF type	Wave	Ref.	Initial wave	Remark
Uniform, isotropic	2D, Gauss	Scalar	[6]	Cylindrical	Analytical
	2D, von Karman		[18]	Cylindrical	
	2D, Gauss	Vector	[8]	Plane	Analytical
			[26]	Cylindrical	Analytical Nonisotropic source
	3D, Gauss	Scalar	[36]	Plane	
			[31]	Spherical	Analytical
			[32]	Plane	Analytical
		Vector	[22]	Plane	Analytical
			[23]	Spherical	Analytical
	3D, von Karman	Scalar	[11]	Plane	
			[17]	Spherical	
			[25]		Analytical
		Vector	[33]	Plane, Spherical	
			[29]	Spherical	Nonisotropic source
Uniform, anisotropic	2D, Gauss	Scalar	[16]	Cylindrical	
	3D, Gauss	Vector	[24]	Plane, Spherical	Analytical
Nonuniform	2D, Gauss	Scalar	[20]	Plane	
	3D, von Karman	Scalar	[34]	Spherical	
Free surface	2D, 3D, Gauss	Vector	[2]	Plane	
Velocity discontinuity	2D, Gauss	Vector	[4]	Plane	
	2D, Exponential	Vector	[5]	Cylindrical	

The "Analytical" in the column of the remark indicate the existence of the analytical solution of the master equation, Eq. (11). The other cases, the master equation is solved numerically

where $k_0' = \omega'/\alpha_0$ and $k_0'' = \omega''/\alpha_0$. To evaluate the third term of Eq. (7), we divide the medium into multiple thin layers with a thickness of Δz, which is larger than a but smaller than the variation of U. This is a stochastic extension of the phase screen method [7]. According to Eq. (5), we can express $U(x, z, \omega)$ using $U(x, z - \Delta z, \omega)$ as

$$U(x,z,\omega)=U(x,z-\Delta z,\omega)+\frac{i}{2k_0}\int_{z-\Delta z}^{z}dz'\left[\partial_x^2U(x,z',\omega)-2k_0^2\xi(x,z')U(x,z',\omega)\right]$$
$$\approx U(x,z-\Delta z,\omega)+\frac{i\Delta z}{2k_0}\partial_x^2U(x,z-\Delta z,\omega)-ik_0U(x,z-\Delta z,\omega)\int_{z-\Delta z}^{z}\xi(x,z')dz', \tag{8}$$

Using Eq. (8), we can evaluate the third term of Eq. (7) as

$$\langle\xi(x',z)U(x',z,\omega')U^*(x'',z,\omega'')\rangle$$
$$\approx -\frac{ik_0}{2}\Gamma_2(x',x'',z-\Delta z,\omega',\omega'')A(0)+\frac{ik_0''}{2}\Gamma_2(x',x'',z-\Delta z,\omega',\omega'')A(x'-x''), \tag{9}$$

where we neglect the back-scattered waves propagating in the negative z-direction. Therefore, there is no correlation between $\xi(x,z)$ and $U(x,z-\Delta z,\omega)$. This derivation is called the Markov approximation because the wavefield at z depends only on that at $z-\Delta z$, as in the Markov process [10]. We define the longitudinal integral of the ACF as $A(x)\equiv\int_{-\infty}^{\infty}R(x,z)dz$ and we assume that the variation of Γ_2 in the z-direction is also small, as is that of U. Thus Eq. (7) can be written as

$$2i\partial_z\Gamma_2+\left(\frac{\partial_x'^2}{k_0'}-\frac{\partial_x''^2}{k_0''}\right)\Gamma_2+i\left[(k_0'^2+k_0''^2)A(0)-2k_0'k_0''A(x'-x'')\right]\Gamma_2=0. \tag{10}$$

We introduce the differential and center of mass coordinates as $k_c\equiv\omega_c/\alpha_0\equiv(\omega'+\omega'')/2\alpha_0$ and $k_d\equiv\omega_d/\alpha_0\equiv(\omega'-\omega'')/\alpha_0$ for the wavenumber and $x_c\equiv(x'+x'')/2$ and $x_d\equiv x'-x''$ for the transverse coordinate. Considering quasi-monochromatic waves, i.e., $\omega_d\ll\omega_c$, we can simplify Eq. (10) to

$$2i\partial_z\Gamma_2-\frac{k_d}{k_c{}^2}\partial_{x_d}{}^2\Gamma_2+2ik_c{}^2\left[A(0)-A(x_d)\right]\Gamma_2+\frac{ik_d{}^2}{2}\left[A(0)+A(x_d)\right]\Gamma_2=0$$
$$\partial_{z\,0}\Gamma_2+i\frac{\omega_d}{2k_c{}^2\alpha_0}\partial_{x_d}{}^2{}_0\Gamma_2+k_c{}^2\left[A(0)-A(x_d)\right]{}_0\Gamma_2=0, \tag{11}$$

where we assume that $A(x_d)\sim A(0)$ and

$$\Gamma_2={}_0\Gamma_2\exp\left[-\frac{\omega_d{}^2A(0)z}{2\alpha_0{}^2}\right]={}_0\Gamma_2w(\omega_d). \tag{12}$$

The term $w(\omega_d)$ is called the wandering term and takes into account the travel time fluctuations due to the long-scale components of the heterogeneity in the process of ensemble averaging [10]. The final equation of (11) is the master equation of ${}_0\Gamma_2$.

The intensity of the x-component is written as

$$
\begin{aligned}
\langle|u_x(x,z,t)|^2\rangle &= \langle \partial_{x'}\phi(x',z,t)\partial_{x''}\phi^*(x'',z,t)\rangle_{x'=x''} \\
&= \frac{1}{2\pi}\int_{-\infty}^{\infty} d\omega' e^{ik_0'z-i\omega' t} \\
&\cdot \frac{1}{2\pi}\int_{-\infty}^{\infty} d\omega'' e^{-ik_0''z+i\omega'' t}\frac{1}{k_0'k_0''}\langle \partial_{x'}U(x',z,\omega')\partial_{x''}U^*(x'',z,\omega'')\rangle_{x'=x''} \\
&= \frac{1}{2\pi}\int_{-\infty}^{\infty} d\omega_c \widehat{I}_x(z,t;\omega_c),
\end{aligned}
\tag{13}
$$

where we define

$$
\widehat{I}_x(z,t;\omega_c) \equiv \frac{1}{2\pi}\int_{-\infty}^{\infty} d\omega_d e^{-i\omega_d(t-z/\alpha_0)}\left[-\frac{1}{{k_c}^2}{\partial_{x_d}}^2 \Gamma_2(x_d,z,\omega_c,\omega_d)\right]_{x_d=0}. \tag{14}
$$

This represents the x-component of MS envelope. We can also define the MS envelope without the wandering effect as

$$
\widehat{I}_{x0}(z,t;\omega_c) \equiv \frac{1}{2\pi}\int_{-\infty}^{\infty} d\omega_d e^{-i\omega_d(t-z/\alpha_0)}\left[-\frac{1}{{k_c}^2}{\partial_{x_d}}^2 {}_0\Gamma_2(x_d,z,\omega_c,\omega_d)\right]_{x_d=0}. \tag{15}
$$

In the same way, we can describe the z-component of the MS envelope without the wandering effect as

$$
\widehat{I}_{z0}(z,t;\omega_c) = \frac{1}{2\pi}\int_{-\infty}^{\infty} d\omega_d e^{-i\omega_d(t-z/\alpha_0)}\left[\left(1+\frac{1}{{k_c}^2}{\partial_{x_d}}^2\right) {}_0\Gamma_2(x_d,z,\omega_c,\omega_d)\right]_{x_d=0}. \tag{16}
$$

In the Markov approximation, we can calculate the MS envelopes by solving Eq. (11) and taking the inverse Fourier transform of ${}_0\Gamma_2$ as (15) and (16). Then, we can obtain the MS envelopes with the wandering effect by convolving the envelopes with the wandering term. For the case of Gaussian ACF type random media, the analytical solution of ${}_0\Gamma_2$ was derived by [8]. The MS envelopes can be scaled by the characteristic time $t_M \equiv \sqrt{\pi}\varepsilon^2 z^2/(2\alpha_0 a)$. We show an example of MS envelopes in Fig. 3a assuming the vertical incidence of an impulsive plane P-wavelet, that is, $\widehat{I}_z(x,z,t) = \delta(t - z/\alpha_0)$ for $z < 0$. For Gaussian ACF type random media, the MS envelope is independent of the frequency. Conversely, for von Karman type random media, the MS envelope becomes broader with increasing frequency [17].

The validity of the Markov approximation has been verified by comparing the Markov envelope with the envelope calculated by a finite difference (FD) simulation [4, 6, 8, 13, 18]. In FD simulations, we numerically solve the wave equation including the effect of the wide-angle scattering, which is neglected in the Markov approximation. For the Markov approximation, the following conditions are necessary: $ak_c \gg 1$ and $a_\perp k_c \gg 1$, where $a_\perp$ is a coherent radius that satisfies $(A(0) - A(a_\perp)){k_c}^2 z = 1$ [15]. The first condition is required for the forward-scattering approximation and to neglect conversion scattering. The second condition can be interpreted as t_M which represents the envelope duration is shorter than half the travel time [25].

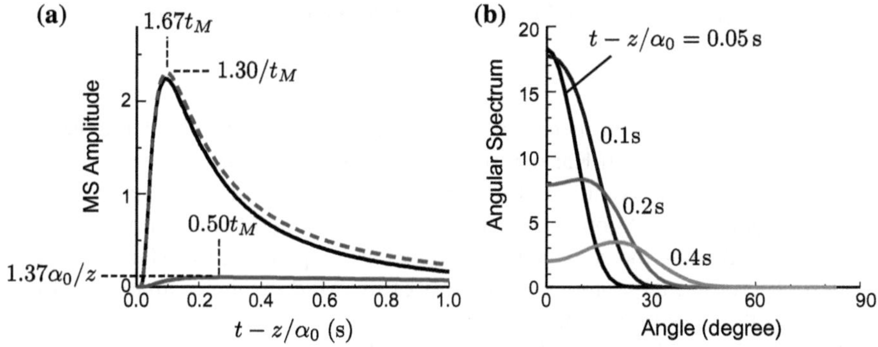

Fig. 3 **a** MS envelopes at $z = 100$ km for the vertical incidence of a plane P-wavelet propagating from $z < 0$. The random medium is characterized by the Gaussian ACF where $a = 5$ km, $\varepsilon = 5\,\%$, and $\alpha_0 = 7.8$ km/s. The *solid black*, *solid gray*, and *gray dashed lines* indicate the z-component, the x-component and the sum of the x- and z-components of the MS envelopes, respectively. **b** Snapshots of the angular spectrum for the case in (**a**)

From the comparison of the FD and Markov envelopes in 2D Gaussian ACF type random media, the Markov approximation is seen to be applicable when $ak_c \geq 8$ and $\varepsilon^2 \leq (0.1 \sim 0.2)a/z$ [4].

5 Extension of the Markov Approximation

To overcome the disadvantages of the Markov approximation, several improvements to the model have been proposed. In this section, we describe a few extensions of the Markov approximation.

5.1 The Hybrid Method

We can adequately describe the early part of the envelope using the Markov approximation; however, the late coda part of the envelope is underestimated because we neglect wide-angle scattering. Conversely, the radiative transfer theory can model the late part of the envelope; however, it cannot adequately describe the early part of the envelope [12]. To compensate for these disadvantages and to model the entire envelope, a hybrid method combining the Markov approximation and the radiative transfer theory has been proposed [18, 27]. Saito et al. [18] replaced the direct term in the radiative transfer equation with the envelope derived using the Markov approximation. However, this approach assumes that the propagator between the scatters is a delta function. Sato et al. [27] used the envelope derived from the Markov approximation as the propagator in the radiative transfer equation and succeeded in synthesizing the entire wave envelope from the onset until the late coda.

5.2 An Analytical Solution for von Karman Type Random Media

For Gaussian type random media, an analytical solution of the TFMCF exists (see Table 1). However, for von Karman type random media, we are required to numerically solve the master equation of the TFMCF using a method such as the Crank–Nicolson method [17]. As opposed to the FD simulation of the wave propagation, we can directly calculate the statistically averaged envelope using the Markov approximation. Therefore, the computational cost of the Markov approximation is lower than that of the FD simulation.

Recently, Sato [25] proposed a new approach to the Markov approximation. In this approach, the contributions of the large-scale and small-sale components of the fluctuation are evaluated separately. The large-scale component affects the envelope broadening, which can be modeled using the Markov approximation. Conversely, the small-scale component contributes to the scattering attenuation. By subtracting the small-scale component of the PSDF and applying the Markov approximation to the subtracted PSDF, we can analytically obtain the MS envelope, even in the case of von Karman type random media.

5.3 Velocity Discontinuity in the Markov Approximation

So far, we have considered a uniformly random inhomogeneous media. To estimate the regional difference and depth dependence of the statistical properties of small-scale heterogeneities, the Markov approximation in a layered random media has been proposed, where the statistical properties are different for different layers, but the background velocities are the same [29, 34]. This approach is based on the stochastic ray path method [36]. We apply the Markov approximation to the mutual coherence function to obtain the probability density function of the ray angles of scattered waves at each layer boundary. According to the probability density function, we numerically simulate the propagation of the intensity particle based on the Monte Carlo method to obtain the MS envelopes.

In the above method, we considered the discontinuity of statistical properties such as a, ε, and κ in Eq. (2); however, we assumed that the background velocities were uniform for all layers. There are several clear velocity discontinuities in the Earth such as the Moho and at slabs. We need to evaluate the effect of these velocity discontinuities to correctly estimate the heterogeneities in the medium. The effect of the velocity discontinuity depends on the incident angle; therefore, we need to know the ray angle distribution of the scattered waves that construct the envelope. We can obtain the distribution of the propagation directions of scattered waves by calculating the angular spectrum, as defined in [8]:

$$ {}_0\breve{\Gamma}_2(k_x, z, \omega_d, \omega_c) \equiv \int_{-\infty}^{\infty} dx_d e^{-ik_x x_d} {}_0\Gamma_2(x_d, z, \omega_d, \omega_c). \tag{17} $$

Using the angular spectrum, we can rewrite the MS envelopes, Eqs. (15) and (16), as

$$ \widehat{I}_{x0}(z, t; \omega_c) = \frac{1}{2\pi} \int_{-k_c}^{k_c} dk_x \frac{{k_x}^2}{{k_c}^2} \widehat{I}_0^A(k_x, z, t; \omega_c), \tag{18} $$

and

$$ \widehat{I}_{z0}(z, t; \omega_c) = \frac{1}{2\pi} \int_{-k_c}^{k_c} dk_x \left(1 - \frac{{k_x}^2}{{k_c}^2}\right) \widehat{I}_0^A(k_x, z, t; \omega_c), \tag{19} $$

where we define the angular spectrum in the time domain as

$$ \widehat{I}_0^A(k_x, z, t; \omega_c) \equiv \frac{1}{2\pi} \int_{-\infty}^{\infty} d\omega_d e^{-i\omega_d(t - z/\alpha_0)} {}_0\breve{\Gamma}_2(k_x, z, \omega_d, \omega_c). \tag{20} $$

The factors appearing in Eqs. (18) and (19) can be expressed using $\theta = \sin^{-1}(k_x/k_c)$ as ${k_x}^2/{k_c}^2 = \sin^2\theta$ and $1 - {k_x}^2/{k_c}^2 = \cos^2\theta$. Therefore, Eqs. (18) and (19) can be thought of as the projection of the angular spectrum onto the x- and z-directions, respectively. The angular spectrum is interpreted as the ray angle distribution of the scattered waves. In Fig. 3b, we show an example of snapshots of the angular spectrum in the time domain. Soon after the onset, the angular spectrum has a peak at 0° and a sharp distribution. This indicates that the envelope around the onset to the peak arrival consists of narrow-angle scattered waves. Conversely, the angular spectrum becomes broad and the peak angle shifts to larger angle. This means that wide-angle scattered waves are dominant in the coda part of the envelope.

By replacing the projection factors with the amplification factors at the free surface, which depend on the incident angle to the free surface, we can obtain the MS envelopes on the free surface [2]. When the medium is homogeneous, the amplitude of the vertical incidence of the plane wave doubles on the free surface; therefore, the energy becomes four times larger than that without a free surface. The primary part of the envelope consists of narrow-angle scattered waves; therefore, the amplification of the envelopes can be roughly assumed to be four for both the vertical and the horizontal components. Strictly speaking, however, because the scattered waves constructing the envelope vary with the lapse time, as shown in Fig. 3b, the amplification of the envelope on the free surface also varies with the lapse time and is not exactly four. For the vertical incidence, in a plane P-wavelet case, the amplification of the horizontal component is larger than that of the vertical component [2].

Similar to the envelopes on a free surface, we can evaluate the effects of the velocity discontinuity, such as transmission, reflection, and conversion [4]. For example, to calculate the MS envelope of the transmitted wave, we calculate the wave propagation in the first layer based on the Markov approximation to obtain the angular spectrum

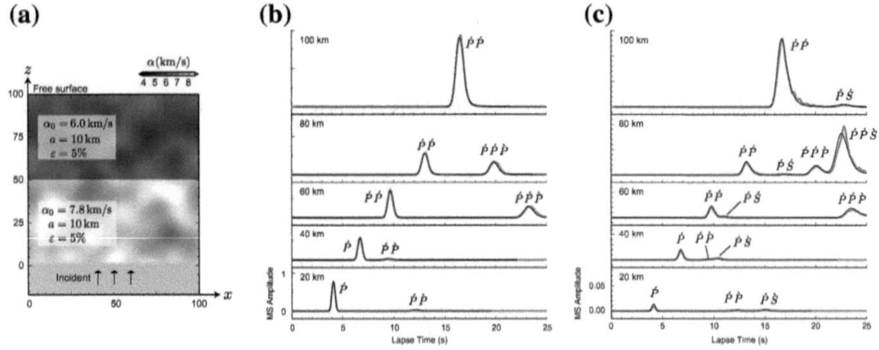

Fig. 4 **a** The layered random medium. Both layers have the same random fluctuation; however, their background velocities are different. **b** z-component of the MS envelopes at propagation distances of 20, 40, 60, 80, and 100 km (the free surface) of the layered random medium shown in (**a**). *Thick gray* and *thin black lines* indicate the MS envelopes calculated using the FD simulation and those derived based on the Markov approximation, respectively. For the FD envelopes, we stacked 2000 traces of different random media generated using different seeds. **c** Same as (**b**) but for the x-component of the MS envelopes. Symbols in the figure indicate the wave type of the propagation (P or S) and the direction ($\acute{}$: up or $\grave{}$: down). For example, $\acute{P}\grave{P}$ indicates the reflected P-wave from the layer boundary, and $\acute{P}\acute{S}$ indicates the P–S converted wave at the layer boundary

at the boundary. Then, we multiply it by the transmission coefficient and calculate the propagation in the second layer using the modified angular spectrum as the initial condition. For the reflected wave from the boundary, we multiply the angular spectrum by the reflection coefficient and propagate it backward in the same layer. Even though we neglect conversion scattering in the propagation process within each layer, we can take into account the conversion at the boundary by multiplying the angular spectrum by the conversion coefficient. For the case of a spherical or cylindrical incident wavelet, we need to modify the geometrical spreading factors at the boundary [5].

We show an example of the envelope synthesis in a layered random medium in Fig. 4. We considered a two-layer random medium where the upper layer has a lower background velocity than the bottom layer. We assumed a vertical incidence of a plane P-wavelet from the bottom of the lower layer (Fig. 4a). Here, we show the z- and x-components of the MS envelopes at several propagation distances in Fig. 4b, c, respectively. We also show the MS envelopes derived using an FD simulation of the elastic wave propagation to check the validity of the Markov approximation in the layered random medium. We can see that the Markov approximation properly models the envelopes calculated by the FD simulation at all propagation distances, even on the free surface.

6 Application of the Markov Approximation

The statistical properties of the random inhomogeneities in the lithosphere have been estimated based on the Markov approximation. Most studies analyze the envelope broadening effect, which is insensitive to intrinsic attenuation and can be aptly modeled using the Markov approximation. Sato [21] was the first to analyze observed seismograms based on the Markov approximation. He analyzed the peak delay time of local earthquakes in the Kanto region and found that the peak delay time increases with increasing epicentral distance. By modeling this observed result based on the Markov approximation, the statistical parameters of the lithosphere were estimated. Saito et al. [17, 19] analyzed the envelope width and the decay of the maximum amplitude of the envelope to estimate the PSDF of the velocity fluctuations in northeastern Honshu. These studies used the Markov approximation in a uniform random medium to evaluate the envelope broadening and estimate the average statistical properties of the random velocity fluctuations in the target region. Takahashi et al. [35] proposed a recursive formula to calculate the peak delay time on the basis of the Markov approximation in a nonuniform random media where the statistical parameters are nonuniform but the background velocity is uniform [34]. They succeeded in mapping the statistical properties in the Tohoku region and fond strong fluctuations beneath the volcanoes (Fig. 5). We can also estimate the random inhomogeneities in the mantle by analyzing the envelope broadenings of the S- and ScS-wave envelopes [3]. The ScS-wave is an S-wave reflected at the core-mantle boundary. We show stacked S-and ScS-wave envelopes in Fig. 6. We cannot see strong envelope broadening of the ScS-wave compared to the S-wave. This means that small-scale heterogeneities in the mantle are weak, if they even exist.

Kubanza et al. [9] focused on analyzing the excitation of the transverse component of the envelope of the teleseismic P-wave based on the Markov approximation and found strong scattering properties in tectonically active regions of the world.

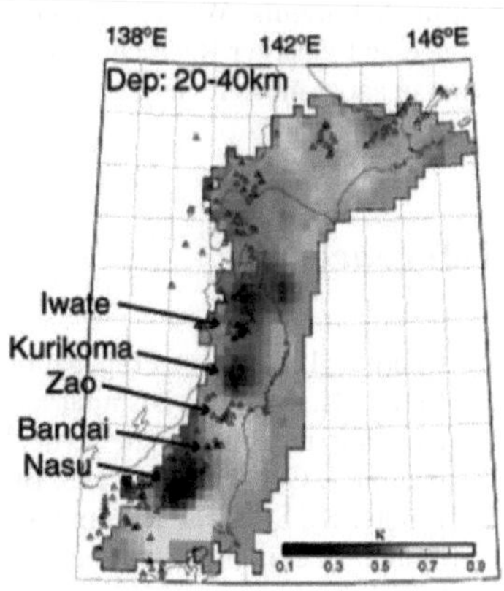

Fig. 5 Spatial variation of the κ value of von Karman type random heterogeneities estimated from an analysis of the peak delay effect at depth of 20–40 km. *Open triangles* denote the Quaternary volcanoes [35, Fig. 11b]

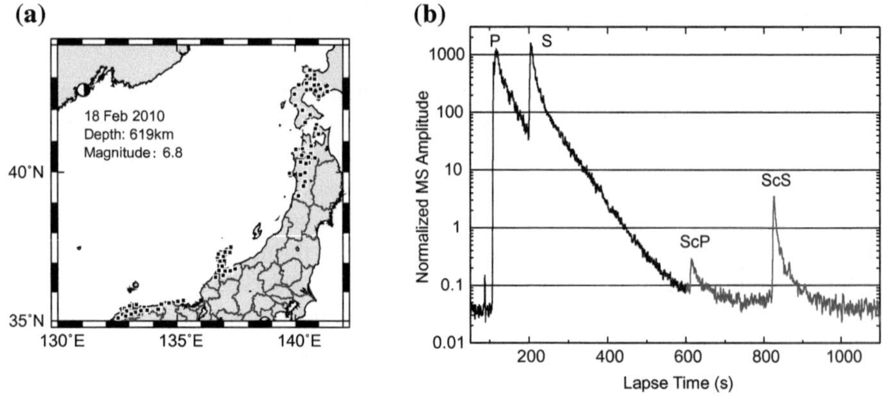

Fig. 6 **a** Map of the epicenter and Hi-net stations (*black squares*). **b** Stacked MS envelopes at 1 Hz. For the *black* and *gray* parts of the envelope, we stacked the sum of two horizontal component velocity seismograms by aligning the onsets of the S- and ScS-waves, respectively

7 Summary

We reviewed the development of the Markov approximation in the field of seismology, where it is used as a multiple forward-scattering approximation for envelope modeling. Using the Markov approximation, we can model the early part of the envelope from the onset to near the peak arrival time. We presented equations to calculate the standard MS envelopes based on the Markov approximation. In addition, we described a method to calculate the MS envelopes in a layered random media. The Markov approximation was improved with the aim of making precise estimations of small-scale heterogeneities in the Earth. Since the Markov approximation was introduced in the seismology in 1989, improvements of the theory are ongoing.

Acknowledgments We are very grateful to the staff of the National Research Institute for Earth Science and Disaster Resilience enable Hi-net to continue functioning. The author would like to thank an anonymous reviewer for his valuable comments to improve this manuscript.

References

1. Chernov, L.A.: Wave Propagation in a Random Medium. McGraw-Hill, New York (1960)
2. Emoto, K., Sato, H., Nishimura, T.: Sysnthesis of vector wave envelopes on the free surface of a random medium for the vertical incidence of a plane wavelet based on the Markov approximation. J. Geophys. Res. **115**, B08,306 (2010)
3. Emoto, K., Sato, H., Nishimura, T.: Stochastic synthesis of vector wave envelopes in layered random media based on the improved Markov approximation. In: EGU General Assembly (2012)
4. Emoto, K., Sato, H., Nishimura, T.: Synthesis and applicable condition of vector wave envelopes in layered random elastic media with anisotropic autocorrelation function based on the markov

approximation. Geophys. J. Int. **188**(1), 325–335 (2012). doi:10.1111/j.1365-246X.2011.05264.x
5. Emoto, K., Sato, H., Nishimura, T.: Envelope synthesis of a cylindrical outgoing wavelet in layered random elastic media based on the Markov approximation. Geophys. J. Int. **194**, 899–910 (2013)
6. Fehler, M., Sato, H., Huang, L.: Envelope broadening of outgoing waves in 2D random media: a comparison between the Markov approximation and numerical simulations. Bull. Seism. Soc. Am. **90**(4), 914–928 (2000)
7. Fisk, M.D., McCartor, G.D.: The phase screen method for vector elastic waves. J. Geophys. Res. **96**, 5985–6010 (1991)
8. Korn, M., Sato, H.: Synthesis of plane vector-wave envelopes in 2-D random elastic media based on the Markov approximation and comparison with finite difference simulations. Geophys. J. Int. **161**(3), 839–848 (2005)
9. Kubanza, M., Nishimura, T., Sato, H.: Evaluation of strength of heterogeneity in the lithosphere from peak amplitude analyses of teleseismic short-period vector P waves. Geophys. J. Int. **171**(1), 390–398 (2007)
10. Lee, L., Jokipii, J.: Strong scintillations in astrophysics. I. the Markov approximation, its validity and application to angular broadening. Astrophys. J. **196**, 695–707 (1975)
11. Lee, L., Jokipii, J.: Strong scintillations in astrophysics. II. a theory of temporal broadening of pulses. Astrophys. J. **201**, 532–543 (1975)
12. Margerin, L.: Introduction to radiative transfer of seismic waves. In: L.A. Nolet (ed.) Seismic Earth: Array Analysis of Broad-band Seismograms, chap. 14, pp. 229–252. AGU (2005)
13. Przybilla, J., Korn, M.: Monte Carlo simulation of radiative energy transfer in continuous elastic random media-three-component envelopes and numerical validation. Geophys. J. Int. **173**(2), 566–576 (2008)
14. Przybilla, J., Korn, M., Wegler, U.: Radiative transfer of elastic waves versus finite difference simulations in two-dimensional random media. J. Geophys. Res. **111**, B04,305 (2006). doi:10.1029/2005JB003952
15. Rytov, S.M., Kravtsov, Y.A., Tatarskii, V.I.: Principles of Statistical Radiophysics 4: Wave Propagation Through Random Media. Springer (1989)
16. Saito, T.: Synthesis of scalar-wave envelopes in two-dimensional weakly anisotropic random media by using the Markov approximation. Geophys. J. Int. **165**, 501–515 (2006)
17. Saito, T., Sato, H., Ohtake, M.: Envelope broadening of spherically outgoing waves in three-dimensional random media having power-law spectra. J. Geophys. Res. **107**(10.1029) (2002)
18. Saito, T., Sato, H., Fehler, M., Ohtake, M.: Simulating the envelope of scalar waves in 2D random media having power-law spectra of velocity fluctuation. Bull. Seismol. Soc. Am. **93**(1), 240–252 (2003)
19. Saito, T., Sato, H., Ohtake, M., Obara, K.: Unified explanation of evnvelope broadening and maximum-amplitude decay of high-frequency seismograms based on the envelope simulation usign the Markov approximation: forearc side of the volcanic front in northeastern Honshu, Japan. J. Geophys. Res. **110**, B01,304 (2005)
20. Saito, T., Sato, H., Takahashi, T.: Direct simulation methods for scalar-save envelopes in two-dimensional layered random media based on the small-angle scattering approximation. Commun. Comput. Phys. **3**(1), 63–84 (2008)
21. Sato, H.: Broadening of seismogram envelopes in the randomly inhomogeneous lithosphere based on the parabolic approximation: Southeastern Honshu, Japan J. Geophys. Res. **94**(B12), 17735–17747 (1989)
22. Sato, H.: Synthesis of vector wave envelopes in three-dimensional random elastic media characterized by a Gaussian autocorrelation function based on the Markov approximation: plane wave case. J. Geophys. Res. **111**(B6), B06,306 (2006)
23. Sato, H.: Synthesis of vector wave envelopes in three-dimensional random elastic media characterized by a Gaussian autocorrelation function based on the Markov approximation: spherical wave case. J. Geophys. Res. **112**(B1), B01,301 (2007)

24. Sato, H.: Synthesis of vector-wave envelopes in 3-D random media characterized by a non-isotropic Gaussian ACF based on the Markov approximation. J. Geophys. Res. **113**, B08,304 (2008)
25. Sato, H.: Envelope broadening and scattering attenuation of a scalar wavelet in random media having power-law spectra. Geophys. J. Int. **204**, 386–398 (2016)
26. Sato, H., Korn, M.: Envelope syntheses of cylindrical vector-waves in 2-D random elastic media based ont he Markov approximation. Earth Planets Space **59**, 209–219 (2007)
27. Sato, H., Fehler, M., Saito, T.: Hybrid synthesis of scalar wave envelopes in two-dimensional random media having rich short-wavelength spectra. J. Geophys. Res. **109**, B06,303 (2004)
28. Sato, H., Fehler, M., Maeda, T.: Seismic Wave Propagation and Scattering in the Heterogeneous Earth, 2nd edn. Springer (2012)
29. Sawazaki, K., Sato, H., Nishimura, T.: Envelope synthesis of short-period seismograms in 3-D random media for a point shear dislocation source based on the forward scattering approximation: application to small strike-slip earthquakes in southwestern Japan. J. Geophys. Res. **116**, B08,305 (2011)
30. Shiomi, K., Sato, H., Ohtake, M.: Broad-band power-law spectra of well-log data in Japan. Geophys. J. Int. **130**(1), 57–64 (1997)
31. Shishov, V.: Effect of refraction on scintillation characteristics and average pulse shape of pulsars. Sov. Astron. **17**, 598–602 (1974)
32. Sreenivasiah, I., Ishimaru, A., Hong, S.: Two-frequency mutual coherence function and pulse propagation in a random medium: an analytic solution to the plane wave case. Radio Sci. **11**(10), 775–778 (1976)
33. Suzaki, A.: Study on vector-wave envelopes in von Karman-type random media. Master's thesis, Tohoku University, Japan (2007)
34. Takahashi, T., Sato, H., Nishimura, T.: Recursive formula for the peak delay time with travel distance in von Karman type non-uniform random media on the basis of the Markov approximation. Geophys. J. Int. **173**(2), 534–545 (2008)
35. Takahashi, T., Sato, H., Nishimura, T., Obara, K.: Tomographic inversion of the peak delay times to reveal random velocity fluctuations in the lithosphere: method and application to northeastern Japan. Geophys. J. Int. **178**, 1437–1455 (2009)
36. Williamson, I.: Pulse broadening due to multiple scattering in the interstellar medium. Mon. Not. R. Astr. Soc. **157**, 55–71 (1972)

Propagation Velocity of Pulse-Like Rupture Along Earthquake Faults

Shiro Hirano

Abstract During earthquakes, the rupture front propagates along faults at approximately 40–90 % of the shear or Rayleigh wave velocity, with slip rate often concentrated in a narrow region behind the front. Past studies have considered this phenomenon using a steady-state pulse-like rupture model and a slip-weakening friction law; however, the results included a trade-off between rupture velocity and the scale of the pulse, which prevents the rupture velocity from being uniquely determined. In this study, we explore this issue and develop a model to determine rupture velocity by considering a friction law based on a numerical simulation of a past study for a slipping plane with its microscopic structure. We combine two models from past studies to construct a relationship between rupture velocity and some tectonophysical/geological parameters.

1 Introduction

Earthquakes are dynamic shear fractures along faults embedded in the Earth's crust. The spatial extension of faults covers a broad range. Small-scale faults extend ∼1 mm in laboratory settings, and are equivalent to mineral grain boundaries, or ∼10 cm in situ features detectable in mines. Large-scale faults extend ∼1000 km along plate interfaces, and can host earthquakes of magnitude ∼9. Geological observations have suggested that fault thicknesses are negligible with respect to fault length; therefore, regardless of scale, faulting is regarded as slip along a crack and is treated using a fracture mechanics framework in earthquake source physics

Using seismological waveform inversions under such a framework, investigations into the spatio-temporal distribution of fault rupture and slip have shown that, in some cases, entire fault surfaces do not go on slipping as the rupture propagates. Instead, the slip rate concentrates in a narrow region behind the rupture front [9].

S. Hirano (✉)
Ritsumeikan University, 1-1-1, Noji Higashi,
Kusatsu, Shiga 525-0058, Japan
e-mail: s-hrn@fc.ritsumei.ac.jp

H. Itou et al. (eds.), *Mathematical Analysis of Continuum Mechanics and Industrial Applications*, Mathematics for Industry 26,
DOI 10.1007/978-981-10-2633-1_10

This phenomenon, which is often referred to as pulse-like rupture, is distinct from crack-like rupture, during which the slip rate is sustained along the entire fault surface as the rupture extends (e.g., [2, 3, 9]).

Freund [5] theoretically simplified pulse-like rupture in 2-D by considering mode-II dynamic rupture propagation of the pulse with constant length and velocity. In his framework, the displacement field caused by the pulse was steady-state within a moving coordinate with velocity equal to that of pulse propagation. This resulted in a singular integral equation (the Airfoil equation) for slip rate inside of the pulse.

At the same time, frictional forces acting on the slipping plane are not negligible because faults are located in the Earth's crust, which is strongly compressed. Therefore, a term representing friction must be introduced in the integral equation. A simplified and classic way to model friction is to assume that constant dynamic friction works uniformly inside of a slipping region. Referred to as Coulomb's friction law, this has been accepted as an efficient approximation of friction. However, slip rates and stress near the leading edge of the pulse diverge, even though the energy release rate is finite [5]. This unphysical divergence can be removed by employing a slip-weakening friction law proposed by Ida [11] and Palmer and Rice [14]. The friction model, originally based on the results of stick-slip laboratory experiments, states that friction is a monotonically decreasing function of slip amount. The model takes a constant value equivalent to dynamic friction after the slip amount exceeds a threshold (i.e., the critical slip distance).

Using the slip-weakening friction model, Rice et al. ([15] hereinafter, RSP) provided a solution of finite slip rate; however, their solution included a trade-off between propagation velocity (rupture velocity) and the length of the pulse, meaning that these two could not be determined uniquely even after assuming values of maximum static friction, dynamic friction, initial stress acting on the fault, and surface energy. In general, rupture velocity is a significant parameter when determining kinetics on and off the fault; for example, maximum slip rate [6], stress distribution around the fault (e.g., RSP), and the dominant frequency of far-field seismic wave [4]. From an observational point of view, rupture velocity is not negligibly small, but approximately 40–90 % of the shear or Rayleigh wave velocity [7, 20]. However, because of the trade-off, even the qualitative characteristic of rupture velocity cannot be explained. For this reason, the theoretical determination of the propagation velocity of pulse-like rupture remains problematic in earthquake source physics.

In this study, we show that the lack of determinability in the RSP framework can be solved using the friction model of Hatano [8], which was developed using a numerical experiment focused on observed microscopic fault structures. We first introduce the singular integral equation for slip rate to describe the balance of traction inside of the pulse. Next, we review the theoretical model of RSP and expose the trade-off between rupture velocity and length of the pulse. After introducing the microscopic structure geologically observed in actual faults and numerically modeled by Hatano [8], we finally show that the Hatano's result contributes to solving the trade-off problem. Moreover, we consider how rupture velocity and length of the pulse depend on tectonophysical or geological parameters, and these are affected by uncertainties in the parameters.

2 Formulation

2.1 PDE and Boundary Conditions

We here define a 3-D coordinate, flat fault, and some physical quantities that are essentially the same as those of Freund [5] and RSP. First, as shown in Fig. 1, we define domains in the x-z plane: $\Gamma = \{(x,z)\,|x \in (-\infty, v_r t), z \in \mathbb{R}\}$ and $\Gamma^* = \{(x,z)\,|x \in (v_r t - L, v_r t)\,, z \in \mathbb{R}\}\,(\subset \Gamma)$, where $t \in \mathbb{R}$ is time, and v_r is the sub-sonic propagation velocity of the moving rupture front toward the positive direction of the x-axis. Physically, we regard the x-z plane as the fault, Γ as an already ruptured region, and Γ^* as a pulse (i.e., a temporarily slipping region). Next, we define the y-axis perpendicular to the x-z plane and the displacement field as $\boldsymbol{u} : \left(\mathbb{R}^3\backslash\Gamma\right) \times \mathbb{R} \ni (x,y,z,t) \mapsto (u_x, u_y, u_z) \in \mathbb{R}^3$ due to slip on Γ. A support of the gap of displacement (i.e., slip) $[\boldsymbol{u}] := \boldsymbol{u}^+ - \boldsymbol{u}^-$ is Γ, where the superscript $\pm$ indicates $\lim_{y\to\pm 0}$. In addition, a support of the slip rate $\boldsymbol{v} := \partial_t\,[\boldsymbol{u}]$ is Γ^*. We assume linear elasticity for a uniform domain corresponding to $\mathbb{R}^3\backslash\Gamma$, so our governing equation corresponds to Navier's equation:

$$\begin{cases} \partial_t^2 \boldsymbol{u} = c_d^2 \nabla(\nabla\cdot\boldsymbol{u}) - c_s^2 \nabla\times(\nabla\times\boldsymbol{u}) & \text{in } \mathbb{R}^3\backslash\Gamma, \quad (1a) \\ [\boldsymbol{u}]\cdot\boldsymbol{n} = [u_y] = 0 & \text{in } \Gamma, \quad (1b) \\ (\sigma\boldsymbol{n})^+ = (\sigma\boldsymbol{n})^- & \text{in } \Gamma, \quad (1c) \\ (\sigma\boldsymbol{n})^\pm - \left\{(\sigma\boldsymbol{n})^\pm\cdot\boldsymbol{n}\right\}\boldsymbol{n} = \boldsymbol{f}(x,z,t) - \boldsymbol{T}_0(x,z) & \text{in } \Gamma^*, \quad (1d) \\ \sigma = 0 & \text{in } |x|\to\infty, |y|\to\infty, \quad (1e) \end{cases}$$

where c_d and c_s indicate the dilatational and shear wave velocities, respectively; note that $c_d = \sqrt{(\lambda + 2\mu)/\rho}$ and $c_s = \sqrt{\mu/\rho}$ hold, where λ and μ are the Lamé constants of rock, and ρ is the density of rock. Furthermore, $\boldsymbol{n}$ indicates a unit normal to Γ, so that Eq. (1b) physically represents a non-opening (i.e., slip) condition on the fault. The tensor σ is the stress perturbation defined as

$$\sigma := \lambda(\nabla\cdot\boldsymbol{u})\boldsymbol{I} + \mu\{\nabla\boldsymbol{u} + (\nabla\boldsymbol{u})^T\}, \tag{2}$$

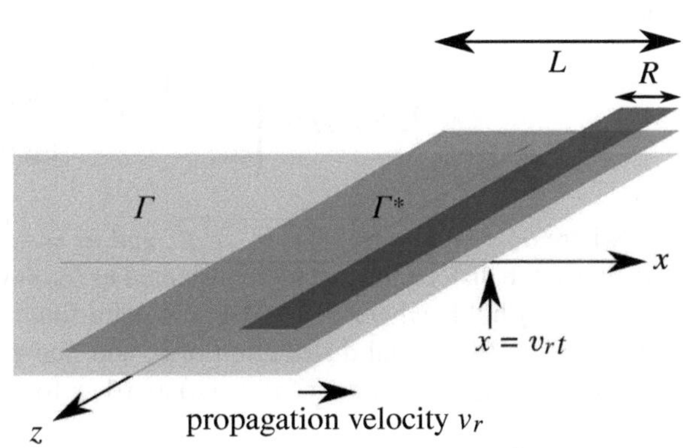

Fig. 1 Domains in the x-z plane. The *light gray* region indicates the ruptured region Γ, and the intermediate *gray* region with width L indicates the temporarily slipping region Γ^*. The *dark gray* region with width R indicates the process zone of slip-weakening friction defined in Eq. (6). All are propagating rightward with velocity v_r

where $\boldsymbol{I}$ indicates the identity, so the boundary condition (1c) physically represents continuity of traction on the fault. Moreover, LHS of Eq. (1d) is the shear component of traction, and $\boldsymbol{f}$ and $\boldsymbol{T}_0$ in RHS of Eq. (1d) are considered to be friction and initial traction applied to Γ, respectively. We assume that $\boldsymbol{T}_0$ is uniform on Γ.

However, the PDE still lacks a boundary condition for $|z| \to \infty$ and an initial condition. Instead of assuming them explicitly, previous studies assumed a uniformity of the system along the z-axis and a steady-state. We assume that the displacement field does not depend on a position in the z-axis, which is formally equivalent to $\boldsymbol{u} = \boldsymbol{u}(x, y, t)$ and $\partial_z \boldsymbol{u} = 0$. Thus, the system no longer requires a boundary condition for $|z| \to \infty$. This assumption splits the governing equation into an in-plane problem for u_x, u_y and an anti-plane problem for u_z. Next, we assume that the displacement field depends only on $(x - v_r t, y)$, as with Γ and Γ^*. This assumption allows us to apply Galilean transformation $x - v_r t \mapsto x'$ to the system, which yields $\partial_t = -v_r \partial_{x'}$. Thanks to the assumption of steady-state, we do not require an initial condition. Subsequently, we once again note x' as x; therefore, the rupture front is located at the origin of the x-y plane (i.e., $\Gamma = (-\infty, 0) \times \mathbb{R}$ and $\Gamma^* = (-L, 0) \times \mathbb{R}$).

2.2 Fundamental Solution and Integral Equation

Regarding the representation of shear traction perturbations due to steady-state propagation of the pulse, one of the simplest cases is to apply Galilean transformation to the moving dislocation model. In the dislocation model, $[u](x) = bH(-x)$ holds when $u = u_x$ for a climbing edge dislocation (mode-II), or $u = u_z$ for a screw edge dislocation (mode-III), and where b is the magnitude of the Burgers vector, and $H(\cdot)$ is the Heaviside function. Noting $(\sigma \boldsymbol{n})^{\pm} = \left(T_{xy}, T_{yy}, T_{zy}\right)^T$, these three components due to the dislocations are obtained as:

$$\left(T_{xy}, T_{yy}, T_{zy}\right)^T = \frac{\mu F(v_r/c_s)}{2\pi} \frac{b}{x} \times \begin{cases} (1, 0, 0)^T & \text{for mode-II,} \\ (0, 0, 1)^T & \text{for mode-III,} \end{cases} \tag{3}$$

where

$$F(v_r/c_s) = \begin{cases} \frac{4\alpha_d \alpha_s - \left(1+\alpha_s^2\right)^2}{\alpha_s\left(1-\alpha_s^2\right)} & \text{for mode-II,} \\ \alpha_s & \text{for mode-III,} \end{cases} \tag{4}$$

and where $\alpha_d = \sqrt{1 - (c_s/c_d)^2 (v_r/c_s)^2}$, and $\alpha_s = \sqrt{1 - (v_r/c_s)^2}$ (e.g., [22]). Slip rate due to the dislocation can be represented as $bv_r\delta(x)$ because $v(x) = \partial_t[u](x) = -v_r \partial_x [u](x)$ holds where $\delta(\cdot)$ is Dirac's delta function. This means that $v_r^{-1} v(x)$ is identical to the spatial distribution of b, and non-zero components in (3) can be represented as a convolution of $v_r^{-1} v(x)$ and $1/x$ for arbitrary v as:

$$T(x) = -\frac{\mu F(v_r/c_s)}{2v_r}\int_{\Gamma^*}\frac{v(\xi)}{\xi - x}\frac{d\xi}{\pi}, \tag{5}$$

where $T = T_{xy}$ for mode-II or $T = T_{zy}$ for mode-III. According to the boundary condition (1d), $(\sigma \boldsymbol{n})^{\pm}$ is equivalent to $f(x) - T_0$ (the x component of $\boldsymbol{f} - \boldsymbol{T}_0$ for mode-II or the z component of $\boldsymbol{f} - \boldsymbol{T}_0$ for mode-III). Therefore, $T(x)$ is a function related to friction and initial traction, and $v(x)$ is an unknown function. The Airfoil equation, a singular integral equation (5), can be turned into the Riemann-Hilbert problem, and its general solution was obtained by [13].

3 Friction Modeling

3.1 RSP Modeling Revisited

In general, a solution of Eq. (5) has a square-root singularity behind the rupture front (i.e., $v(-\varepsilon) \propto 1/\sqrt{\varepsilon}$ for a sufficiently small positive value of ε). As discussed in Sect. 1, this singularity can be removed under slip-weakening friction. RSP considered $f(x)$ and $T(x)$ at $x \in \Gamma^* = (-L, 0)$ as follows:

$$T(x) + T_0 = f(x) = \begin{cases} T_s + \frac{T_s - T_d}{R}x & \text{for } -R < x \le 0, \\ T_d & \text{for } -L \le x \le -R, \end{cases} \tag{6}$$

where T_s and T_d are the maximum static friction and the dynamic friction, respectively, and should satisfy the relationship $0 \le T_d \le T_0 < T_s$. Note that Eq. (6) is an actual form of Eq. (1d), and the LHS of Eq. (6) physically represents the total shear traction temporarily acting on the pulse. According to Eq. (6), the maximum static friction occurs at the rupture front ($x = 0$), and constant dynamic friction occurs when $-L < x < -R$. Moreover, there exists a process zone for the weakening of friction at $-R < x \le 0$. Under this setting, $[u]$ increases and f decreases as x decreases in Γ^* if $v \ge 0$. Therefore, friction is a monotonically decreasing function of the slip amount. Thus, RSP considered that Eq. (6) corresponds to the slip-weakening friction law.

By solving the Airfoil equation, which consists of Eqs. (5) and (6), RSP obtained a condition to remove the square-root singularity as follows:

$$\frac{1}{1+S} = \frac{\theta'}{\pi} - \frac{\theta' - \sin\theta'}{2\pi\sin^2(\theta'/2)}, \tag{7}$$

where $S = (T_s - T_0)\,/\,(T_0 - T_d)$, and $\theta' = 2\arcsin\sqrt{R/L}$. This condition shows that only R/L is determined uniquely after S is given. In other words, the absolute values of L and R cannot be determined uniquely even after S is given.

RSP also considered the energy release rate due to the propagation of the pulse, which can be calculated via:

$$G = \int_0^D \{f([u]) - T_d\}\, d[u], \tag{8}$$

where $D := \lim_{t\to\infty}[u]$ is the total locked-in displacement [14]. As a result, they concluded that:

$$G = \frac{(T_s - T_d)^2 R}{\mu F(v_r/c_s)} \left(\frac{\theta'}{\pi} - \frac{\theta' - \sin\theta'}{2\pi \sin^2(\theta'/2)}\right)\left(\frac{\theta' - \sin\theta' \cos\theta'}{4\sin^4(\theta'/2)}\right). \tag{9}$$

Combining Eqs. (7) and (9), we can write Ψ as a relationship between some parameters:

$$\Psi\,(T_s - T_d, S, G, R, v_r) = 0. \tag{10}$$

In Eq. (10), a frictional parameter $T_s - T_d$, the stress ratio S and a material quantity G can be regarded as prescribed mechanical parameters independent of kinematics. On the other hand, R and v_r are kinematic parameters that describes an apparent movement of the system. Hence, in view of mechanics, R and v_r are determined after the mechanical parameters are given. However, Eq. (10) means that there exists a trade-off between R and v_r, which cannot be determined uniquely on the basis of this single relationship. Therefore, another relationship between R and v_r needs to be identified.

3.2 Modeling the Microscopic Structure of Fault Planes

During modeling, prior studies neglected fault thickness; although, geologists have observed small but finite thicknesses of localized shear layers in fault outcrops. These layers are full of a finely crushed matrix (or fault gouge) and have thicknesses of 10^{-4}–10^{2} m, roughly proportional to 10^{-4} times fault length (e.g., [16, 18]). Therefore, the regions Γ and Γ^* in our notation are extended to $\Gamma = \{(x, y, z)\,|x < 0, y \in B, z \in \mathbb{R}\}$ and $\Gamma^* = \{(x, y, z)\,|x \in (-L, 0), y \in B, z \in \mathbb{R}\}$, respectively, where $B := (-H_0/2, +H_0/2)$, and a small constant H_0 is the thickness of the gouge layer. The fault gouge likely plays a significant role in the energy dissipation processes of faulting. Hatano [8] modeled each particle of a granular layer as an elastic sphere and numerically simulated the macroscopic frictional behavior of the gouge layer using a discrete element method. In his model, the granular layer was sandwiched between two rigid walls, one of which was forced to slide at a given rate. When the rate was changed in a step-like manner, frictional resistance between the granular layer and the wall varied continuously; on this basis, Hatano [8] obtained a relaxation time τ for the transient variation, which was independent of the slip rate and was given as:

$$\tau = cH_0\sqrt{\frac{\rho_p}{P}}, \tag{11}$$

where $c \sim 1.0$ is a numerical constant, ρ_p is density of the particle, and P is pressure acting on the layer. This means that the relaxation time of friction is prescribed prior to slippage. Moreover, Hatano [8] suggested that τ can roughly represent the scale of transient time from the maximum static friction to dynamic friction. Therefore, these findings presented the possibility for a model of rupture propagation that incorporated the weakening process of friction.

3.3 Determination of Kinematic Parameters

As discussed, our x was originally $x - v_r t$ before the Galilean transformation was applied, while the relaxation time is Eq. (11) if the slipping plane consists of a granular layer. By returning x into $x - v_r t$ and considering Eq. (6), we find that the slip-weakening friction assumed by RSP has a relaxation time of R/v_r. Hence,

$$\tau = \frac{R}{v_r} \tag{12}$$

can be regarded as a constant. It is clear that Eq. (12) represents another trade-off between R and v_r; therefore, the two parameters are uniquely determined after substituting Eq. (12) into (10). Moreover, by taking Eq. (11) into account, we obtain a relationship between the parameters $T_s - T_d$, S, G, H_0, ρ_p, P, and v_r. Furthermore, considering Eqs. (5), (7) and (12), we are also able to uniquely determine slip rate $v(x)$, slip $[u](x) = -v_r^{-1} \int_0^x v(\xi)d\xi$, and pulse length L.

4 Discussion

4.1 Analysis with Huge Uncertainties of Parameters

In general, the tectonophysical and geological parameters T_s, T_d, T_0, G, H_0 and P contain huge uncertainties, of which we often know only the order. Therefore, we propose only a rough determination of kinematic parameters (e.g., v_r, R and L) depending on the order. By defining $R_0 := \lim_{v_r \to 0} R$ and regarding Eq. (9), RSP obtained the following:

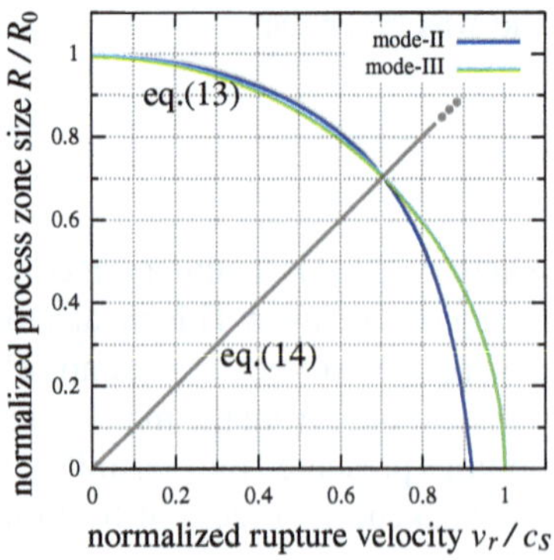

Fig. 2 Plot of Eq. (13), where *blue* denotes mode-II and *green* denotes mode-III, and Eq. (14). For mode-II, $c_d/c_s = \sqrt{3}$ is assumed, which is equivalent to Poisson's ratio of 1/4 and an approximate value for crustal rocks. Zero of the curve for mode-II is identical to the Rayleigh wave velocity

$$\frac{R}{R_0} = \frac{F(v_r/c_s)}{F(0)}. \tag{13}$$

On the other hand, Eq. (12) is equivalent to the following:

$$\frac{R}{R_0} = \frac{\tau c_s}{R_0} \frac{v_r}{c_s}. \tag{14}$$

Figure 2 shows a curve and line based on Eqs. (13) and (14), respectively, where their intersection provides actual values of v_r/c_s and R/R_0. The slope of the line, which corresponds to the coefficient $\tau c_s/R_0$ in Eq. (14), depends on the aforementioned uncertain parameters, while Eq. (13) depends only on v_r/c_s. Therefore, we are only able to consider the order of the coefficient. With respect to Fig. 2, we make the rough divisions:

$$\begin{cases} v_r/c_{\text{lim}} \ll 1 & \text{if } \tau c_s/R_0 \gg 1, \\ v_r/c_{\text{lim}} \sim 0.7 & \text{if } \tau c_s/R_0 \sim 1, \\ v_r/c_{\text{lim}} \sim 1 & \text{if } \tau c_s/R_0 \ll 1, \end{cases} \tag{15}$$

where c_{lim} is the Rayleigh wave velocity for mode-II, or the shear wave velocity for mode-III. As discussed in Sect. 1, v_r/c_{lim} is usually approximately 40–90 %. Hence, we conclude that: (1) rupture velocity of several tens of percent for c_{lim} is possible when $\tau c_s/R_0$ is near unity; (2) rupture velocity approximately equivalent to c_{lim} is possible when $\tau c_s/R_0$ is negligible; and (3) slow rupture velocity is rarely observed, likely because $\tau c_s/R_0$ is not sufficiently greater than unity.

4.2 Applicability to Interfacial Problem

Faults commonly lie along material interfaces (e.g., plate boundaries and other tectonic lines), and the theoretical modeling discussed in Sects. 2 and 3.1 has been extended to such interfacial problems [1, 10, 21]. Especially, in the case of mode-II, normal stress perturbation inside of an interfacial pulse, the second component of Eq. (3), is not zero, but rather is Dirac's delta function with the sign dependent on the direction of rupture propagation [21]. Hence, many studies have predicted asymmetric or unidirectional rupture along material interfaces, while others have actually observed [12, 17]; however, a theoretical understanding of them is still needed.

Hirano and Yamashita [10] extended the method of RSP and derived a Carleman-type singular integral equation (c.f., [19]) instead of Eq. (5). On solving the equation, they found a relationship between some parameters (i.e., generalization of Eq. (7)), which removed divergence of the slip rate near the leading edge of an interfacial pulse. This generalized relationship included not only S and R/L, but also the direction and velocity of pulse propagation. Moreover, G for the interfacial pulse can be determined by using their solution for slip rate and Eq. (8). Therefore, we can now extend Eq. (10) and our modeling to determine propagation velocity of the interfacial pulse, which can be asymmetric or unilateral.

5 Conclusions

We reviewed the methods of RSP for modeling the propagation of the pulse-like rupture with the slip-weakening friction law and highlighted their difficulties in the determination of the propagation velocity of pulse-like rupture. Considering the frictional model of Hatano [8], we proposed determinability of the propagation velocity. The propagation velocity depends on some tectonophysical/geological parameters, which usually have huge uncertainty, and we discussed the velocity only on the basis of orders of the parameters. Even with such uncertainty, we found that we could roughly estimate the propagation velocity and discuss why propagation velocity is not negligibly low. Moreover, we showed that our method can be extended to a problem of interfacial rupture propagation, which is important and controversial in earthquake source physics.

Acknowledgments The author is grateful to T. Yamashita and T. Hatano for helpful discussions.

References

1. Adda-Bedia, M., Ben Amar, M.: Self-sustained slip pulses of finite size between dissimilar materials. J. Mech. Phys. Solids **51**, 1849–1861 (2003)

2. Ampuero, J.-P., Ben-Zion, Y.: Cracks, pulses and macroscopic asymmetry of dynamic rupture on a bimaterial interface with velocity-weakening friction. Geophys. J. Int. **173**, 674–692 (2008)
3. Bizzarri, A.: On the deterministic description of earthquakes. Rev. Geophys. **49**, RG3002 (2011)
4. Dong, G., Papageorgiou, A.S.: Seismic radiation from a unidirectional asymmetrical circular crack model, part I: constant rupture velocity. Bull. Seism. Soc. Am. **92**(3), 945–961 (2002)
5. Freund, L.B.: Dynamic Fracture Mechanics. Cambridge University Press, New York (1990)
6. Gabriel, A.-A., Ampuero, J.-P., Dalguer, L.A., Mai, P.M.: Source properties of dynamic rupture pulses with off-fault plasticity. J. Geophys. Res. Solid Earth **118**, 4117–4126 (2013)
7. Geller, R.J.: Scaling relations for earthquake source parameters and magnitudes. Bull. Seism. Soc. Am. **66**(6), 1501–1523 (1976)
8. Hatano, T.: Scaling of the critical slip distance in granular layers. Geophys. Res. Lett. **36**, L18304 (2009)
9. Heaton. T.H.: Evidence for and implications of self-healing pulses of slip in earthquake rupture. Phys. Earth Planet. Int. **64**, 1–20 (1990)
10. Hirano, S., Yamashita, T.: Modeling of interfacial dynamic slip pulses with slip-weakening friction. Bull. Seism. Soc. Am. **106**, 1628–1636 (2016)
11. Ida, Y.: Cohesive force across the tip of a longitudinal shear crack and Griffith·s crack specific surface energy. J. Geophys. Res. **77**, 3796–3805 (1972)
12. Kane, D.L., Shearer, P.M., Goertz-Allman, B.P., Vernon, F.L.: Rupture directivity of small earthquakes at Parkfield. J. Geophys. Res. **118**, 1–10 (2013)
13. Muskhelishvili, N.I.: Singular Integral Equations. P. Noordhoff Ltd., Groningen (1953)
14. Palmer, A.C., Rice, J.R.: The growth of slip surfaces in the progressive failure of over-consolidated clay. Proc. R. Soc. Lond. A **322**, 527–548 (1973)
15. Rice, J.R., Sammis, C.G., Parsons, R.: Off-fault secondary failure induced by a dynamic slip pulse. Bull. Seism. Soc. Am. **95**(1), 109–134 (2005)
16. Robertson, E.C.: Continuous formation of gouge and breccia during fault displacement. In: Goodman, R.E., Hulse, F. (eds.) Issues in Rock Mechanics, Proceedings of 23rd Symposium of Rock Mechanics, pp. 397–404. American Institute of Mining and Metallurgical Engineers, New York, NY (1982)
17. Rubin, A., Gillard, D.: Aftershock asymmetry/rupture directivity among central San Andreas fault microearthquakes. J. Geophys. Res. **105**, 19095–19109 (2000)
18. Scholz, C.H.: Wear and gouge formation in brittle faulting. Geology **15**, 493–495 (1987)
19. Tricomi, F.G.: Integral Equations. Dover, New York (1957)
20. Venkataraman, A., Kanamori, H.: Observational constraints on the fracture energy of subduction zone earthquakes. J. Geophys. Res. **109**, B05302 (2004)
21. Weertman, J.: Unstable slippage across a fault that separates elastic media of different elastic constants. J. Geophys. Res. **85**, 1455–1461 (1980)
22. Weertman, J., Weertman, J.R.: Moving dislocations. In: Nabarro, F.R.N. (ed.) Dislocations in Solids, vol. 3, chap. 8, pp. 1–59, North-Holland Publishing Co., Amsterdam (1980)

Inverse Source Problem for a Wave Equation with Final Observation Data

Daijun Jiang, Yikan Liu and Masahiro Yamamoto

Abstract In this chapter, we study the inverse problem on recovering a spatial component of the source term in a wave equation by the final observation data. Employing the analytic Fredholm theory, we establish a generic well-posedness result concerning the uniqueness of our inverse source problem. Numerically, by treating a corresponding minimization problem, we investigate the variational equation for the minimizer and develop an iterative thresholding algorithm. One- and two-dimensional numerical experiments are implemented to demonstrate the robustness and accuracy of the proposed algorithm.

1 Introduction

Let $T > 0$ and Ω be an open-bounded domain in $\mathbb{R}^d$ $(d = 1, 2, \dots)$ whose boundary is of C^2 class. Consider an initial-boundary value problem for a hyperbolic equation with the homogeneous Neumann boundary condition:

$$u(f)\begin{cases}\partial_t^2 u(x,t) - \Delta u(x,t) = f(x)R(x,t) & (x \in \Omega,\ 0 < t \le T),\\ u(x,0) = \partial_t u(x,0) = 0 & (x \in \Omega),\\ \partial_\nu u(x,t) = 0 & (x \in \partial\Omega,\ 0 < t \le T),\end{cases} \tag{1}$$

D. Jiang
School of Mathematics and Statistics & Hubei Key Laboratory of Mathematical Sciences, Central China Normal University, Wuhan 430079, People's Republic of China
e-mail: jiangdaijun@mail.ccnu.edu.cn

Y. Liu (✉) · M. Yamamoto
Graduate School of Mathematical Sciences, The University of Tokyo, 3-8-1 Komaba, Tokyo, Meguro-ku 153-8914, Japan
e-mail: ykliu@ms.u-tokyo.ac.jp

M. Yamamoto
e-mail: myama@ms.u-tokyo.ac.jp

H. Itou et al. (eds.), *Mathematical Analysis of Continuum Mechanics and Industrial Applications*, Mathematics for Industry 26,
DOI 10.1007/978-981-10-2633-1_11

where $\nu = \nu(x)$ is the unit outward normal vector at $x \in \partial\Omega$ and $\partial_\nu u := \nabla u \cdot \nu$ denotes the normal derivative. Here, we write the solution as $u(f)$ to emphasize its dependency upon the time-independent function f.

In this chapter, we investigate the following inverse source problem:

Problem 1 Let u be the solution to (1). Provided that R is known, determine the spatial component $f(x)$ $(x \in \Omega)$ of the source term by the final observation data $u(\,\cdot\,, T)$.

Within the past few decades, inverse hyperbolic problems have been investigated extensively from both theoretical and numerical aspects (see [1, 4, 5, 9, 12, 14–16] and the references therein). However, it reveals that the majority of the existing literature deals with partial boundary or interior observation data except for [15]. Regardless of the practical importance, publications treating the final observation data are quite limited to the authors' best knowledge. Indeed, such a formulation of Problem 1 is applicable in recovering the spatial distribution of the external force in the elastic wave model or the nucleation rate in the one-dimensional time cone model (see [13]), where the final state of the process is observed. As the first attempt, this chapter aims at understanding the ill-posedness and solvability of Problem 1, as well as developing a numerical reconstruction method for applications.

The rest of this chapter is organized as follows. Motivated by the methodology of [14], in Sect. 2 we establish a generic well-posedness result concerning the uniqueness of Problem 1. In Sect. 3, we purpose an iteration thresholding algorithm for the numerical reconstruction. Finally, one- and two-dimensional numerical experiments are implemented in Sect. 4.

2 Generic Well-Posedness for Problem 1 and the Proof

In this section, we give an affirmative answer to Problem 1, asserting that the problem is almost well-posed under certain conditions. To this end, we define the admissible set of the known component R in the source term as

$$\mathscr{R} = \{R \in C^2([0, T]; L^\infty(\Omega));\ R(\,\cdot\,, 0) = 0,\ R(\,\cdot\,, T) \not\equiv 0 \text{ in } \overline{\Omega}\}. \tag{2}$$

First we state the main theoretical result in this chapter.

Theorem 1 (Generic well-posedness) *Let $T > 0$ and $R \in \mathscr{R}$ be arbitrarily given, where $\mathscr{R}$ is defined in* (2). *Then for any $\varepsilon > 0$, there exists $R_\varepsilon \in \mathscr{R}$ such that $\|R - R_\varepsilon\|_{C^2([0,T];L^\infty(\Omega))} < \varepsilon$ and Problem 1 with R_ε admits a unique solution.*

The above conclusion is significant in the sense that the subset of $\mathscr{R}$ guaranteeing the uniqueness of Problem 1 is dense in $\mathscr{R}$ by the norm of $C^2([0, T]; L^\infty(\Omega))$. Meanwhile, we emphasize that the observation time T is allowed to be arbitrary, while it is not the case in most inverse hyperbolic problems due to the finite propagation speed of wave (see, e.g., [4, 5]).

Proof Fix $R \in \mathscr{R}$ arbitrarily. We divide the proof of Theorem 1 into three steps. The main ingredient is the application of the analytic Fredholm theory.

Step 1 First, we reduce the problem to a Fredholm equation of the second kind. Set $v = \partial_t^2 u$ and $g = u(\,\cdot\,, T)$. Since $R(\,\cdot\,, 0) = 0$, we see that v satisfies

$$\begin{cases} \partial_t^2 v - \Delta v(x,t) = f(x)\partial_t^2 R(x,t) & (x \in \Omega,\ 0 < t \le T), \\ v(x,0) = 0,\ \partial_t v(x,0) = f(x)\partial_t R(x,0) & (x \in \Omega), \\ \partial_\nu v(x,t) = 0 & (x \in \partial\Omega,\ 0 < t \le T), \end{cases} \tag{3}$$

and

$$f = \frac{v(\,\cdot\,, T) - \Delta g}{R(\,\cdot\,, T)}.$$

Next we define an R-valued operator

$$K_R f := \frac{v(\,\cdot\,, T)}{R(\,\cdot\,, T)}, \quad f \in L^2(\Omega).$$

Then $R(\,\cdot\,, T)K_R : L^2(\Omega) \longrightarrow H^1(\Omega)$ is a bounded linear operator (see Lions and Magenes [10]), and thus K_R is a compact operator for $R \in \mathscr{R}$.

Step 2 Fix $R_1, R_2 \in \mathscr{R}$ arbitrarily and write $H_1 := R_1(\,\cdot\,, T)$, $H_2 := R_2(\,\cdot\,, T)$. For $0 < \theta < 1$, we show that $K(\theta) := K_{\theta R_1 + (1-\theta)R_2}$ is analytic for $\theta \in (0, 1)$. For any fixed $f \in L^2(\Omega)$, denote the solution to (1) at $t = T$ with R_1 and R_2 by w_1 and w_2, respectively. Then we have

$$K(\theta) f = \frac{\theta w_1 + (1-\theta)w_2}{\theta H_1 + (1-\theta)H_2}.$$

Therefore, for any $\theta_0 \in (0, 1)$, we calculate

$$\begin{aligned} K(\theta)f - K(\theta_0)f &= \frac{\theta w_1 + (1-\theta)w_2}{\theta H_1 + (1-\theta)H_2} - \frac{\theta_0 w_1 + (1-\theta_0)w_2}{\theta_0 H_1 + (1-\theta_0)H_2} \\ &= \frac{(\theta-\theta_0)(w_1 H_2 - w_2 H_1)}{(\theta H_1 + (1-\theta)H_2)(\theta_0 H_1 + (1-\theta_0)H_2)}. \end{aligned}$$

Passing $\theta \to \theta_0$, we see that the limit

$$\lim_{\theta \to \theta_0} \frac{K(\theta)f - K(\theta_0)f}{\theta - \theta_0} = \frac{w_1 H_2 - w_2 H_1}{(\theta_0 H_1 + (1-\theta_0)H_2)^2} \in L^2(\Omega)$$

exists. In other words, $K(\theta)$ is analytic for $\theta \in (0, 1)$.

Step 3 We search $R(x,t) =: r(t) \in \mathscr{R}$ guaranteeing the well-posedness of the inverse problem. To this end, we employ the eigensystem $\{(\lambda_n, \varphi_n)\}_{n=0}^{\infty}$ of $-\Delta$ with the homogeneous Neumann boundary condition. Then $0 = \lambda_0 < \lambda_1 \le \lambda_2 \le \cdots$, $\lambda_n \to \infty$ as $n \to \infty$, and $\{\varphi_n\} \subset H^2(\Omega)$ forms a complete orthonormal basis of

$L^2(\Omega)$. According to the Fourier method, the solution to (3) with $R(x,t) = r(t)$ reads

$$v(\cdot,t) = (f,\varphi_0)r(t)\varphi_0 + \sum_{n=1}^{\infty} \frac{(f,\varphi_n)}{\sqrt{\lambda_n}} \int_0^t \sin\left(\sqrt{\lambda_n}\,(t-s)\right) r''(s)\,\mathrm{d}s\,\varphi_n,$$

where $(\cdot\,,\,\cdot)$ denotes the inner product in $L^2(\Omega)$. Taking $t = T$ in the above expression and using the identity

$$v(\cdot\,,T) = \partial_t^2 u(\cdot\,,T) = \Delta u(\cdot\,,T) + r(T)\,f = \Delta g + r(T)\,f,$$

we obtain

$$\Delta g + \sum_{n=1}^{\infty}(f,\varphi_n)r(T)\varphi_n = \sum_{n=1}^{\infty} \frac{(f,\varphi_n)}{\sqrt{\lambda_n}} \int_0^T \sin\left(\sqrt{\lambda_k}\,(T-s)\right) r''(s)\,\mathrm{d}s\,\varphi_n.$$

Hence, if

$$r(T) > \frac{1}{\sqrt{\lambda_n}} \int_0^T \sin\left(\sqrt{\lambda_k}\,(t-s)\right) r''(s)\,\mathrm{d}s, \quad \forall n = 1,2,\ldots, \tag{4}$$

then $g = 0$ implies $f = 0$, and thus the Fredholm alternative yields that the inverse problem is well-posed with $R = r$. For simplicity, we further restrict r as

$$r(t) = \mathrm{e}^{\delta t} - \delta t - 1 \quad (\delta > 0).$$

Then we see that (4) is equivalent to

$$\mathrm{e}^{\delta T} - \delta T - 1 > \frac{\delta^2 I}{\sqrt{\lambda_n}}\ (\forall n = 1,2,\ldots), \quad I := \int_0^T \sin\left(\sqrt{\lambda_n}\,(T-t)\right) \mathrm{e}^{\delta t}\,\mathrm{d}t.$$

Direct calculations yield

$$\begin{aligned} I &= \frac{1}{\delta}\left(-\sin\left(\sqrt{\lambda_n}\,T\right) + \sqrt{\lambda_n}\int_0^T \cos\left(\sqrt{\lambda_n}\,(T-t)\right)\mathrm{e}^{\delta t}\,\mathrm{d}t\right) \\ &= \frac{1}{\delta}\left\{-\sin\left(\sqrt{\lambda_n}\,T\right) + \frac{\sqrt{\lambda_n}}{\delta}\left(\mathrm{e}^{\delta T} - \cos\left(\sqrt{\lambda_n}\,T\right) - \sqrt{\lambda_n}\int_0^T \sin\left(\sqrt{\lambda_n}\,(T-t)\right)\mathrm{e}^{\delta t}\,\mathrm{d}t\right)\right\} \\ &= -\frac{\lambda_n I}{\delta^2} + \frac{1}{\delta^2}\left\{\sqrt{\lambda_n}\,\mathrm{e}^{\delta T} - \left(\delta\sin\left(\sqrt{\lambda_n}\,T\right) + \sqrt{\lambda_n}\cos\left(\sqrt{\lambda_n}\,T\right)\right)\right\} \end{aligned}$$

and thus

$$\begin{aligned} I &= \frac{1}{\delta^2+\lambda_n}\left\{\sqrt{\lambda_n}\,\mathrm{e}^{\delta T}-\left(\delta\sin\left(\sqrt{\lambda_n}\,T\right)+\sqrt{\lambda_n}\cos\left(\sqrt{\lambda_n}\,T\right)\right)\right\} \\ &\le \frac{\sqrt{\lambda_n}}{\delta^2+\lambda_n}\mathrm{e}^{\delta T}+\frac{1}{\sqrt{\delta^2+\lambda_n}}. \end{aligned}$$

Therefore, it suffices to choose $\delta=\delta(T)>0$ such that

$$\mathrm{e}^{\delta T}-\delta\,T-1>\frac{\delta^2}{\delta^2+\lambda_n}\mathrm{e}^{\delta T}+\frac{\delta^2}{\sqrt{\lambda_n(\delta^2+\lambda_n)}}\quad (\forall n=1,2,\dots),$$

which is guaranteed by

$$\mathrm{e}^{\delta T}-\delta\,T-1>\frac{\delta^2}{\delta^2+\lambda_1}\mathrm{e}^{\delta T}+\frac{\delta^2}{\sqrt{\lambda_1(\delta^2+\lambda_1)}}$$

or equivalently

$$\lambda_1\,\mathrm{e}^{\delta T}>(\delta^2+\lambda_1)\left(\frac{\delta^2}{\sqrt{\lambda_1(\delta^2+\lambda_1)}}+\delta\,T+1\right). \tag{5}$$

Since the right-hand side of (5) is polynomial with respect to δ, we can choose a sufficiently large $\delta=\delta(T)>0$ such that (5) holds and thus (4) is satisfied.

Consequently, since we have constructed a special $r\in\mathscr{R}$ ensuring the uniqueness of Problem 1, the application of Kato [8, Theorem 1.9] completes the proof. □

3 Iterative Thresholding Algorithm

Based on the generic well-posedness obtained in the previous section, this section is devoted to the development of an effective algorithm for the numerical reconstruction of the source term in the same manner as that in [7, 11]. To this end, first we recall the following definition of the weak solution.

Definition 1 *(see, e.g.,* [6]) Let $F\in L^2(\Omega\times(0,T))$, $u_0\in H^1(\Omega)$ and $u_1\in L^2(\Omega)$. We say that $u\in H^1(\Omega\times(0,T))$ is a weak solution to

$$\begin{cases} \partial_t^2 u-\Delta u=F & \text{in } \Omega\times(0,T],\\ u=u_0,\ \partial_t u=u_1 & \text{in } \Omega\times\{0\},\\ \partial_\nu u=0 & \text{on } \partial\Omega\times(0,T] \end{cases}$$

if $u = u_0$ in $\Omega \times \{0\}$ and

$$\int_0^T \int_\Omega (\nabla u \cdot \nabla z - \partial_t u\, \partial_t z)\, \mathrm{d}x\mathrm{d}t = \int_0^T \int_\Omega F\, z\, \mathrm{d}x\mathrm{d}t + \int_\Omega u_1\, z(\,\cdot\,, 0)\, \mathrm{d}x$$

holds for any $z \in H^1(\Omega \times (0, T))$ with $z(\,\cdot\,, T) = 0$.

Parallel, the weak solution to a backward wave equation can be defined in the same manner.

Numerically, we treat Problem 1 by interpreting its solution as a minimizer of the minimization problem

$$\min_{f \in L^2(\Omega)} J(f), \quad J(f) := \|u(f)(\,\cdot\,, T) - g^\delta\|^2_{L^2(\Omega)} + \alpha \|f\|^2_{L^2(\Omega)}, \tag{6}$$

where $g^\delta \in L^2(\Omega)$ is the observation data with noise, and $\alpha > 0$ stands for the regularization parameter. Taking Fréchet derivative yields

$$J'(f)\widetilde{f} = 2 \int_\Omega \left(u(f)(\,\cdot\,, T) - g^\delta\right) \left(u'(f)\widetilde{f}\right)(\,\cdot\,, T)\, \mathrm{d}x + 2\alpha \int_\Omega f\, \widetilde{f}\, \mathrm{d}x,$$

where

$$u'(f)\widetilde{f} = \lim_{\varepsilon \to 0} \frac{u(f + \varepsilon \widetilde{f}) - u(\widetilde{f})}{\varepsilon} = u(\widetilde{f}), \quad \widetilde{f} \in L^2(\Omega).$$

This indicates

$$\frac{1}{2} J'(f)\widetilde{f} = \int_\Omega \left\{\left(u(f)(\,\cdot\,, T) - g^\delta\right) u(\widetilde{f})(\,\cdot\,, T) + \alpha\, f\, \widetilde{f}\right\} \mathrm{d}x.$$

To characterize the minimizer $f^* \in L^2(\Omega)$ to (6), we introduce the corresponding adjoint system

$$z(f) \begin{cases} \partial_t^2 z - \Delta z = 0 & \text{in } \Omega \times [0, T), \\ z = 0,\ \partial_t z = u(f)(\,\cdot\,, T) - g^\delta & \text{in } \Omega \times \{T\}, \\ \partial_\nu z = 0 & \text{on } \partial\Omega \times [0, T), \end{cases} \tag{7}$$

whose weak solution can be defined similarly as that in Definition 1. Now for any $f, \widetilde{f} \in L^2(\Omega)$, we take $z(f)$ and $u(\widetilde{f})$ as mutual test functions for $u(\widetilde{f})$ and $z(f)$ to deduce

$$
\begin{aligned}
&\int_\Omega \left(u(f)(\cdot\,,T) - g^\delta\right) u(\widetilde{f})(\cdot\,,T)\,\mathrm{d}x \\
&= -\int_0^T\!\!\int_\Omega \left(\nabla u(\widetilde{f}) \cdot \nabla z(f) - \partial_t u(\widetilde{f})\, \partial_t z(f)\right) \mathrm{d}x\mathrm{d}t \\
&= -\int_0^T\!\!\int_\Omega \widetilde{f}\, R\, z(f)\,\mathrm{d}x\mathrm{d}t + \int_\Omega u(\widetilde{f})(\cdot\,,0)\, z(f)(\cdot\,,0)\,\mathrm{d}x \\
&= -\int_\Omega \widetilde{f} \int_0^T R\, z(f)\,\mathrm{d}t\mathrm{d}x,
\end{aligned}
$$

implying

$$
\frac{1}{2} J'(f)\widetilde{f} = \int_\Omega \widetilde{f} \left(\alpha\, f - \int_0^T R\, z(f)\,\mathrm{d}t\right) \mathrm{d}x, \quad \forall\, f, \widetilde{f} \in L^2(\Omega).
$$

This immediately gives a characterization of the solution to the minimization problem (6).

Proposition 1 *A function $f^* \in L^2(\Omega)$ is a minimizer of the functional $J(f)$ in* (6) *only if it satisfies the variational equation*

$$
\alpha\, f^* = \int_0^T R\, z(f^*)\,\mathrm{d}t, \tag{8}
$$

where $z(f^*)$ solves the backward system (7) *with the coefficient f^*.*

Adding Mf^* ($M > 0$) to both sides of (8) and rearranging in view of the iteration, we are led to the iterative thresholding algorithm

$$
f_{k+1} = \frac{M}{M+\alpha} f_k + \frac{1}{M+\alpha} \int_0^T R\, z(f_k)\,\mathrm{d}t \quad (k = 0, 1, \ldots), \tag{9}
$$

where $M > 0$ is a tuning parameter for the convergence. Relating the minimization problem (6) with that of the corresponding surrogate functional (see [2]), we see that the convergence is guaranteed by choosing

$$
M \ge \|\mathscr{A}\|_{\mathrm{op}}^2, \quad \text{where} \quad \begin{aligned} \mathscr{A} : L^2(\Omega) &\to L^2(\Omega), \\ f &\mapsto u(f)(\cdot\,,T). \end{aligned} \tag{10}
$$

Now we are well prepared to state the main algorithm for the reconstruction.

Algorithm Choose a tolerance $\varepsilon > 0$, a regularization parameter $\alpha > 0$ and a tuning constant $M > 0$ according to (10). Give an initial guess f_0, and set $k = 0$.

1. Compute f_{k+1} by the iterative update (9).

2. If $\|f_{k+1} - f_k\|_{L^2(\Omega)} / \|f_k\|_{L^2(\Omega)} \le \varepsilon$, then stop the iteration. Otherwise, update $k \leftarrow k+1$ and return to Step 1.

We conclude this section with stating the convergence result of the above Algorithm, which is a direct application of [2, Theorem 3.1].

Lemma 1 *Let $M > 0$ be a constant satisfying condition* (10). *Then for any $f_0 \in L^2(\Omega)$, the sequence $\{f_k\}_{k=1}^{\infty}$ produced by the iteration* (9) *converges in $L^2(\Omega)$ to the solution to the minimization problem* (6).

It turns out that for the iteration (9), we need not the condition $R(\,\cdot\,, 0) = 0$ in (2). In other words, this restriction is only technically obligatory in proving the theoretical uniqueness.

4 Numerical Experiments

In this section, we shall apply the iterative thresholding algorithm developed in the previous section to numerically identify the spatial component $f(x)$ in the source term of Eq. (1). The general settings of the numerical reconstruction are assigned as follows. We simply take $T = 1$, $\Omega = (0, 1)^d$ $(d = 1, 2)$, and denote the true solution by f_{true}. Although the difference between the noiseless data $u(f_{\mathrm{true}})(\,\cdot\,, T)$ and the actually observed data g^δ was evaluated in the $L^2(\Omega)$-norm, here for simplicity we produce g^δ by adding uniform random noises to the noiseless data, i.e.,

$$g^\delta(x) = u(f_{\mathrm{true}})(x, T) + \delta\,\mathrm{rand}(-1, 1), \quad \forall x \in \Omega,$$

where $\mathrm{rand}(-1, 1)$ denotes the uniformly distributed random number in $[-1, 1]$, and $\delta \ge 0$ is the noise level. Here we choose δ as a certain portion of the amplitude of the exact solution, i.e.

$$\delta := \delta_0 \max_{\overline{\Omega}} |u(f_{\mathrm{true}})(\,\cdot\,, T)|, \quad 0 < \delta_0 < 1.$$

We shall demonstrate the reconstruction method by several test examples in one and two spatial dimensions. Other than the illustrative figures, we mainly evaluate the numerical performance by the number K of iterations, the relative L^2-norm error

$$\mathrm{err} := \frac{\|f_K - f_{\mathrm{true}}\|_{L^2(\Omega)}}{\|f_{\mathrm{true}}\|_{L^2(\Omega)}},$$

where f_K is regarded as the numerical result. In our implementations, all the forward and backward wave equations involved in the iterative thresholding algorithm are solved by the von Neumann scheme for the one-dimensional case and the alternating direction implicit method (see [3]) for the two-dimensional case.

4.1 One-Dimensional Examples

In case of $d = 1$, we divide the space-time region $\overline{\Omega} \times [0, T] = [0, 1] \times [0, 1]$ as 51×51 equidistant meshes, and fix the known component of the source term as $R(x, t) = x + \pi t + \pi^2$. For the parameters involved in algorithm (9), we set $M = 20$ and $\alpha = 10^{-5}$. The stopping criteria is chosen as $\varepsilon = 2\,\%\,\delta_0$.

Now we are ready to present the numerical tests for the identification of the spatial component $f(x)$ in the source term of system (1).

Example 1 We choose $f_{\text{true}}(x) = \cos(\pi x) + 2$ and the initial guess $f_0 \equiv 2$. We will change noise levels as $\delta_0 = 1, 2, 4, 8\,\%$ to see the stability of the proposed algorithm with respect to a noise level. Figure 1 compares the exact solutions with the reconstructions with noise level $\delta_0 = 1\,\%$ (left) and $\delta_0 = 8\,\%$ (right). Table 1 lists the iteration number K and relative errors with different noise levels.

Example 2 We will select various true solutions f_{true} to show the accuracy and robustness of the proposed algorithm. More precisely, we take (a) $f_{\text{true}}(x) = \sin(\frac{\pi}{2}x) + 2$, (b) $f_{\text{true}}(x) = -x + 2$ and (c) $f_{\text{true}}(x) = 4 - e^x$. In all the cases, we fix the noise level $\delta_0 = 5\,\%$ and the initial guess $f_0 \equiv 2$. The iteration numbers K and the relative errors for different f_{true} are listed in Table 2.

We can see from the above examples that the numerical reconstructions appear to be quite satisfactory in spite of the ill-posedness of the inverse problem, even

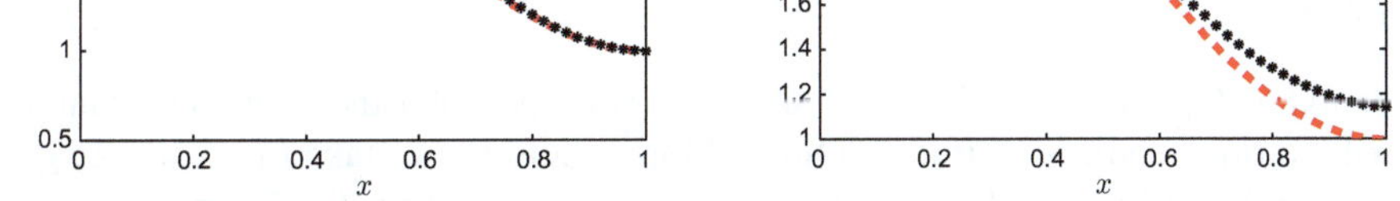

Fig. 1 Exact and reconstructed solutions in Example 1 with noise level $\delta_0 = 1\,\%$ (*left*) and $\delta_0 = 8\,\%$ (*right*)

Table 1 Iteration numbers K and relative errors with different noise levels in Example 1

δ_0 (%)	1	2	4	8
K	13	11	9	7
err (%)	0.86	1.94	2.02	4.45

Table 2 Iteration numbers K and relative errors with different true solutions in Example 2

$f_{\text{true}}(x)$	K	err (%)
$\sin(\frac{\pi}{2}x)+2$	9	3.14
$-x+2$	11	3.06
$4-e^x$	8	4.34

Table 3 Iteration numbers K and relative errors with different noise levels in Example 3

δ_0 (%)	1	2	4	8
K	36	32	28	24
err (%)	0.21	0.36	0.90	1.74

with very bad initial constant guesses and noise levels up to 8 %. Moreover, Table 1 indicates that the proposed algorithm is very stable against the noise level, while Table 2 further shows the accuracy of the proposed algorithm.

4.2 *Two-Dimensional Examples*

Similarly, in the two-dimensional case we divide $\overline{\Omega} \times [0, T] = [0, 1]^2 \times [0, 1]$ as $51^2 \times 51$ equidistant meshes. For simplicity, we choose a space-independent known component $R(x, t) = 2 + \pi^2 t^2$. We set $M = 2$ and $\alpha = 10^{-5}$ in algorithm (9), and fix the stopping criteria as $\varepsilon = 1\,\%\,\delta_0$.

Example 3 Similarly to Example 1, we first fix $f_{\text{true}}(x_1, x_2) = \cos(\pi x_1)\cos(\pi x_2) + 1$ and the initial guess $f_0 \equiv 1$ to test the robustness of algorithm (9) against the noise by changing the noise levels as $\delta_0 = 1, 2, 4, 8\,\%$. The resulting iteration numbers and relative errors with different noise levels are concluded in Table 3.

Example 4 Again, we shall test the accuracy and robustness of algorithm (9) with various choices of true solutions. More precisely, we take (a) $f_{\text{true}}(x_1, x_2) = \cos(\pi x_1)\cos(2\pi x_2) + 1$, (b) $f_{\text{true}}(x_1, x_2) = \sin x_1 \cos(\pi x_2) + 1$ and (c) $f_{\text{true}}(x_1, x_2) = x_1(x_1 - 1)\cos(2\pi x_2) + 1$. In all the cases, we fix the noise level $\delta_0 = 5\,\%$ and the initial guess $f_0 \equiv 1$. As a representative, Fig. 2 illustrates the true solution and the recovered one for the choice of (a). In Table 4, we list the iteration numbers K and relative errors for different choices of f_{true}.

Similarly to the one-dimensional case, the above two-dimensional examples demonstrate the robustness and accuracy of the proposed iterative thresholding algorithm as that observed in the one-dimensional examples. Nevertheless, it turns out that the two-dimensional reconstructions require more iterations until convergence, due to the increasing complexity with respect to the spatial dimensions.

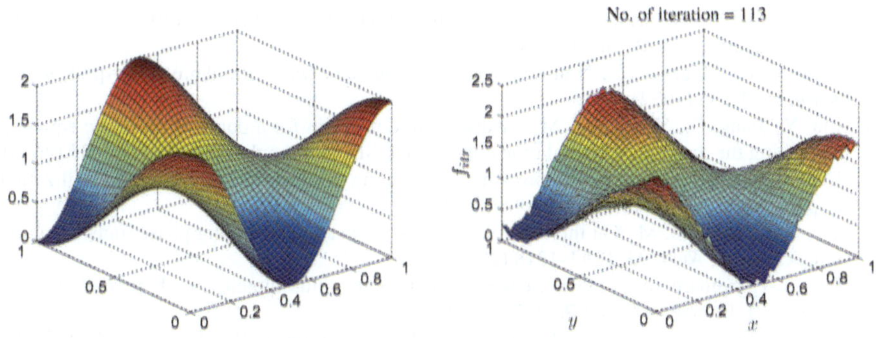

Fig. 2 Exact (*left*) and reconstructed (*right*) source terms for the choice of **a** $f_{\text{true}}(x_1, x_2) = \cos(\pi x_1)\cos(2\pi x_2) + 1$ in Example 4

Table 4 Iteration numbers and relative errors with different true solutions in Example 4

$f_{\text{true}}(x_1, x_2)$	K	err (%)
$\cos(\pi x_1)\cos(2\pi x_2) + 1$	113	3.86
$\sin x_1 \cos(\pi x_2) + 1$	20	2.27
$\cos(2\pi x_1)\cos(2\pi x_2) + 2$	48	4.30

Acknowledgements The work was supported by A3 Foresight Program "Modeling and Computation of Applied Inverse Problems," Japan Society for the Promotion of Science (JSPS). The first author was financially supported by self-determined research funds of Central China Normal University from the colleges' basic research and operation of MOE (No. CCNU14A05039), National Natural Science Foundation of China (Nos. 11326233, 11401241 and 11571265). The second and the third authors are partially supported by Grant-in-Aid for Scientific Research (S) 15H05740, JSPS.

References

1. Bukhgeim, A.L., Klibanov, M.V.: Global uniqueness of a class of multidimensional inverse problems. Sov. Math. Dokl. **24**, 244–247 (1981)
2. Daubechies, I., Defrise, M., De Mol, C.: An iterative thresholding algorithm for linear inverse problems with a sparsity constraint. Comm. Pure Appl. Math. **57**, 1413–1457 (2004)
3. Fairweather, G., Mitchell, A.R.: A high accuracy alternating direction method for the wave equation. IMA J. Appl. Math. **1**, 309–316 (1965)
4. Imanuvilov, O.Y., Yamamoto, M.: Global uniqueness and stability in determining coefficients of wave equations. Comm. Partial Differ. Equ. **26**, 1409–1425 (2001)
5. Imanuvilov, O.Y., Yamamoto, M.: Global Lipschitz stability in an inverse hyperbolic problem by interior observation. Inverse Prob. **17**, 717–728 (2001)
6. Isakov, V.: Inverse Problems for Partial Differential Equations. Springer, New York (2006)
7. Jiang, D., Liu, Y., Yamamoto, M.: Inverse source problem for the hyperbolic equations with a time-dependent principal part. J. Differ. Equation. (in press)
8. Kato, T.: Perturbation Theory for Linear Operators. Springer, Berlin (1980)

9. Klibanov, M.V., Timonov, A.: Carleman Estimates for Coefficient Inverse Problems and Numerical Applications. VSP, Utrecht (2004)
10. Lions, J.-L., Magenes, E.: Non-homogeneous Boundary Value Problems and Applications. Springer, Berlin (1972)
11. Liu, Y., Jiang, D., Yamamoto, M.: Inverse source problem for a double hyperbolic equation describing the three-dimensional time cone model. SIAM J. Appl. Math. **75**, 2610–35 (2015)
12. Liu, Y., Xu, X., Yamamoto, M.: Growth rate modeling and identification in the crystallization of polymers. Inverse Prob. **28**, 095008 (2012)
13. Liu, Y., Yamamoto, M.: On the multiple hyperbolic systems modelling phase transformation kinetics. Appl. Anal. **93**, 1297–1318 (2014)
14. Puel, J.-P., Yamamoto, M.: Generic well-posedness in a multidimensional hyperbolic inverse problem. J. Inverse Ill-Posed Prob. **5**, 55–84 (1997)
15. Yakhno, V.G.: A converse problem for a hyperbolic equation (in Russian). Mat. Zametki **26**, 39–44 (1979)
16. Yamamoto, M.: Stability, reconstruction formula and regularization for an inverse source hyperbolic problem by a control method. Inverse Prob. **11**, 481–496 (1995)

Part IV
Fluid Mechanics and Interface Dynamics

The Contribution of Kawada to the Analytical Solution for the Velocity Induced by a Helical Vortex Filament and Modern Applications of Helical Vortices

Yasuhide Fukumoto, Valery L. Okulov and David H. Wood

Abstract Currently, the analytical form of the velocity field induced by a helical vortex filament is well known as Hardin's solution (1982). But essentially the same result had been obtained by a Japanese scientist Sandi Kawada, which predates Hardin by as long as 46 years. Kawada (1936) provided a comprehensive treatment of deriving the induced velocity by helical vortices with a view to applying it to the propeller theory. This paper recollects Kawada's contribution, together with his life devoted to lead the Japanese aeronautical engineering in the time of its dawning.

1 Introduction

The Mitsubishi Regional Jet (MRJ for short) is a twin-engine regional jet aircraft manufactured by Mitsubishi Aircraft Corporation, and successfully made its first flight on November 11, 2015. This is the first airliner designed and produced in Japan since the NAMC YS-11 of the 1960. But before the World War II, the Japanese aeronautical engineers and scientists had developed a worldwide top-level technology for producing propeller aircrafts. Sandi Kawada was a member of the Japanese pioneering team. He was an expert on motion of propellers and flows around them. He published

Y. Fukumoto (✉)
Institute of Mathematics for Industry, Kyushu University,
744 Motooka, Nishi-ku, Fukuoka 819-0395, Japan
e-mail: yasuhide@imi.kyushu-u.ac.jp

V.L. Okulov
Wind Energy Department, Technical University of Denmark, 2800 Lyngby, Denmark
e-mail: vaok@dtu.dk

V.L. Okulov
Institute of Thermophysics, Siberian Branch of the Russian Academy of Sciences,
630090 Novosibirsk, Russia

D.H. Wood
Schulich School of Engineering, University of Calgary,
2500 University Dr NW, Calgary T2N 1N4, Canada
e-mail: dhwood@ucalgary.ca

H. Itou et al. (eds.), *Mathematical Analysis of Continuum Mechanics and Industrial Applications*, Mathematics for Industry 26,
DOI 10.1007/978-981-10-2633-1_12

a comprehensive theory [1, 2] for calculating flow field around a helicoidal vortex sheet [3], which serves as a model for the propeller wake flow. Unfortunately his paper had gone unnoticed till today in the community of fluid dynamics, before the last author of Ref. [4] rediscovered it.

Helical vortices play an important role in swirling flows, for instance, in tornadoes, in vortex devices and rotating tanks, in pipes, *etc*. These vortices are mathematical models for tip vortices in wakes behind a screw, propeller, or wind turbine. Since a helicoidal vortex structure in the wake can be viewed as a superposition of helical vortices, the latter is the fundamental vortex structure of any rotor wake. The helical vortex theory is of special importance in rotor aerodynamics.

Among all helical structures, a regular infinite helical vortex filament is an elementary and fundamental object of vortex dynamics. This constitutes a basic element for an assembly of helical vortices. In contrast to cases of a straight line vortex and a vortex ring, the Biot–Savart integral for the flow induced by a helical filament does not have a closed form; it cannot be represented as a simple pole like singularity, or expressed with use of the complete elliptic integrals, as in the case of an infinitely thin vortex ring [5]. For this reason, any analytical representation for the flow around a helix filament is desired for the further development of the helical vortex theory. Since the past decades of the twentieth century, it has wrongly been assumed that Jay Hardin [6] was the first who has derived this analytical solution for the helical filament. However, this is not the case. We have found that in 1936 Sandi Kawada [1, 2] was the first to establish this result [4]. The results obtained by the both scientists met the needs, though not being the same, of their times.

Kawada's solution was concerned exclusively with the propeller lifting-line theory. Lerbs [7], using Kawada's results, presented calculations for 3-, 4-, and 5-bladed propellers and a design method using the induction factor, introduced by Moriya [8], along a single radial direction. He also investigated the equivalence between the use of the Biot–Savart low and the velocity potential. Following this result, Morgan and Wrench [9] showed that the use of the Biot-Savart law in the vortex theory for a moderately loaded propeller gives a better physical insight into the problem than the use of the velocity potential for the induction factor of the fixed radial direction. The difficulty in using Kawada's solution for approximating the induction factor restricted application of the velocity potential in favor of direct simulation with use of the Goldstein function. In 1964, high-precision tables of the Goldstein function were completed by direct calculation of the Goldstein circulation along the blade [10]. Possibly, these tables replaced Kawada's solution and the interest in it has been lost.

The second interest to this solution, which is traced to Hardin's solution [6], has led to the current extensive investigation of helical vortices [5, 11–21]. Our recent paper [4] made an attempt to fill the gap between the two streams of the helical vortex theory.

This paper gives an essence of Kawada's bibliography (Sect. 2) and contribution to derive the velocity field induced by a helocal vortex filament (Sect. 3). This paper closes with von Karman's remark [22] on the contributions by Kawada and Moriya (Sect. 4).

2 Bibliography of Sandi Kawada [23]

Sandi Kawada was a pioneer of the aeronautical engineering in Japan, and played a leading role in raising this field in Japan. He was born on May 26, 1899 and died on July 16, 1970. Kawada was one of the first graduates, in March 1923, of Department of Aeronautics, Faculty of Engineering, the Imperial University of Tokyo, currently being known as the University of Tokyo. Tomijiro Moriya was his classmate. After working in Aviation Laboratory as a part-time employee and in the Japanese Army as an aviation soldier for a few years, he was appointed as an associate professor of Tokyo University in 1926, and in parallel, he worked as a researcher in Aviation Laboratory, Tokyo University. In 1939, he was promoted as a professor of Tokyo University. Kawada was widely engaged in aerodynamic research with his fields ranging over propeller theory, axial blower, turbojet, and high-speed aerodynamics (sub-sonic, transonic, and super-sonic regimes). He committed in building a number of wind tunnels of the first generation in Japan. Among others, his theory of propellers reached the frontier of the aerodynamics [24].

Aviation Laboratory was established, as a research institute attached to the Imperial University of Tokyo in 1918. The laboratory, with only a couple of specialists in airplanes, was not active, and, to be worse, its building was destroyed by the Great Kanto earthquake, calamity attacking Tokyo region in 1923. When the Shohwa era began in 1926, Chuzaburou Shiba, the Director of Laboratory, embarked on a project of making an airplane which was able to establish a world record, with a view to publicize the work of the Laboratory, and thereby to convince the laboratory staff of their high ability. In those days, long-distance flight was a challenge of attracting attention in the aviation world. In 1931, Shiba's proposal "Project of design and construction of long-distance flying airplane" to the Ministry of Education was approved by the Imperial Members of Parliament, and an unusually large amount of budget at that time, 250 thousands JPY (=2,500 USD), was allocated to it. This project was launched in the summer of 1931, collecting all the members of the Laboratory. Kawada was a chief researcher responsible for propellers. Ichiro Tani took charge of aerodynamics. At that time, the development of aviation technology was rapid, and world record of long-distance flight was updated year by year. The design of the body was made smoothly, but the cost for manufacturing the real body and the engine, together with those of test airplanes for research and development, exceeded by far the allocated budgets. Getting help from a few companies such as Tokyo Gas and Electric Industry, Kawasaki and Sumitomo Co Ltds., the project proceeded to a stage of planning for construction from 1933, but, with too much unknown, the progress was slow. At last in 1937, the desired airplane for long-distance flight was completed. On May 13, 1938, it took off at the Kisarazu Airport, Chiba Prefecture, for the third flight, continued to fly over 62 h, and flied over the distance of 11,651.011 Km in total. This flight distance was, together with its high-speed 186.192 Km/h, admitted as the new world record by FAI, Federation Aeronautique Internationale. Kawada contributed to improve the efficiency of propellers.

The next project was manufacturing an airplane for commercial flight from Tokyo to New York. Unfortunately, the effort in this direction was suspended by the outbreak of the Second World War in the pacific region in 1941.

The activities of the Japanese aeronautical engineering were completely stopped when Japan surrendered to the Allies in August 1945. The GHQ or the GHQ/SCAP (General Headquarters, the Supreme Commander for the Allied Powers) banned the aircraft industry and the related research in Japan. This policy was applied to all the laboratories associated with aeronautical engineering in Japan. The Japan Society of Aeronautical Science was broken up and was remodeled into the Japan Science Culture Society in 1946. In 1947, this society and other engineering societies were reorganized into the Japan Society of Applied Mechanics, which dealt with mechanical problems shared in common by a variety of fields in engineering. Aviation Laboratory was abolished in December 1945, but was partly revived as the Institute of Science and Engineering in April 1946. Kawada's laboratory ensued in the latter. Prohibition of aeronautics research lasted until 1953. Kawada waved his route off aeronautics, entered into civil engineering and geophysical fluid mechanics, being featured by environment aspect, and opened up a new field of what he called 'Engineering Geophysics'. For instance, one of the first project of the groups of Kawada and Ryouma Kawamura was a problem of sediment transport of sands in a coast. This problem had bearing with windbreaks in forestry. Kawada struggled to amalgamate the aerodynamics, being valid over a smaller scale, and the meteorology, being valid over a large scale. He and Moriya noticed the importance of wind as an energy source, and paid attention to research of wind turbine for gaining energy. They mentioned a pioneering work being made in Institute of High Speed Mechanics, Tohoku University, now being remodeled as Institute of Fluid Science.

Aviation industry was almost extinguished when the research ban in aeronautical engineering was lifted in 1953. In this year, the Japan Society of Aeronautical Science got back together. But the number of members was small and the society suffered from a serious financial deficiency. In 1954, Kawada was elected as the president of the society, and devoted himself to reconstructing the society. In 1970, the society has been remodeled as the Japan Society of Aeronautical and Space Sciences (JSASS), augmented with a scope for space enterprises and development of space science. Wind turbine and wind energy are also included in the scope of the JSASS.

Sandi Kawada was among the first graduates of the Department of Aeronautical Engineering, the Imperial University of Tokyo, and achieved the first class research in propeller theory. He made a basis, through his original research and education as well, for the JSASS, currently being appreciated worldwidely. He opened up the field of environmental science of including both local and global scales, currently being intensively studied as multi-scales and multi-physics. It is not by chance that Kawada had reached the same formulas for the velocity field induced by a helical vortex as those derived by Hardin. These formulas now stand as a milestone for analyzing the helical vortical structure.

3 Kawada's Potential Theory for a Helical Vortex Filament

In this section, we trace Kawada's derivation [1, 2] for the induced velocity by a helical vortex filament. His method is an adaptation of Goldstein's model with a regular helicoidal vortex sheet behind a blade [3].

We denote a to be the radius of a cylinder on which the helical vortex filament is wound and $h = 2\pi l$ to be the helical pitch. He introduced the helical variable $\chi = \theta + z/l$ (in our notation) for which the velocity has periodicity in accordance with number p of blades. For simplicity, we use one blade only ($p = 1$). Then the potential Φ of the flow due to the helical vortices is a function of r and χ only.

Putting $\mu = r/l$, the Laplacian equation takes the form:

$$\frac{\partial^2 \Phi}{\partial \mu^2} + \frac{1}{\mu}\frac{\partial \Phi}{\partial \mu} + \left(1 + \frac{1}{\mu^2}\right)\frac{\partial^2 \Phi}{\partial \chi^2} = 0. \tag{1}$$

The boundary conditions to be satisfied are (i) Φ is a discontinuous function on the blade with the vortex position at $\chi = 0$ and 2π, when $\mu \le a/l$, and the discontinuity must be equal to the circulation Γ; (ii) Φ is a single-valued function of μ and χ which must be odd and periodic in χ and zero at limit $\mu \to \infty$, when $\mu \ge a/l$.

A general solution of (1) satisfying the boundary conditions (i) and (ii) is represented in a form of the Fourier–Bessel series

$$\Phi = -\left\{ \begin{matrix} \Gamma(\pi - \chi)/2\pi \\ 0 \end{matrix} \right\} + \sum_{m=1}^{\infty} \left\{ \begin{matrix} a_m I_m(m\mu) \\ b_m K_m(m\mu) \end{matrix} \right\} \sin m\chi, \tag{2}$$

where a_m and b_m are constants, and $I_m(\cdot)$ and $K_m(\cdot)$ are the modified Bessel functions, with order m, of the first and the second kinds. The upper in braces corresponds to the case $\mu \le a/l$, and the lower is to $\mu \ge a/l$. Kawada determined the constants a_m and b_m by the continuity of the solution (2) at $\mu = a/l$, except at the singular point $\chi = 0$ and 2π where he used the following Fourier series form of the first term of (2):

$$\frac{\pi - \chi}{2} = \sum_{m=1}^{\infty} \frac{1}{m} \sin m\chi. \tag{3}$$

The two solutions must be, together with their derivatives, continuous at $\mu = a/l$ so that the coefficients of the same harmonic of the velocity potential (2) fulfill

$$\begin{aligned} -\frac{\Gamma}{\pi m} + a_m I_m\left(\frac{ma}{l}\right) &= b_m K_m\left(\frac{ma}{l}\right), \\ a_m I_m'\left(\frac{ma}{l}\right) &= b_m K_m'\left(\frac{ma}{l}\right). \end{aligned} \tag{4}$$

With the aid of the identity

$$I'_m\left(\frac{ma}{l}\right)K_m\left(\frac{ma}{l}\right)-I_m\left(\frac{ma}{l}\right)K'_m\left(\frac{ma}{l}\right)=\frac{l}{am}, \tag{5}$$

Kawada determined the coefficients as

$$a_m=-\frac{\Gamma a}{\pi l}K'_m\left(\frac{ma}{l}\right),\quad b_m=-\frac{\Gamma a}{\pi l}I'_m\left(\frac{ma}{l}\right). \tag{6}$$

The expression for the velocity potential from (2) takes a final form

$$\Phi=-\left\{\begin{matrix}\Gamma(\pi-\chi)/2\pi\\ 0\end{matrix}\right\}+\frac{\Gamma a}{\pi l}\sum_{m=1}^{\infty}\left\{\begin{matrix}K'_m(ma/l)\,I_m(mr/l)\\ I'_m(ma/l)\,K_m(mr/l)\end{matrix}\right\}\sin m\chi. \tag{7}$$

The upper entry in braces corresponds to the case $r<a$, and the lower to $r>a$.

Kawada then introduced "the tangential velocity at the immediate neighborhood of the blade" by

$$w_\theta=\frac{1}{2}\left(\frac{1}{r}\frac{\partial\Phi}{\partial\chi}\right)_{\chi=0}, \tag{8}$$

using

$$\frac{\partial\Phi}{\partial\theta}=\frac{\partial\Phi}{\partial\chi}\frac{\partial\chi}{\partial\theta}, \tag{9}$$

supplemented by $\partial\chi/\partial\theta=1$. With this form, he could produce the formulas for the radial distribution of tangential velocity at $\chi=0$ and a relation between the axial (w_z) and the azimuthal (w_θ) velocity:

$$\begin{aligned}w_\theta&=\frac{\Gamma}{4\pi r}\left\{\begin{matrix}1\\0\end{matrix}\right\}-\frac{\Gamma a}{2\pi rl}\sum_{m=1}^{\infty}m\left\{\begin{matrix}K'_m(ma/l)\,I_m(mr/l)\\ I'_m(ma/l)\,K_m(mr/l)\end{matrix}\right\},\\ w_z&=\frac{\Gamma}{4\pi l}\left\{\begin{matrix}1\\0\end{matrix}\right\}-\frac{\Gamma a}{2\pi l^2}\sum_{m=1}^{\infty}m\left\{\begin{matrix}K'_m(ma/l)\,I_m(mr/l)\\ I'_m(ma/l)\,K_m(mr/l)\end{matrix}\right\},\end{aligned} \tag{10}$$

so that $w_z=rw_\theta/l$. Kawada did not derive the radial component of the induced velocity.

On the other hand, Hardin [6] derived the induced velocity field by way of the Biot–Savart law. Suppose that the filament curve of a very thin helical vortex is represented as $\boldsymbol{x}=\boldsymbol{X}(s)$ in terms of the arclength parameter s. Then the induced velocity field $\boldsymbol{u}(\boldsymbol{x})$ by the helical vortex filament, with the vorticity field $\boldsymbol{\omega}(\boldsymbol{x})$ concentrated on it, is calculated through

$$\boldsymbol{u}(\boldsymbol{x})=\frac{\Gamma}{4\pi}\int\frac{\boldsymbol{t}(s)\times(\boldsymbol{x}-\boldsymbol{X}(s))}{|\boldsymbol{x}-\boldsymbol{X}(s)|^3}ds, \tag{11}$$

where $\boldsymbol{t} = d\boldsymbol{X}/ds$ is the unit tangent vector to the filament curve. Hardin obtained the velocity field valid at any point outside a single helical filament of infinite extent. Kawada's solution (10) had, though valid only in the neighborhood of a helicoidal blade, reached the same velocity field. It is readily verified that the induced velocities at a point on a blade equals half the induced velocity at a corresponding point of the full helix system [20]. The velocity from (10) should be doubled for the correct comparison with the formula based on (11). We mention that the vortex system used by Kawada was not only semi-infinite but it included the vortex sheet with nonuniform vortex strength along blades [2]. He further made a refined modeling of this vortex sheet by a finite number of helical vortices of different strengths to be determined self-consistently.

4 von Kármán's Recognition

It is worth noting, as a closing remark, that Theodore von Kármán recognized Kawada's contribution to development of the vortex model for the wake of rotors [22].

"The second step in the development constitutes a direct application of the Lanchester-Prandtl ideas to rotating bound vortices representing the propeller blades. Helicoidal vortex sheets now replace the free vortex sheet of the Prandtl's theory. This idea was first carried out mathematically by Sydney Goldstein in his doctor's thesis at Göttingen University [3]. Goldstein became one of the leading aerodynamicists in England .... Two Japanese aerodynamicists, Moriya and Kawada, continued the work of Goldstein ...."

Acknowledgements YF is grateful to Prof. Kojiro Suzuki of University of Tokyo for providing us with documents on Sandi Kawada and Tomijiro Moriya. This work has been carried out with a support of the Danish Council for Strategic Research for the project COMWIND—Center for Computational Wind Turbine Aerodynamics and Atmospheric Turbulence: grant 2104_09_067216/DSF and the Russian Science Foundation (grant no. 14-29-00093). DHW acknowledges the support of the Canadian Natural Science and Engineering Research Council through its Industrial Research Chair program in conjunction with the ENMAX Corporation.

References

1. Kawada, S.: Induced velocity by helical vortices. J. Aeronaut. Sci. **3**, 86–87 (1936)
2. Kawada, S.: Calculation of induced velocity by helical vortices and its application to propeller theory. Rep. Aeronaut. Res. Inst. Tokyo Imperial Univ. **14**, 2–57 (1939)
3. Goldstein, S.: On the vortex theory of screw propellers. Proc. R. Soc. Lond. A **123**, 440–465 (1929)
4. Fukumoto, Y., Okulov, V.L., Wood, D.H.: The contribution of Kawada to the analytical solution for the velocity induced by a helical vortex filament. Appl. Mech. Rev. **67**, 060801 (2015)
5. Alekseenko, S.V., Kuibin, P.A., Okulov, V.L.: Theory of Concentrated Vortices: An Introduction. Springer, Berlin (2007)

6. Hardin, J.C.: The velocity field induced by a helical vortex filament. Phys. Fluids **25**, 1949–1952 (1982)
7. Lerbs, H.: Moderately loaded propeller with a finite number of blades and an arbitrary distribution of circulation. Trans. SNAME **60**, 73–123 (1952)
8. Moriya, T.: On the induced velocity and characteristics of a propeller. J. Eng. **20**, 147–162 (1933). Tokyo Imperial University
9. Morgan, B.M., Wrench Jr, J.W.: Some computation aspects of propeller design. Methods Comput. Phys. **4**, 301–331 (1965)
10. Tibery, C.L., Wrench Jr, J.W.: Tables of Goldstein factor. Report 1534 Department of Navy Washington, pp. 1–69 (1964)
11. Ricca, R.L.: The effect of torsion on the motion of a helical vortex filament. J. Fluid Mech. **273**, 241–259 (1994)
12. Kuibin, P.A., Okulov, V.L.: Self-induced motion and asymptotic expansion of the velocity field in the vicinity of helical vortex filament. Phys. Fluids **10**, 607–614 (1998)
13. Boersma, J., Wood, D.H.: On the self-induced motion of a helical vortex. J. Fluid Mech. **384**, 263–280 (1999)
14. Alekseenko, S.V., Kuibin, P.A., Okulov, V.L.: Helical vortices in swirl flow. J. Fluid Mech. **382**, 195–243 (1999)
15. Wood, D.H., Boersma, J.: On the motion of multiple helical vortices. J. Fluid Mech. **447**, 149–171 (2001)
16. Okulov, V.L.: On the stability of multiple helical vortices. J. Fluid Mech. **521**, 319–342 (2004)
17. Fukumoto, Y., Okulov, V.L.: The velocity field induced by a helical vortex tube. Phys. Fluids **17**, 107101 (2005)
18. Okulov, V.L., Sørensen, J.N.: Stability of helical tip vortices in rotor far wake. J. Fluid Mech. **576**, 1–25 (2007)
19. Okulov, V.L., Sørensen, J.N.: Refined Betz limit for rotors with a finite number of blades. Wind Energy **11**, 415–426 (2008)
20. Okulov, V.L., Sørensen, J.N.: Maximum efficiency of wind turbine rotors using Joukowsky and Betz approaches. J. Fluid Mech. **649**, 497–508 (2010)
21. Okulov, V.L., Sørensen, J.N.: Applications of 2D helical vortex dynamics. Theor. Comput. Fluid Dyn. **24**, 395–401 (2010)
22. von Kármán, T.: Aerodynamics: Selected Topics in the Light of their Historical Development. Dover Publications, Mineola (2004)
23. Hayashi, T.: Obituary of professor Sandi Kawada [in Japanese]. J. Jpn. Soc. Aeronaut. Eng. 200 (1970)
24. Takenaka, K.: Classical airplane museum [in Japanese]. http://www.ne.jp/asahi/airplane/museum/kouken.html

A New Model for Fungal Hyphae Growth Using the Thin Viscous Sheet Equations

Thomas de Jong, Georg Prokert and Joost Hulshof

Abstract In this paper, we model the growth of single nonbranching fungal hypha cell. The growth proceeds as an elongating expansion in a single direction. Modelling of hyphae growth consists out of two parts: transport of cell wall building material to the cell wall and growth of the cell wall as new cell wall building material arrives. In this paper we present a new model for hyphae growth using the work of Barnicki-Garcia et al. (1989), which assumes that cell wall building material is transported in straight lines by an isotropic point source, and the work of Campas and Mahadevan (2009), which assumes that the cell wall is a thin viscous sheet. Furthermore, we include a novel equation which models the hardening of the cell wall with age. We show numerically that these governing equations have solutions corresponding to hyphae growth. We also compute asymptotic expansions near the apex and the base of the cell.

1 Introduction

Biological cells typically have a spherical-like shape. This makes sense because the pressure inside the cell is high compared to the outside. Hyphae cells exhibit an extreme lengthwise growth while maintaining a high pressurized environment inside

T. de Jong (✉)
Department of Mathematics and Computer Science,
Vrije Universiteit Amsterdam and Technische Universiteit Eindhoven,
P.O. Box 513, 5600 MB Eindhoven, The Netherlands
e-mail: t.g.d.jong@tue.nl

G. Prokert
Department of Mathematics and Computer Science, Technische Universiteit Eindhoven,
P.O. Box 513, 5600 MB Eindhoven, The Netherlands
e-mail: g.prokert@tue.nl

J. Hulshof
Department of Mathematics, Vrije Universiteit Amsterdam,
De Boelelaan 1081a, 1081 HV Amsterdam, The Netherlands
e-mail: j.hulshof@vu.nl

H. Itou et al. (eds.), *Mathematical Analysis of Continuum Mechanics and Industrial Applications*, Mathematics for Industry 26,
DOI 10.1007/978-981-10-2633-1_13

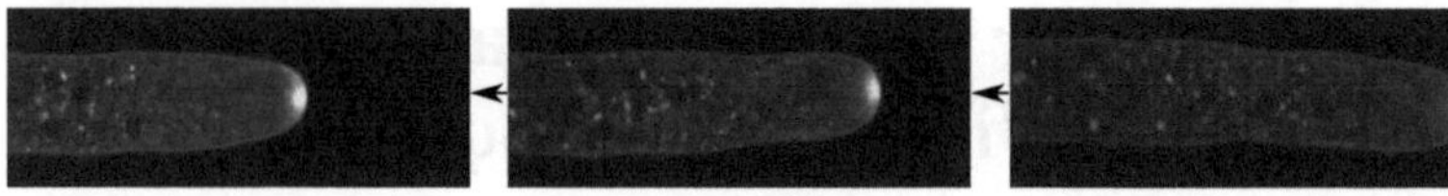

Fig. 1 Hyphae growth [6]. These pictures are cross sections of a single hypha cell. In the full spatial dimensions it has a cigar-like shape. The pictures where taken at equal time steps. The hyphae cell moves from right to left. The white spot close to the tip of the cell is the Spitzenkörper. It produces vesicles which contain cell wall building material. These vesicles are transported to the cell wall which leads to the growth of the cell

the cell, see Fig. 1. Hence the evolution of their cell shape is completely different from spherical-like cells.

In this paper, we construct a new model which explains how the type of growth in Fig. 1 can occur in hyphae. We will call the hyphae growth in Fig. 1 steady tip growth. More specifically, this steady growth is described by a single-celled nonbranching hypha which grows at a constant speed while preserving its overall tube-like form.

This steady tip growth is controlled by an organelle called the Spitzenkörper located at an approximately fixed distance from the tip. The Spitzenkörper produces packages of cell wall building material, vesicles, see Fig. 1. These vesicles are transported to the cell wall. Once these vesicles hit the cell wall they get absorbed and thereby the cell wall grows. Due to the closeness of the Spitzenkörper to the tip this leads to the characteristic tip growth of hyphae. Consequently, in terms of modelling we need to model two biological components: the Spitzenkörper and the evolution of the cell wall as vesicles get absorbed.

In Sect. 2 we construct our new model. We model the Spitzenkörper as an isotropic point source which transports cell wall building material in straight lines to the cell as done in the work of Bartnicki-Garcia et al. [1] and we model the cell wall as a thin viscous sheet as done by Campas and Mahadevan [2]. However, it will turn out that solutions corresponding to Fig. 1 will not exist. By introducing a novel equation for the age of the cell wall and by letting the cell wall harden with age we can overcome this problem. In Sect. 3 we formulate the abstract mathematical problem. In Sect. 4 we numerically compute a hyphae growth solution. Furthermore, we also compute asymptotic expansions close to the tip and the base of the cell. Finally, in Sect. 5 we present our conclusions and discuss topics for future work.

2 The New Fungal Hyphae Growth Model

We will first argue that steady tip growth, see Fig. 1, can be mathematically described by travelling waves with a specific shape. Then we set up governing equations for general travelling wave solutions by modelling the Spitzenkörper and the evolution of the cell wall as vesicles get absorbed. The aim is then to show that these governing equations admit travelling wave solutions which correspond to steady tip growth.

2.1 Steady Tip Growth

Let us first mathematically describe the steady tip growth in Fig. 1. We assume that the cell shape is axially symmetric. Thus, at a fixed time we describe the cell shape by cylindrical coordinates $(r(s), z(s), \phi)$ where s is the arc length to the apex, $s = 0$. Observe that we have the equality

$$(r')^2 + (z')^2 = 1, \tag{1}$$

where the prime mark denotes the derivative with respect to s. From Fig. 1 we see that the cell grows in the direction of the normal at the apex of the hyphae cell while preserving its overall shape. Hence, we model steady tip growth by $(r(s), z(s) + ct, \phi)$ where c is a positive constant, see Fig. 2. Observe that this corresponds to travelling waves. Finally, we need to give conditions on the cell shape for steady tip growth. Based on Fig. 1 we require that

$$\begin{aligned} &r(s) > 0\ \forall s \in (0,\infty),\ \lim_{s\to 0} r(s) = 0,\ \lim_{s\to\infty} r(s) = r_\infty > 0,\\ &r'(s) > 0\ \forall s \in (0,\infty),\ \lim_{s\to 0} r'(s) = 1,\ \lim_{s\to\infty} r'(s) = 0,\\ &z'(s) > 0\ \forall s \in (0,\infty),\ \lim_{s\to 0} z(s) = z_0,\ \lim_{s\to\infty} z(s) = \infty,\\ &r''(s) < 0\ \forall s \in (0,\infty). \end{aligned} \tag{2}$$

Furthermore, based on the smoothness and the axial symmetry at the apex we expect for $s \to 0$ that z and r are a smooth even function and a smooth odd functions, respectively.

2.2 The Cell Wall as a Thin Viscous Sheet

Following Campas and Mahadevan [2] we model the cell wall as a thin viscous sheet. From a geometrical viewpoint this makes sense because the cell wall is a shell-like structure whose thickness is much smaller than the observed radii of curvature of the shell [2]. Given the shape of the sheet the principal curvatures are then given by

$$\kappa_s = -r''/z', \quad \kappa_\phi = z'/r. \tag{3}$$

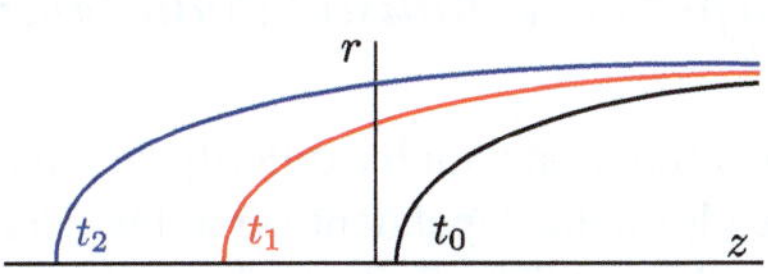

Fig. 2 Steady tip growth: We assumed that the cell shape is axially symmetric hence we can represent the shape in the (r, z)-plane. The travelling wave moves from right to left. It is drawn at times t_0, t_1, t_2 with $t_0 < t_1 < t_2$

Following Campas and Mahadevan the stress resultants σ_{ss} and $\sigma_{\phi\phi}$ in the shell are subject to a pressure difference P between the inside and outside of the cell which yields the equations:

$$\kappa_s \sigma_{ss} + \kappa_\phi \sigma_{\phi\phi} = P, \quad \kappa_\phi \sigma_{ss} = P/2. \tag{4}$$

Campas and Mahadevan consider the simplest stress-strain relation by assuming that the cell wall behaves as a incompressible viscous fluid where the stress-strain rates are linear:

$$\sigma_{ss} = 4\mu h(u' + ur'/2), \quad \sigma_{\phi\phi} = 4\mu h(u'/2 + ur'), \tag{5}$$

where μ is the viscosity and u, h are s-dependent variables with u the velocity and h the cell wall thickness. Finally, we have an equation for mass conservation given by

$$(urh)' = r\gamma/\rho_w, \tag{6}$$

where ρ_w is the density of the cell wall which we take constant and γ is the local rate of inflow of new cell wall material over an area element $dsd\phi$. First observe that Eqs. (4) and (6) describe the travelling wave in stationary co-ordinates. However, the travelling wave velocity can be retrieved by the limit $\lim_{s\to\infty} u(s)$. In terms of steady tip growth we require for u, h that

$$\begin{aligned} u(s) > 0 \ \forall s \in (0, \infty), \ \lim_{s\to 0} u(s) = u_0 > 0, \ \lim_{s\to\infty} u(s) = u_\infty > 0, \\ h(s) > 0 \ \forall s \in (0, \infty), \ \lim_{s\to 0} h(s) = h_0 > 0, \ \lim_{s\to\infty} h(s) = h_\infty > 0. \end{aligned} \tag{7}$$

Furthermore, based on the smoothness and the axial symmetry at the apex we expect for $s \to 0$ that u, h are smooth even functions.

The Eqs. (1), (4) and (6) can also be obtained in a more rigorous way. By assuming that the thickness of the cell wall is small compared to the length scales we can write the variables in their corresponding scale and collect terms of the same scale to obtain the thin viscous sheet equations. This more rigorous approach can be found in the work of Howell [7, 8].

The Eqs. (1), (4) and (6) yield a complete set of equations for the dependent variables (r, z, u, h) once we have chosen γ [2].

2.3 The Spitzenkörper as a Ballistic Point Source

In [2] $\gamma(s)$ is taken to be a function which explicitly depends on s. However, we will consider a γ which depends on the dependent variables. To set up equations for γ we need to model the Spitzenkörper. The Spitzenkörper is located at an approximately fixed distance from the tip. It produces at an approximately constant rate packages of cell wall building material, vesicles. These vesicles are transported to the cell wall where they get absorbed. We follow Bartnicki-Garcia et al. [1] which models the

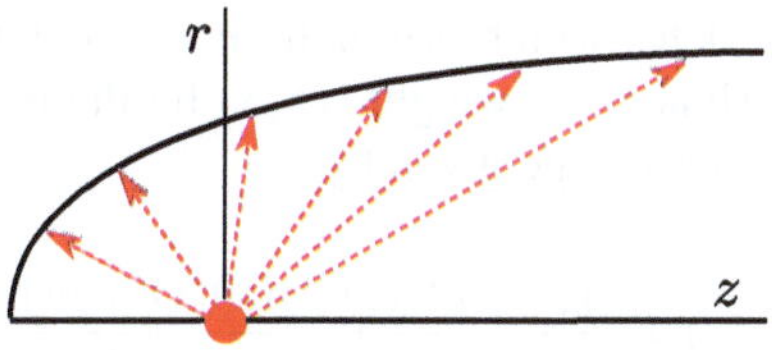

Fig. 3 Ballistic vesicle supply center: a point source on the z-axis at a fixed distance from the apex sends vesicles at a constant rate and in a straight path to the cell wall

Spitzenkörper as a point source which emits isotropically vesicles at a constant rate. Hence, this point source is called the vesicle supply center (VSC). We fix the VSC on the z-axis at a constant distance from the tip of the cell. We assume that the VSC sends cell wall building material ballistically to the cell wall, which means that the trajectories are straight lines, see Fig. 3. We fix this point source at $(r, z) = (0, 0)$. Consequently, we assume that $z_0 < 0$ in (2). The γ will then be given by

$$\gamma(r, z, r') := C\frac{r\sqrt{1-(r')^2} - zr'}{2(r^2+z^2)^{3/2}}, \tag{8}$$

where C is the rate of cell wall material emitted by the point source.

2.4 Age Function

Contrary to [2] let us assume that μ is constant. Suppose that steady tip growth solutions exist. This means that (2) and (7) are satisfied. Then it follows by (4) that $\lim_{s\to\infty} du/ds > 0$. But this contradicts the condition $\lim_{s\to\infty} u(s) = u_\infty > 0$ in (7). In [2] Campas and Mahadevan take $\mu(s)$ to be a function which explicitly depends on s such that $\lim_{s\to\infty} \mu(s) = \infty$. Then if steady tip growth solutions exist it follows by (4) that $\lim_{s\to\infty} du/ds = 0$.

We will consider a μ which increases with the age of the cell wall material. Biologically this models cell wall loosening enzymes, we refer to [3–5]. In terms of modelling we need to set up an age equation and we need to define a coupling between the viscosity and the age. Let us first set up the age equation. We can write the tangential velocity as

$$u = \frac{ds}{dt}, \tag{9}$$

where t is the time variable. Denote by $t(\tau, s)$ the time it takes to travel from τ to s then using (9) we have that

$$t(\tau, s) = \int_\tau^s \frac{1}{u(\sigma)} d\sigma. \tag{10}$$

At a point s there are particles which originally entered at τ and have been part of the cell wall for $t(\tau, s)$. Hence, we weight $t(\tau, s)$ by the particles which enter at τ to construct a weighted average age given by

$$\Psi(s) = \frac{\int_0^s (r\gamma(r, z, r'))(\tau)t(\tau, s)d\tau}{\int_0^s (r\gamma(r, z, r'))(\tau)d\tau}. \tag{11}$$

We can rewrite (11) as an ODE:

$$\Psi' = \frac{1}{u} - \frac{r\gamma(r, z, r')}{rhu}\Psi. \tag{12}$$

In terms of the steady tip growth solution we require

$$\Psi(s) > 0 \;\forall s \in (0, \infty),\; \lim_{s\to 0} \Psi(s) = \Psi_0 > 0,\; \lim_{s\to\infty} \Psi = \infty. \tag{13}$$

Furthermore, based on the smoothness and the axial symmetry at the apex we expect for $s \to 0$ that Ψ is a smooth even function.

For the coupling between viscosity and age we introduce the function $G(\Psi)$:

$$\mu(\Psi) = 1/G(\Psi) \text{ with } (dG/dy)(y) < 0\,,\;\; G(y) > 0 \;\;\forall y \in \mathbb{R},\;\; \lim_{y\to\infty} G(y) = 0. \tag{14}$$

3 Governing Equations: A First Order 5-Dimensional ODE

We will rewrite the governing equations as a 5-dimensional first order ODE. There exists a transformation of the variables such that the ODE becomes independent of the parameters P, ρ_w, C. We will not introduce this transformation and set $P = \rho_w = C/2 = 1$.

We consider solutions satisfying the following 5-dimensional first order ODE:

$$\begin{aligned}
\rho' &= \frac{3}{2}\frac{1-\rho^2}{r}\left(-1 + \frac{\rho\Gamma(r, z)\sqrt{1-\rho^2}}{G(\Psi)r^3}\right),\\
r' &= \rho,\\
h' &= \left(\frac{r\gamma(r, z, \rho)}{\Gamma(r, z)} - \frac{G(\Psi)r^2}{2\Gamma(r, z)\sqrt{1-\rho^2}} - \frac{\rho}{2r}\right)h,\\
\Psi' &= \frac{rh}{\Gamma(r, z)} - \frac{r\gamma(r, z, \rho)}{\Gamma(r, z)}\Psi,\\
z' &= \sqrt{1-\rho^2},
\end{aligned} \tag{15}$$

where $G(\Psi)$ satisfies (14) and where

$$\Gamma(r,z) := 1 + z/\sqrt{r^2+z^2}. \tag{16}$$

The phase space is given by

$$M = \{(\rho, r, h, \Psi, z) \in (-1,1) \times \mathbb{R}_+ \times \mathbb{R} \times \mathbb{R} \times \mathbb{R}\}. \tag{17}$$

The conditions (2), (7), (13) for steady tip growth solutions become

$$\begin{aligned}
&r(s) > 0\ \forall s \in (0,\infty),\ \lim_{s\to 0} r(s) = 0,\ \ \lim_{s\to\infty} r(s) = r_\infty > 0,\\
&\rho(s) > 0\ \forall s \in (0,\infty),\ \lim_{s\to 0} \rho(s) = 1,\ \ \lim_{s\to\infty} \rho(s) = 0,\\
&\rho'(s) < 0\ \forall s \in (0,\infty),\ \lim_{s\to 0} z(s) = z_0 < 0,\ \ \lim_{s\to\infty} z(s) = \infty,\\
&\Psi(s) > 0\ \forall s \in (0,\infty),\ \lim_{s\to 0} \Psi(s) = \Psi_0 > 0,\ \lim_{s\to\infty} \Psi(s) = \infty,\\
&h(s) > 0\ \forall s \in (0,\infty),\ \lim_{s\to 0} h(s) = h_0 > 0,\ \ \lim_{s\to\infty} h(s) = h_\infty > 0.
\end{aligned} \tag{18}$$

Furthermore, based on the smoothness and the axial symmetry at the apex we expect for $s \to 0$ that ρ, h, Ψ, z are smooth even function and that r is a smooth odd function.

Given solutions satisfying (18) we have that

$$\Gamma(r,z)(s) = \int_0^s (r\gamma(r,z,\rho))(\sigma)d\sigma \tag{19}$$

Hence, $\Gamma(r,z)(s)$ represents the cumulative flux over a cap from the apex to s. Note that there is no u-equation in (15). Assuming (18) we have eliminated u by using (6) and (19):

$$u = \Gamma(r,z)/(rh). \tag{20}$$

Analysis of the asymptotics at the apex yields: $\lim_{s\to 0}\Psi(s) = h_0 z_0^2$. In combination with (18) this yields two degrees of freedom at the apex: h_0, z_0. Assume that $G(y) \in o(y)$ for large y then analysis of the asymptotics at the base, $s \to \infty$ yields: $\lim_{s\to\infty}\Psi(s) - \frac{r_\infty h_\infty}{2}s = \Psi_c$, $\lim_{s\to\infty} z(s) - s = z_c$ with $\Psi_c, z_c \in \mathbb{R}$. In combination with (18) this yields four degrees of freedom at the base: $r_\infty, h_\infty, \Psi_c, z_c$.

Observe that the vector field is singular at the limits (18). In the next section we compute the local asymptotics close to these limits.

4 Numerical Results

The vector field is singular at the apex and at the base of the steady tip growth solution. Hence, to numerically compute the steady tip growth solution we need to deal with the singularities. We compute asymptotic expansions of the steady tip growth solutions close to both singularities. When we are sufficiently far away from the singularities we use a standard numerical method to approximate the solution. Then by varying at the apex h_0, z_0 and at the base r_∞, h_∞, z_c, Ψ_c we connect the two numerical solutions.

4.1 Asymptotic Expansions at the Apex and Base

Let us consider an explicit $G(\Psi)$ given by

$$G(\Psi) = \frac{1}{1+\Psi^3}. \tag{21}$$

To deal with the singularities we approximate the solution by an asymptotic expansion. We present the two lowest order terms and the two highest order terms at the apex and at the base, respectively:

$$\begin{aligned}
\tilde{r}_{\text{apex}}(s; h_0, z_0) &= s - 2z_0^4 s^3/(27(1+h_0^3 z_0^6)^2) + O(s^5),\\
\tilde{z}_{\text{apex}}(s; h_0, z_0) &= z_0 + z_0^2 s^2/(3+3h_0^3 z_0^6) + O(s^4),\\
\tilde{\rho}_{\text{apex}}(s; h_0, z_0) &= 1 - 2z_0^4 s^2/(9(1+h_0^3 z_0^6)^2) + O(s^4),\\
\tilde{h}_{\text{apex}}(s; h_0, z_0) &= h_0 + h_0(-72 - 64z_0^3 - (8+135h_0^3)z_0^6 - 56h_0^3 z_0^9 + \cdots\\
&\quad (4h_0^3 - 54h_0^6)z_0^{12} + 8h_0^6 z_0^{15} + 9h_0^9 z_0^18)s^2/\cdots\\
&\quad (6(10+h_0^3 z_0^6)(z_0 + h_0^3 z_0^7)^2) + O(s^4),\\
\tilde{\Psi}_{\text{apex}}(s; h_0, z_0) &= h_0 z_0^2 + h_0(27 + 24z_0^3 + (8+54h_0^3)z_0^6 + 24h_0^3 z_0^9 + 27h_0^6 z_0^12)s^2/\cdots\\
&\quad (18(1+h_0^3 z_0^6)(10+h_0^3 z_0^6)) + O(s^4),
\end{aligned} \tag{22}$$

and

$$\begin{aligned}
\tilde{r}_{\text{base}}(s; r_\infty, h_\infty, \Psi_c, z_c) &= r_\infty - 2/(h_\infty^3 s^2) + O(s^{-3}),\\
\tilde{z}_{\text{base}}(s; r_\infty, h_\infty, \Psi_c, z_c) &= s + z_c + O(s^{-1}),\\
\tilde{\rho}_{\text{base}}(s; r_\infty, h_\infty, \Psi_c, z_c) &= 4/(h_\infty^3 s^3) - 24\Psi_c/(h_\infty^4 r_\infty^4 s^4) + O(s^{-5}),\\
\tilde{h}_{\text{base}}(s; r_\infty, h_\infty, \Psi_c, z_c) &= h_\infty + (2/(h_\infty^2 r_\infty) - h_\infty r_\infty^2/4)s^{-2} + O(s^{-3}),\\
\tilde{\Psi}_{\text{base}}(s; r_\infty, h_\infty, \Psi_c, z_c) &= h_\infty r_\infty s/2 + \Psi_c + O(s^{-1}).
\end{aligned} \tag{23}$$

4.2 Connecting the Asymptotic Expansions

We start with the asymptotic expansion at the tip. Recall that h_0, z_0 can be chosen freely. Once we are sufficiently far away from the singularity we can use a standard

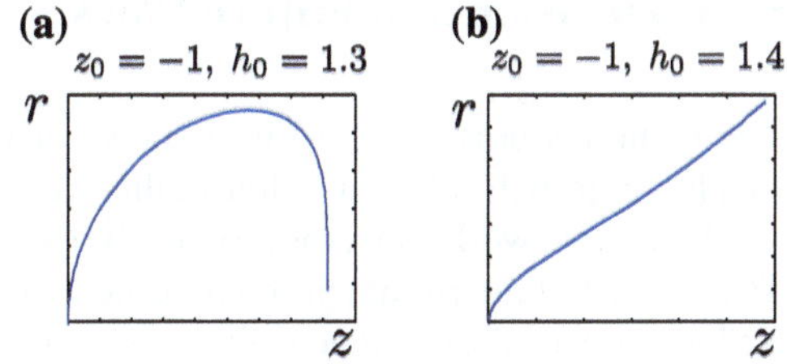

Fig. 4 Numerical plots in the r, z-plane of solutions starting with the asymptotics of the steady tip growth solution at the apex. These solutions do not have the desired global behaviour

numerical solver to continue the solution. It turns out that numerically there exists an h_1 for every z_0 such that for any $h_0 \in (0, h_1)$ there exist $s_0, s_1 \in \mathbb{R}_+$ with $s_0 < s_1$ such that

$$\rho'(s; h_0, z_0) < 0, \quad \rho(s; h_0, z_0) > 0\ \forall s \in (0, s_0), \quad \rho(s_0; h_0, z_0) = 0,$$
$$\rho'(s; h_0, z_0) < 0\ , \quad \rho(s; h_0, z_0) < 0\ \forall s \in (s_0, s_1),$$

while all the other variables remain bounded, see Fig. 4a. Similarly, we see that there exist $h_2, h_3 > h_1$ for every z_0 such that for any $h_0 \in (h_2, h_3)$ there exist $s_0, s_1 \in \mathbb{R}_+$ with $s_0 < s_1$ such that

$$\rho'(s; h_0, z_0) < 0, \quad \rho(s; h_0, z_0) > 0\ \forall s \in (0, s_0), \quad \rho'(s_0; h_0, z_0) = 0,$$
$$\rho'(s; h_0, z_0) > 0\ , \quad \rho(s; h_0, z_0) > 0\ \forall s \in (s_0, s_1),$$

while all the other variables remain bounded, see Fig. 4b. Observe that for the steady tip growth solutions we are only interested in $h_0 \in [h_1, h_2]$. The numerics shows us that h_1 and h_2 are close to each other. Theoretically, it might be that $h_1 = h_2$. Given a $h_0 \in [h_1, h_2]$ we can numerically continue the solution until an $s = s_{\max}$ where $(\rho, r, h, \Psi, z)(s_{\max}; h_0, z_0)$ is close to the singularity at the base. Then by using a numerical root solver we can find $\sigma, r_\infty, h_\infty, \Psi_c, z_c$ such that

$$(\rho, r, h, \Psi, z)(s_{\max}; h_0, z_0) - (\tilde{\rho}_{\text{base}}, \tilde{r}_{\text{base}}, \tilde{h}_{\text{base}}, \tilde{\Psi}_{\text{base}}, \tilde{z}_{\text{base}})(\sigma; r_\infty, h_\infty, \Psi_c, z_c) = 0.$$

Consequently, we can connect the asymptotic expansions, see Fig. 5.

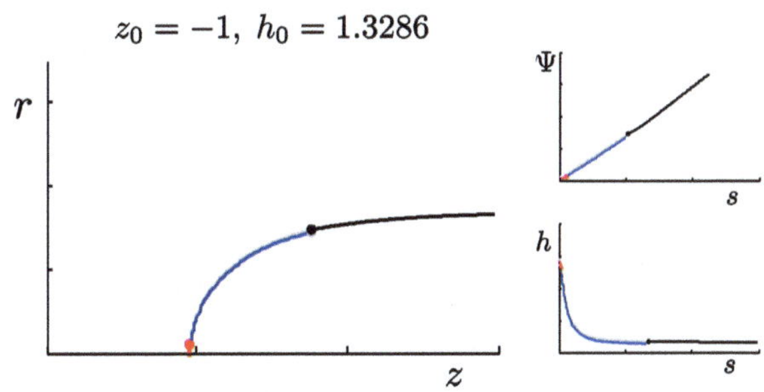

Fig. 5 Steady tip growth solution: The short *red line* is the asymptotic expansion at the apex, Eq. (22), the *blue line* is the continuation of the asymptotic expansion by a standard numerical ODE solver and the *black line* is the asymptotic expansion at the base, Eq. (23)

5 Conclusion and Future Work

The numerics presented in Fig. 5 shows that the governing equations (15) admits steady tip growth solutions. Hence, this validates our modelling.

Our current work concerns proving the existence of the steady tip growth solutions. The whole 5-dimensional problem turned out to difficult to tackle at once. However, we have proven the existence of these solution for a simplified 3-dimensional model. We are currently extending the method to a 5-dimensional model.

References

1. Bartnicki-Garcia, S., Hergert, F., Gierz, G.: Computer simulation of fungal morphogenesis and the mathematical basis for hyphal (tip) growth. Protoplasma **153**, 46–57 (1989)
2. Campas, O., Mahadevan, L.: Shape and dynamics of tip growing cells. Curr. Biol. **19**, 2102–2107 (2009)
3. Cole, R.A., Fowler, J.E.: Polarized growth: maintaining focus on the tip. Curr. Opin. Plant Biol. **9**, 579–588 (2006)
4. Cosgrove, D.J.: Loosening of plant cell walls by expansions. Nature **405**, 312–326 (2000)
5. Cosgrove, D.J.: Growth of the plant cell wall. Nat. Rev. Mol. Cell. Biol. **6**, 850–861 (2005)
6. Fisher-Parton, S., Parton, R.M., Hickey, P.J., Dijksterhuis, J., Atkinson, H.A., Read, N.D.: The University of Edinburgh
7. Howell, P.D.: Extensional thin layer flows, Ph.D. thesis, University of Oxford (1994)
8. Howell, P.D.: Models for thin viscous sheets. Europ. J. Appl. Math. **7**, 321–343 (1996)

On Boundary Conditions for Hele-Shaw Problem

Hisasi Tani

Abstract Weakly nonlinear stability analysis is carried out for a two-dimensional radially growing interface between two immiscible viscous fluids, known as the Hele-Shaw problem. In contrast to the previous studies, here we consider the boundary conditions with the wetting-layer effect and the VNS effect, respectively. The difference in fingering patterns is investigated by employing each boundary condition, and then it is discussed about which one reflects the interfacial phenomena more appropriately.

1 Introduction

Two-dimensional interface growth has been intensively studied because of its wide variety of related phenomena, for instance, the motions of rising bubbles in the water, pattern formation of snow crystals, filtration of the water into the soil, and so on. It is also interested because of its possibilities of applications such as microfluidic devices and injection moulding. The examples above are concerned with a motion of a free boundary in a Hele-Shaw cell, known as the Hele-Shaw problem. Here, Hele-Shaw cell is an experimental apparatus which consists of a pair of parallel plates with narrow gap as depicted in Fig. 1, and enables us to observe two-dimensional hydrodynamical phenomena.

In the pioneering work by P.G. Saffman and G.I. Taylor, it was revealed that a interface in a Hele-Shaw cell becomes unstable if the less viscous fluid expels the more viscous one, otherwise it remains stable [18]. In general, once the interface becomes unstable, a highly ramified finger-like pattern appears, which is referred to as viscous fingering. The investigations of Hele-Shaw problem so far have been categorized into two geometries; the rectangular one as depicted in Fig. 1, and the radial one, in which a circular interface expands from a point source on the cell. In the present paper we focus on viscous fingering patterns for the radial geometry,

H. Tani (✉)
Meiji University, Tama Ku Higashi-mita 1-1-1, Kawasaki-shi, Kanagawa, Japan
e-mail: hisasitani@gmail.com

H. Itou et al. (eds.), *Mathematical Analysis of Continuum Mechanics and Industrial Applications*, Mathematics for Industry 26,
DOI 10.1007/978-981-10-2633-1_14

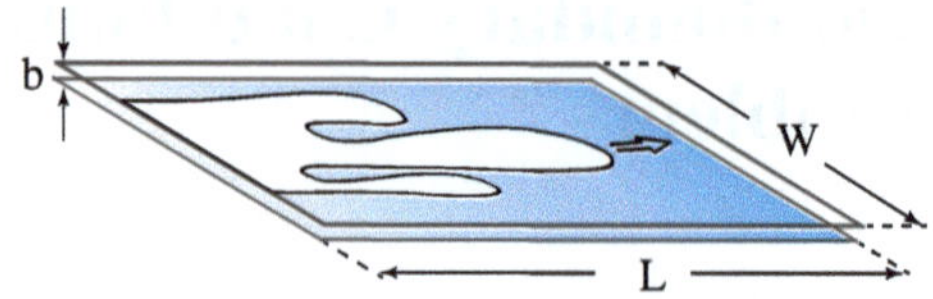

Fig. 1 Hele-Shaw cell (rectangular geometry)

since they seem to be more closely related to the patterns such as the dendrite crystal [3, 4] and the filtration phenomenon into the soil [1, 16], than those for the rectangular ones. Not only in the field of physics, it is also studied from a mathematical point of view, for instance, the existence and uniqueness of the classical solution [6], the complex analysis by making use of the conformal mapping [8] and a mathematical model for the rising bubbles [20]. Thus the Hele-Shaw problem has been investigated for more than 50 years, however, there remains a lot of topics to be clarified.

In particular, boundary conditions for the Hele-Shaw problem have been ceaselessly paid attention and discussed ever. In general, as seen in the next section, Hele-Shaw problem is equivalent to solve the Laplace equation for velocity potential under the kinematic and dynamical boundary conditions. Previously, Young–Laplace relation were employed as the dynamical boundary condition in the almost all of theoretical studies. However, for the case of rectangular geometry, McLean and Saffman [12], and then Park and Homsy [15] indicated that it was not valid since the theoretically predicted behaviours of the stationary finger did not coincide with the experimental results in the low-capillary number region. In [15], Park and Homsy suggested that the reason of the discrepancy was that more viscous fluid wets the cell plate and then affects the boundary condition. They achieved the asymptotic expansion with respect to the capillary number, and proposed the modified Young–Laplace relation which includes the effect of the thin wetting-layer in the Hele-Shaw cell. Marthyshev and Birzina carried out the linear stability analysis by employing this boundary condition [11], and recently almost at the same time, Dias and Miranda, H.T. and Tani independently employed it to the weakly nonlinear analysis [5, 19]. In addition, other approaches to modify the Young–Laplace relation have also been reported. Kim et al. simply employed the balance of the normal components of the normal stress tensor and carried out the linear stability analysis [9]. Reinelt took the thickness of the wetting-layer into consideration and determined the local thickness as a function of the capillary number [17]. In contrast, Weinstein et al. consider the motion of a contact line, at which the interface contacts with the plate of the cell with a certain contact angle [21]. This boundary condition seems to be quite significant now, since it is closely related with the development of devices or materials by making use of the wettability of the cell.

Thus the modifications for Young–Laplace equation have already been investigated, however, it has not enough been understood how these effects modify the theoretical results about the radial viscous fingering patterns. Moreover, it has never been clarified which modifications reflect the phenomena more appropriately.

Under these backgrounds, we focus on the boundary condition with the effect of the wetting-layer [15] and that based on the normal stress balance [9], and compare their influences on the fingering patterns. The contents of the paper are as follows; in section two, the formulation of Hele-Shaw problem is introduced. And then the weakly nonlinear stability analysis and its application to our model are presented in section three. Finally, concluding remarks are addressed in section four.

2 Formulations

2.1 Hele-Shaw Problem

Let us consider the configuration as depicted in Fig. 2. Here, fluid 1 is injected into the Hele-Shaw cell filled with fluid 2, at the origin O with a constant injection rate Q. The viscosity of fluid 1 and 2 are μ_1 and μ_2, respectively, and hereafter they satisfy $\mu_1 < \mu_2$, required for the Saffman–Taylor instability.

Assume that the fluids are incompressible and obey the stationary Stokes equation;

$$0 = -\nabla p_i + \mu_i \Delta \mathbf{v}_i. \tag{1}$$

Here p_i, μ_i and $\mathbf{v}_i$ are the pressure, viscosity and velocity vector of the fluid $i = 1, 2$, respectively. Moreover we assume that the flows are uniform for the thickness of the cell, and then by averaging with respect to z, the Darcy's law is obtained as

$$\mathbf{v}_i = -\frac{b^2}{12\mu_i}\nabla p_i, \tag{2}$$

where b denotes the thickness of the cell. From the Darcy's law and the incompressibility, one can easily find that the motion of fluids and the interface in a Hele-Shaw cell are determined by solving the Laplace equation

$$\Delta p_i = 0 \quad \text{in } \Omega_i, \tag{3}$$

with the kinematic boundary condition

$$\mathbf{v} \cdot \mathbf{n} = V \quad \text{on } \Gamma(t), \tag{4}$$

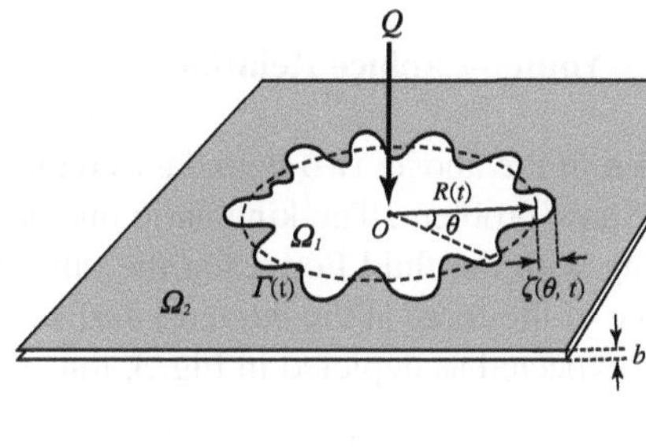

Fig. 2 Hele-Shaw cell and the interface

and the Young–Laplace relation as the dynamical boundary conditions

$$p_1 - p_2 = \sigma\kappa_M \quad \text{on } \Gamma(t), \tag{5}$$

where σ and κ_M denote the surface tension coefficient and the mean curvature of the interface, respectively. It is convenient to introduce the velocity potential $\phi_i = (b^2/12\mu_i)p_i$ such that

$$\mathbf{v}_i = -\nabla\phi_i. \tag{6}$$

Then the above formulations are rewritten as

$$\Delta\phi_i = 0 \quad \text{in } \Omega_i, \tag{7}$$

$$-\nabla\phi_i \cdot \mathbf{n} = V \quad \text{on } \Gamma(t), \tag{8}$$

$$\mu_1\phi_1 - \mu_2\phi_2 = \frac{b^2\sigma}{12}\left(\kappa_\perp + \kappa_\parallel\right) \quad \text{on } \Gamma(t). \tag{9}$$

Here $\kappa_\perp$ and $\kappa_\parallel$ are the curvatures perpendicular and parallel to the plates, respectively. Now we consider the system as depicted in Fig. 2, and assume that the perturbed interface $\mathscr{R}$ is described as $\mathscr{R}(\theta, t) = R(t) + \zeta(\theta, t)$, where $R(t)$ is radius of the unperturbed circular interface, and $\zeta(\theta, t)$ the deviation. Under an assumption that $\zeta(\theta, t)$ does not influence the area of the fluid 1, it holds

$$R(t) = \sqrt{R_0^2 + \frac{Q}{\pi}t},$$

from the conservation of volume of the fluid 1. Here R_0 is the radius of the initial interface. Thus the normal velocity of the interface V is written as

$$V = \dot{\mathscr{R}} = \frac{Q}{2\pi R} + \dot{\zeta}(\theta, t), \tag{10}$$

where the dot represents the derivative with respect to t.

2.2 Boundary Conditions

2.2.1 Young–Laplace Relation

As seen in the above, two boundary conditions are employed in the formulation of Hele-Shaw problem. The kinematic one describes that the interface moves accompanying with the fluid flows. On the other hand, the dynamical one represents the balance of the force at the curved interface. The cross section of the interface in the cell is expected as depicted in Fig. 3, thus $\kappa_\perp$ in Eq. (9) is equivalent to $2/b$.

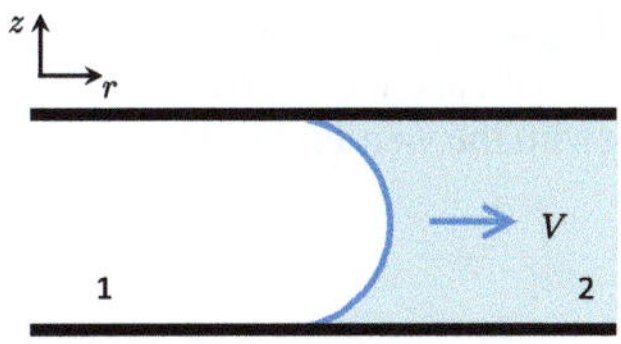

Fig. 3 The cross section of the cell and the interface (Young–Laplace relation)

As mentioned in introduction, the validity of the Young–Laplace relation (5) has been controversial because of the discrepancies between the theoretical and experimental results [12, 15]. In addition, this relation implies the balance of the static pressure, and then, in principle, does not hold for a deforming interface in a nonequilibrium system. These reasons support that the modifications for Young–Laplace relation should be considered. Hereafter, we employ the balance of the normal stress, and the balance including the wetting-layer effect, as the dynamical boundary condition.

2.2.2 Balance of the Normal Stress Tensor

From the momentum conservation law, the balance of the normal components of the normal stress hold as

$$-\mathbf{n}\cdot\mathbf{T}_1\cdot\mathbf{n}+\mathbf{n}\cdot\mathbf{T}_2\cdot\mathbf{n}=\sigma\kappa_M, \tag{11}$$

where $\mathbf{T}_i$ is the stress tensor for fluid i, which is written as $\mathbf{T}_i=-p_i\mathbf{I}+2\mu_i\mathbf{e}_i$ for the Newtonian fluid (for instance, [2]). Here $\mathbf{I}$ is the unit tensor and $\mathbf{e}$ denotes the rate of strain tensor defined as $\mathbf{e}_i=(\nabla\mathbf{v}_i+(\nabla\mathbf{v}_i)^T)/2$. Thus Eq. (11) can be rewritten as

$$p_1-p_2-\mathbf{n}\cdot\mathbf{e}_1\cdot\mathbf{n}+\mathbf{n}\cdot\mathbf{e}_2\cdot\mathbf{n}=\sigma\kappa_M. \tag{12}$$

Kim et al. carried out the linear stability analysis by employing this boundary condition [9]. Following [9], in the present paper we refer to the third and fourth terms in the left-hand side of Eq. (12) as viscous normal stress (VNS) term. This boundary condition is the same as the one in [20], in which VNS term was expressed by Hessian using the Darcy's law. One can easily see that Eq. (12) results in the Young–Laplace relation (5) only if $\mathbf{e}_i=\mathbf{0}$ and then VNS term vanishes, i.e. the interface moves rigidly without any deformation.

2.2.3 Boundary Condition with the Wetting-Layer Effect

As mentioned in introduction, it is considered that there exists a thin wetting- layer of the more viscous fluid in the cell, which is left behind the interface as depicted in

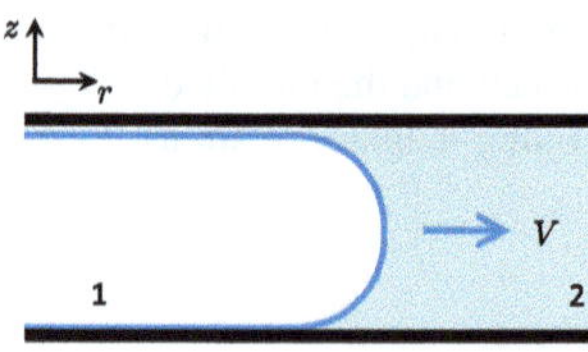

Fig. 4 The cross section of the cell and the interface (with the wetting-layer)

Fig. 4. In [15], it was indicated that the effect of a wetting-layer in a Hele-Shaw cell had to be taken into consideration as

$$p_1 - p_2 = \frac{2\sigma}{b}\left(1 + J Ca^{2/3}\right) + \frac{\pi\sigma}{4}\kappa_{\parallel} \quad \text{on } C. \tag{13}$$

In the above, Ca is the capillary number defined as $Ca = \mu_2 V/\sigma$, and the coefficient $J = 3.8$ is obtained by a numerical integration in [15]. As seen in Fig. 4, $\kappa_{\perp}$ become smaller than $2/b$, and the balance of the pressure at the interface is changed because of the effect of the dragging of the wetting- layer, which is equivalent to the result by Landau and Levich [10].

3 Weakly Nonlinear Stability Analysis

In this section, we show our results based on the boundary conditions including the effect of VNS and the wetting-layer, respectively. In general, linear stability analysis tells us if a certain mode remains stable or not, as its control parameter increases. However, for the case that the perturbation amplitudes become large and are not negligible, we have to carry out the weakly nonlinear stability analysis. It was applied originally for, for instance the stability of the Rayleigh–Benard convection and the Taylor–Couette flow [7, 14], and also for the stability analysis of the interface in the Hele-Shaw cell [13]. Hereafter, our results are presented by following the method in [13].

3.1 Analytical Results

3.1.1 Mode Coupling Equation Based on the Normal Stress Balance

Under the assumption that the solution is written by the Fourier power series, we can obtain the mode coupling equation for the perturbation amplitude. From the Laplace equation (7) the kinematic boundary condition (8) with the normal velocity (10) and the normal stress balance (12), the mode coupling equation is

$$\dot{\zeta}_n(t) = \lambda(n,t)\zeta_n + \sum_{n'} X(n,n',t)\zeta_{n'}\zeta_{n-n'}, \tag{14}$$

up to $(\zeta_n/R)^2$. Here $\zeta_n(t)$ and $\dot{\zeta}_n(t)$ denote the Fourier amplitude of the perturbation with mode $n \in (-\infty, \infty)$ and its time derivative. In the above $\lambda(n,t)$ denotes the linear growth rate defined as

$$\begin{aligned}\lambda(n,t) &= \frac{Q}{2\pi R^2}(A|n|-1) - \frac{\alpha}{R^3}|n|(n^2-1)\\ &\quad + \varepsilon n^2\frac{Q}{\pi R^2}(A|n| - A^2 - 2) + 2\varepsilon\frac{\alpha}{R^3}n^2(n^2-1)(A+|n|),\end{aligned} \tag{15}$$

where $A = (\mu_2 - \mu_1)/(\mu_2 + \mu_1)$ is the viscosity difference ratio, $\alpha = b^2\sigma/12(\mu_1 + \mu_2)$ and $\varepsilon = \varepsilon(t) = b^2/12R(t)^2$ is the thickness parameter of the cell. The third and forth terms in Eq. (15) are derived by considering the normal stress balance, and they are positive since A satisfies $0 \le A \le 1$. The coupling coefficient $X(n,n',t)$ between modes n' and $n-n'$ is as follows;

$$\begin{aligned}X(n,n',t) &= \frac{1}{R}\Bigg[\frac{Q}{2\pi R^2}\left(A|n|\left\{A|n'|(1-\mathrm{sgn}(nn')) - \frac{1}{2}\right\} - (A|n'|-1)\right)\\ &\qquad - \frac{\alpha}{R^3}\left(|n|\left(1 - \frac{1}{2}nn' - \frac{3}{2}n'^2\right) + \{A|n|(1-\mathrm{sgn}(nn')) - 1\}|n'|(n'^2-1)\right)\Bigg]\\ &\quad + \frac{\varepsilon}{R}\Bigg[-\frac{Q}{\pi R^2}\Bigg(An^2|n'|(1-A^2)\mathrm{sgn}(nn') + A|n|\left\{-4n'^2(n'+1) + \frac{1}{2}|n|(A+|n|+4n') + 1\right\}\\ &\qquad + A|n|(A|n'|-1)\left\{|n|(A+|n|) + |n'|(2n'^2 - nn' + 2)\right\}\\ &\qquad - n'^2(A|n'| - A^2 + 2)\{A|n|(1-\mathrm{sgn}(nn')) - 1\}\Bigg)\\ &\quad + 2\frac{\alpha}{R^3}\Bigg(n^2(A+|n|)\left(1 - \frac{1}{2}nn' - \frac{3}{2}n'^2\right) + n^2|n'|(n'^2-1)(1-A^2)\mathrm{sgn}(nn')\\ &\qquad + A|n||n'|(n'^2-1)\left\{|n|(A+|n|) + |n'|(2n'^2 - nn' + 2)\right\}\\ &\quad - |n|n'(n'^2-1)(4n'-n) + n'^2(n'^2-1)(A+|n'|)\{A|n|(1-\mathrm{sgn}(nn')) - 1\}\Bigg)\Bigg].\end{aligned} \tag{16}$$

In addition, $\mathrm{sgn}(nn')$ is a sign function equals to the unity for $nn' > 0$, otherwise -1. By instituting Eqs. (15) and (16) into the nonlinear approximated solution (18), we can obtain the time evolution of the perturbation of mode n including the effect of VNS.

By following the analysis due to [13], as the first step, we neglect the nonlinear term and obtain the linearized solution $\zeta_n^{\mathrm{lin}}(t) = \zeta_n(0)\exp(\int \lambda(n,t)\mathrm{d}t)$. From the linear growth rate (15), it is easily seen that, for a fixed mode n, $\lambda(n,t)$ is a monotonically increasing function with respect to t, which has a zero-point $t_c(n) > 0$ such that $\lambda(n, t_c(n)) = 0$. This indicates that $\lambda(n,t)$ remains negative for $R < R_c(n)$, which means the interface is stable and vice versa, where $R_c(n) = R(t_c(n))$ is the critical radius. Moreover, we assume that the nonlinear term in Eq. (14) can be approximately replaced by $\zeta_n^{\mathrm{lin}}(t)$ as

$$\dot{\zeta}_n(t) = \lambda(n,t)\zeta_n + \Xi(n,t), \\ \Xi(n,t) = \sum_{n'} X(n,n',t)\zeta_{n'}^{\text{lin}}\zeta_{n-n'}^{\text{lin}}. \tag{17}$$

Thus it is easy to show that Eq. (17) can be approximately solved as

$$\zeta_n(t) = \begin{cases} \zeta_n(0) & (R(t) < R_c(n)), \\ \zeta_n^{\text{lin}}(t)\left\{1 + \int_{t_c(n)}^{t} \frac{\Xi(n,t')}{\zeta_n^{\text{lin}}(t')}\mathrm{d}t'\right\} & (R(t) > R_c(n)), \end{cases} \tag{18}$$

up to the first order of ε. This solution (18) is equivalent to that derived in [13] as ε tends to zero.

3.1.2 Mode Coupling Equation with the Effect of the Wetting-Layer

As a next, we consider the mode coupling equation with the wetting-layer effect. It is to be noted that the mode coupling equation and its nonlinear approximated solution are the same as (14) and (18), however the linear growth rate is different as follows;

$$\lambda(n,t) = \frac{Q}{2\pi R^2}(A|n|-1) - \frac{\pi}{4}\frac{\alpha}{R^3}|n|(n^2-1) - \varepsilon\, Ca^{2/3}\,\tilde{J}\,\tilde{\alpha}|n|(A|n|-1), \tag{19}$$

where the coefficients are defined as $\tilde{J} = (3.8\cdot 4)/(3\cdot 12^{1/3})$ and $\tilde{\alpha} = b^{1/3}\sigma/(12(\mu_1+\mu_2)\mu_2^{2/3})$. The coupling coefficient are also different from Eq. (16) as

$$\begin{aligned} X(n,n',t) = \frac{1}{R}\Bigg[&\frac{Q}{2\pi R^2}\left(A|n|\left\{A|n'|(1-\operatorname{sgn}(nn')) - \frac{1}{2}\right\} - (A|n'|-1)\right) \\ &- \frac{\pi}{4}\frac{\alpha}{R^3}\left(|n|\left(1-\frac{1}{2}nn' - \frac{3}{2}n'^2\right) + \left\{A|n|(1-\operatorname{sgn}(nn')) - 1\right\}|n'|(n'^2-1)\right)\Bigg] \\ -\varepsilon\frac{|n|}{R}Ca^{2/3}\frac{\tilde{J}\alpha}{(b\mu_2)^{2/3}}\Bigg[&A|n||n'|(\operatorname{sgn}(nn')-1) + \frac{A|n|}{b}\left(\frac{1}{2} - \operatorname{sgn}(nn')\right) + \frac{A}{2}|n'| - \frac{11}{6} \\ &- \frac{1}{2}\left(3A - 1 - \frac{1}{|n'|}\right)n'(n-n') - \frac{1}{3R^2}(2A|n'| - 5 + 3A|n-n'|(A|n'|-1)) \\ &+ \frac{1}{b}(A|n'|-1)(|n|+|n'|)\left(A(1-\operatorname{sgn}(nn')) - \frac{1}{|n|}\right) \\ &+ \frac{2\pi}{Q}\frac{\pi}{4}\frac{\alpha}{R^3}\left\{|n'|(n'^2-1)(A|n-n'|-1) + (A|n'|-1)|n-n'|\left((n-n')^2-1\right)\right\}\Bigg]. \end{aligned} \tag{20}$$

Here ε and A are already defined above. As ε tends to zero, Eqs. (19) and (20) are also equivalent to those derived in [13].

3.1.3 Numerical Results

From the nonlinear solution (18) with the linear growth rate (15) or (19) and the coupling coefficient (16) or (20), we can obtain the time evolution of the perturbation of mode n explicitly. Here the parameter values are chosen the same as those in Paterson's experiments [16], and the numerical results by Miranda and Widom [13] as $b = 0.15\,\text{cm}$, $Q = 9.3\,\text{cm}^2/\text{s}$, $\sigma = 63\,\text{dyn/cm}$, $\mu_1 = 0.1 \times 10^{-4}\,\text{g/(cm s)}$, $\mu_2 = 5.21\,\text{g/(cm s)}$ and the initial condition: the slightly perturbed circle $R = R_0 + (R_0/500)\cos(n\theta)$, with $R_0 = 0.05\,\text{cm}$.

To calculate the time evolution, we focus on the tip-splitting of the fingers, which the tip of a finger split into two or three and grow independently. This phenomenon seems to be an essential process in the growth of the viscous fingering patterns. In order to reveal how the effect of VNS and the wetting-layer influence the tip-splitting of the fingers, we consider the simplest case that the mode is limited to n, $2n$ and $3n$. For instance, for the case of $n = 6$, their amplitude are determined by Eq. (18), and then the perturbed interface $\mathscr{R}$ is calculated as shown in Fig. 5.

In Fig. 5, the curves represent the interfaces per second from the centre outward. As seen in Fig. 5a, the interface calculated by Young–Laplace relation becomes unstable, however, tip-splitting does not appear even though it is frequently observed in the experiment in [16]. In contrast, for the case of the VNS effect in Fig. 5b, the interface as a whole remains the six-fold symmetry, and the tip of each finger becomes blunt and finally split, which is similar to the patterns in [16]. Finally, for the wetting-layer effect in Fig. 5c, the interface becomes more unstable than the case of (a), however, tip-splitting of fingers are not seen. The similar results can be obtained in the case of $n = 5$ and 6. These results suggest that the interface calculated by including the VNS effect qualitatively agrees with the experimental results. Therefore it is indicated that the normal stress balance, including the VNS effect, is more appropriate for Hele-Shaw problem than the other boundary condition.

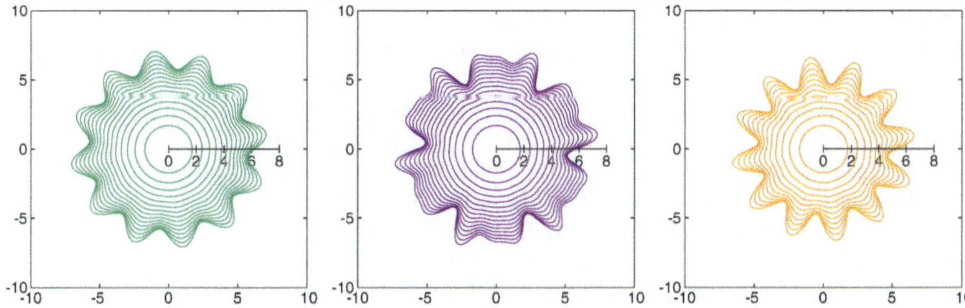

(a) Young–Laplace relation (b) The VNS effect (c) The wetting-layer effect

Fig. 5 The time evolution of the interfaces

4 Concluding Remarks

In the present paper, we carry out the weakly nonlinear stability analysis for the Hele-Shaw problem in radial geometry. As the dynamical boundary condition, Young–Laplace relation seems not to be valid and the modified conditions have been proposed. Among them, we focus on the boundary condition with the VNS effect and the one with the effect of the wetting-layer. By the weakly nonlinear analysis, we can obtain the nonlinear approximated solution for the perturbation amplitude, and then calculated the interfacial patterns for the simplest case whose the modes n, $2n$ and $3n$. As a result, the tip-splitting phenomena of the fingers are reproduced only for the case of the VNS effect, which qualitatively agree with the experimental results. These results support that the normal stress balance including the VNS effect is the more appropriate as the dynamical boundary condition, than Young–Laplace relation and the condition with the wetting-layer effect. However, the results obtained here are only for the case of the simplest modes. It has to be examined for the case of including the more modes, and also the dependence of the patterns on the initial condition. The analysis based on the other boundary conditions, the one including the local thickness of the wetting-layer and the effect of the contact line, should be also investigated. These are the open problem to be solved soon.

References

1. Bataille, J.: Rev. Inst. Pétrole **23**, 1349 (1968)
2. Batchelor, G.: An Introduction to Fluid Dynamics. Cambridge University Press (2000)
3. Ben-Jacob, E., Godbey, R., Goldenfeld, N.D., Koplik, J., Levine, H., Mueller, T., Sander, L.M.: Phys. Rev. Lett. **55**, 1315–1318 (1985)
4. Chen, J.-D.: Exp. Fluids **5**, 363–371 (1987)
5. Dias, E.O., Miranda, J.A.: Phys. Rev. E **88**, 013016 (2013)
6. Escher, J., Simonett, G.: Adv. Differ. Equ. **2**, 619–642 (1997)
7. Fujimura, K., Mizushima, J.: Euro. J. Mech. B **10**, 25–30 (1991)
8. Gustafsson, B., Vasil'ev, A.: Conformal and Potential Analysis in Hele-Shaw Cells, Birkhäuser (2006)
9. Kim, H., Funada, T., Joseph, D.D., Homsy, G.M.: Phys. Fluids **21**, 074106 (2009)
10. Landau, L., Levich, B.: Acta Physicochim. U.R.S.S. **17**, 42–54 (1942)
11. Martyushev, L.M., Birzina, A.I.: J. Phys. Condens. Matter **20**, 045201 (2008)
12. McLean, J.W., Saffman, P.G.: J. Fluid Mech. **102**, 455–469 (1981)
13. Miranda, J.A., Widom, M.: Phys. D **120**, 315–328 (1998)
14. Mizushima, J., Fujimura, K.: J. Fluid Mech. **234**, 651–667 (1992)
15. Park, C.W., Homsy, G.M.: J. Fluid Mech. **139**, 291–308 (1984)
16. Paterson, L.: J. Fluid Mech. **113**, 513–529 (1981)
17. Reinelt, D.A.: J. Fluid Mech. **183**, 219–234 (1987)
18. Saffman, P.G., Taylor, G.I.: Proc. R. Soc. Lond. A **245**, 312–329 (1958)
19. Tani, H., Tani, A.: J. Phys. Soc. Jpn. **83**, 034401 (2014)
20. Ushijima, T., Yazaki, S.: Proc. Czech-Jpn. Semin. Appl. Math. 146–152 (2005)
21. Weinstein, S.J., Dussan, E.B., Ungar, L.H.: J. Fluid Mech. **221**, 53–76 (1990)

Part V
Industrial Applications

Computer Simulation of the Phase Separation of Polymeric Materials for Industrial Applications

Takeshi Aoyagi

Abstract Mixtures of polymers and polymer solutions are easily separated into two phases in a manner similar to water and oil. Even block copolymers, which have a chain of multiple blocks possessing different chemical structures, exhibit a multiphase structure. Such multiphase structures of polymeric materials exhibit unique forms and properties that cannot be reproduced by a simple homogeneous mixture of materials, and controlling the dynamics of the phase separation is very important for developing highly functional materials for industry.

Computational simulation is a powerful tool for studying the dynamics of phase separation. There are many approaches for studying phase behavior, including particle models such as the lattice and off-lattice Monte Carlo models and molecular dynamics, and continuous models such as the self-consistent field (SCF) [5] and Cahn–Hilliard [2] theories. We introduce two examples of the computational simulation of phase separation dynamics using continuous models.

1 Microphase Separation of Block Copolymers for the Study of Thermoplastic Elastomers

1.1 Introduction

Block copolymers that possess blocks of immiscible chemical units exhibit a microphase-separated structure whose morphology depends upon the structure and properties of the polymers, such as miscibility, and the volume fraction of each block. Figure 1 shows examples of microphase-separated structures of AB-type diblock copolymers of different volume fractions.

Thermoplastic elastomers (TPEs) are an industrial application of block copolymers. A typical TPE consists of ABA type triblock copolymer such as Styrene-

T. Aoyagi (✉)
National Institute of Advanced Industrial Science and Technology (AIST),
Central 2, 1-1-1, Umezono, Tsukuba, Ibaraki 305-8568, Japan
e-mail: aoyagi.t@aist.go.jp

H. Itou et al. (eds.), *Mathematical Analysis of Continuum Mechanics and Industrial Applications*, Mathematics for Industry 26,
DOI 10.1007/978-981-10-2633-1_15

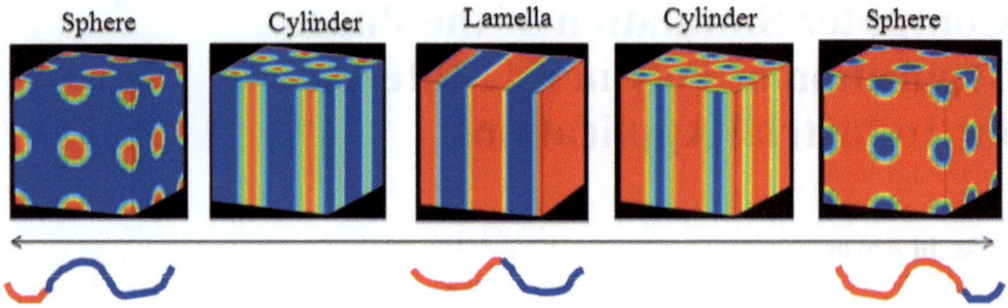

Fig. 1 Microphase-separated structures of diblock copolymers. The morphology changes from a *red block* sphere to a *blue block* sphere by increasing the volume fraction of the *red block*

Butadiene-Styrene, consisting of both glassy blocks and rubber-like blocks. The glassy blocks create a hard domain that acts as a physical crosslink and exhibits rubber elasticity below the glass transition temperature, Tg, while plastic flow is observed above Tg. TPEs are widely used for rubbery materials, plasticizers for hard polymeric materials and pressure-sensitive adhesives, for example, and it is important to control their morphologies to produce the desired properties by changing the fine chemical structure of the polymer chains and the amount of each polymer in the mixture. In addition to the morphology, the chain conformation of the soft block, such as the fraction of bridging chain between different glassy domains, affects elastic behavior. Thus the microstructural analysis of chain conformation is also desired by theoretical calculation.

Self-consistent field (SCF) theory is a powerful method for predicting microphase-separated structures of block copolymers. The chemical structure of a polymer is coarse-grained, and the polymer chain is described by the connection of segments that normally represent a small portion of a real polymer, such as several monomer units. Although SCF theory is based on a continuous model, the effect of polymer architecture is considered in the statistical approach; therefore the SCF calculation can give the spatial distribution of each segment in the phase-separated state in addition to the distribution of each type of segment. This kind of information provides details concerning the conformation of each polymer chain that are important for understanding the physical properties of TPEs.

We analyzed the details of chain conformation of the microphase-separated structure of block copolymers using SCF calculations to study the elastic behavior of TPEs, including realistic industrial materials which have complicated polymer structures and components.

1.2 Calculation Method

The details of SCF theory are described in the literature [5, 6]. SCF theory provides the statistical weight of a chain by solving the following Edwards equation.

$$\frac{\partial}{\partial s}Q_i(s', \mathbf{r}'; s, \mathbf{r}) = \left[\frac{b_K^2}{6}\nabla^2 - \frac{1}{k_B T}\rho_K V_i(\mathbf{r})\right]Q_i(s', \mathbf{r}'; s, \mathbf{r}). \quad (1)$$

where $Q_i(s', \mathbf{r}'; s, \mathbf{r})$ is the path integral describing the equilibrium statistical weight of the ith subchain composed of K-type segments, k_B is Bolzmann's constant, T is the absolute temperature, b_K and ρ_K are the Kuhn statistical length and specific volume of K-type segments respectively, and $V_i(\mathbf{r})$ is the external potential acting on the segments of the ith subchain at position $\mathbf{r}$. $V_i(\mathbf{r})$ is divided into two contributions,

$$V_i(\mathbf{r}) = W_K(\mathbf{r}) - \mu_i(\mathbf{r}). \quad (2)$$

$W_K(\mathbf{r})$ is a segment–segment interaction potential given by

$$W_K(\mathbf{r}) = \sum_{K'} \chi_{KK'}\phi_{K'}(\mathbf{r}), \quad (3)$$

where $\phi_{K'}(\mathbf{r})$ is the sum of the segment densities of the subchains composed of K'-type segments, and $\chi_{KK'}$ is an interaction parameter describing the miscibility between K and K'-type segments. $\mu_i(\mathbf{r})$ is an external potential such as an incompressible condition and an external field.

To study the details of chain conformation, we introduced the labeled fraction of a triblock copolymer where one end of the chain is constrained at a specified domain. The bridge fraction can be determined by calculating the distribution of the free end of the chain, i.e. the fraction of the free end that is distributed in the domain other than that which exists at the constrained end corresponds to the bridge fraction. The details of the method are described in the original paper [1].

1.3 Results

Figure 2 shows an example of the spatial distribution of the volume fractions of A blocks of an ABA triblock copolymer. Figure 2a shows the overall distribution of A blocks. When the end of a small part of one A block is constrained in the center domain (Fig. 2b), the distribution of free ends of the other A block is obtained (Fig. 2c). From this distribution we can obtain the fraction of bridged chain, which bridges between different domains that become an origin of stress during deformation.

Table 1 shows the bridge fraction ϕ_{bridge} of several ABA symmetric triblock copolymers of equilibrium morphologies. Lamellar, cylindrical and spherical domain structures are obtained by 1D, 2D, and 3D calculations with periodic boundary conditions respectively. vf_A is the volume fraction of segment type A, and χ_{AB} is an interaction parameter corresponding to the miscibility of two types of segments, A and B. Larger values of χ correspond to lower miscibility.

It is evident that the bridge fraction increases as the domain morphology changes from lamellar to cylindrical and then to spherical phases.

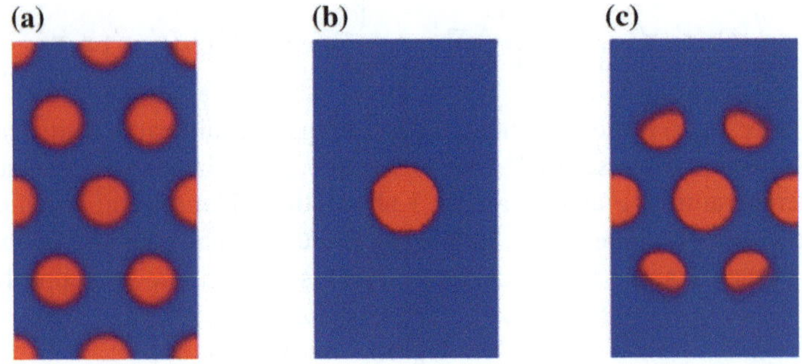

Fig. 2 Two-dimensional spatial distribution of an ABA triblock copolymer illustrating **a** total A block, **b** constrained A block, and **c** free end of the other A block of constrained chains

Table 1 Bridge fraction of various micro domain structures of ABA symmetric triblock copolymers

	Molecule	vf_A	χ_{AB}	ϕ_{bridge}
Lamella	A10B20A10	0.5	2.0	0.45
	A20B40A20	0.5	1.0	0.45
	A40B80A40	0.5	1.0	0.41
Cylinder	A5B40A5	0.20	1.5	0.63
	A6B28A6	0.30	1.25	0.63
Sphere	A3B54A3	0.10	3.0	0.75
	A5B40A5	0.20	1.0	0.78
	A6B28A6	0.30	0.75	0.80

The previous studies for the analysis of bridge fraction using SCF calculation [1, 7] are restricted to ABA symmetric triblock copolymer and equilibrated morphology such as lamellar, hexagonal cylinder and body center cubic phase.

In this study, we applied this method to more complicated systems, which are commonly used for industrial materials. Takayanagi reported a Styrene-Isoprene type TPE produced by blending a symmetric triblock copolymer and an asymmetric triblock copolymer [8]. Since the asymmetric copolymer contains long and short styrene blocks, large and small domains of styrene blocks are observed using transmission electron microscopy. Figure 3 shows a schematic of the blend of symmetric and asymmetric triblock copolymers, and an example of the results of SCF calculation for the blend. Furthermore, the middle block of the asymmetric copolymer is shorter than that of the symmetric copolymer, and it is thought that the asymmetric copolymer cannot easily adopt a bridged conformation.

The bridge fraction of both the symmetric and the asymmetric copolymer can be calculated by constraining one end of the chain. Figure 4 shows the distributions of the constrained end and the free end of symmetric and asymmetric copolymers. In this study, the end of the large block is constrained in the case of asymmetric copolymers.

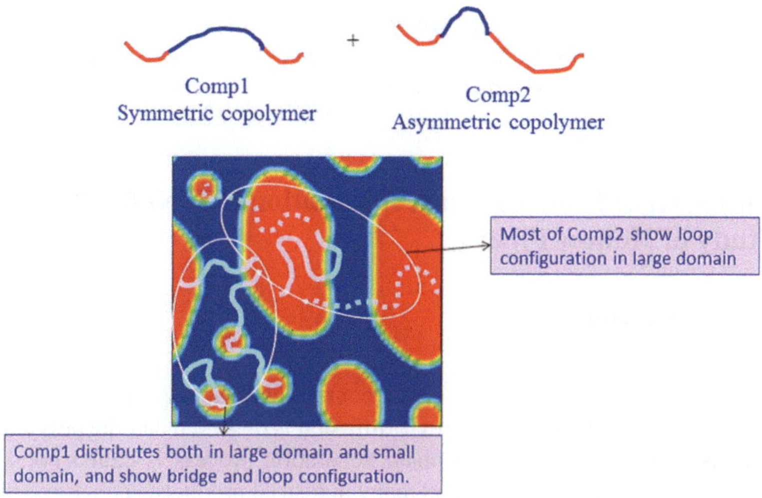

Fig. 3 Schematic of the blend of symmetric and asymmetric triblock copolymers, and an example of the results of SCF calculation for the blend

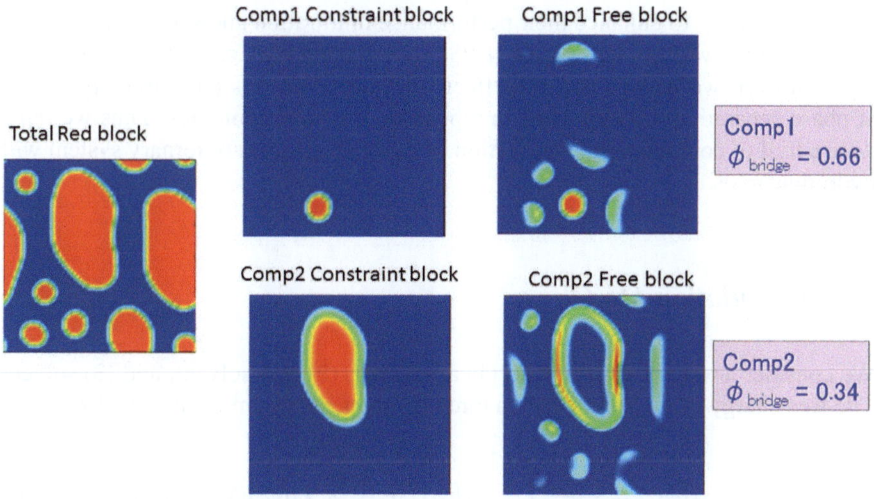

Fig. 4 Distribution of constrained and free blocks and bridge fraction of each chain

We found that the large red block was distributed across the whole of the large domain, while the short red block of the asymmetric copolymer and the symmetric copolymer was distributed only at the small domain and the surface of the large domain. The bridge fractions of symmetric and asymmetric copolymers with the selected constraint positions are 0.66 and 0.34, respectively. The fixed domain is selected arbitrarily and we should note that the distribution and bridge fraction will be changed by the selection of domain for the constrained end. However, it is clear that the bridge fraction of the asymmetric copolymer which has a short middle block is smaller than that of the symmetric copolymer.

2 Non-solvent Induced Phase Separation (NIPS) for the Study of Hollow Fibers

2.1 *Introduction*

When a polymer solution in good solvent contacts with a non-solvent, an exchange between the solvent and non-solvent via diffusion is observed and the polymer solution begins to precipitate. Such a phase separation process is called non-solvent induced phase separation (NIPS), and many industrial products, such as membranes for filtration and separation, are made using the process. Hollow fibers are one popular product of these porous membranes and are used for water purification and artificial kidneys, for example. It is important to control the size and shape of these porous structures to enhance their performance in filtration and separation.

Zhou and Powell studied such a NIPS process using the Cahn–Hilliared equation [9]. However, we believe that the effects of hydrodynamics and viscocity change at phase separation are important to reproduce final morphologies. Thus we introduce a trial of computational simulation of the NIPS process of ternary system with hydrodynamics.

2.2 *Calculation Model*

We consider a ternary system with (1) a polymer, (2) non-solvent, and (3) solvent. The free energy of mixing of such a three component system is given by [9]

$$F_{mix}[\{\phi_i\}] = \frac{k_B T}{b^3} \int \left[f_0(\{\phi_i\}) + \frac{1}{2} \sum_{i=1}^{2} \kappa_i(\phi_i) |\nabla \phi_i|^2 \right] d\mathbf{r}, \tag{4}$$

where ϕ_i is the volume fraction of component i, b is the monomer size and κ_i is the gradient coefficient. The first term f_0 on the right side of Eq. 4 is the local, homogeneous mixing free energy of the system [4]. We use a multicomponent Flory-Huggins type model. In the Flory-Huggins model, the homogeneous term is given by a mixing entropy and an enthalpy of interaction,

$$f_0 = \frac{\phi_1}{N} \ln \phi_1 + \phi_2 \ln \phi_2 + \phi_3 \ln \phi_3 + \chi_{12}\phi_1\phi_2 + \chi_{13}\phi_1\phi_3 + \chi_{23}\phi_2\phi_3, \tag{5}$$

where N is the degree of polymerization of the polymer and the χ_{ij} are the Flory interaction parameters between components i and j. The second term (the sum of square gradients) accounts for inhomogeneity in the solution.

We assume incompressibility, and the mass conservation equation can be written as

$$\frac{\partial \phi_i}{\partial t} = -\nabla \cdot (\phi_i \mathbf{v}_i), \quad (6)$$

where $\mathbf{v}_i$ is the component velocity. Incompressibility of the total solution is expressed by the total mass balance

$$\nabla \cdot \mathbf{v}_i = 0, \quad (7)$$

where

$$\mathbf{v} = \sum_{i=1}^{3} \phi_i \mathbf{v}_i \quad (8)$$

is the solution velocity. By definition, the volume fractions satisfy

$$\sum_{i=1}^{3} \phi_i = 1 \quad (9)$$

and only two of the volume fractions are independent.

The chemical potentials μ_i are obtained by taking the functional derivatives with respect to volume fraction. Neglecting any dependence of the gradient coefficients on concentration gives

$$\mu_1 = \frac{k_B T}{b^3}\left[\ln\left(\frac{\phi_1^{1/N}}{1-\phi_1-\phi_2}\right) + \frac{1-N}{N} + (\chi_{12}-\chi_{23}-\chi_{13})\,\phi_2 + \chi_{13}(1-2\phi_1) - \kappa_1 \nabla^2 \phi_1\right] \quad (10)$$

$$\mu_2 = \frac{k_B T}{b^3}\left[\ln\left(\frac{\phi_2}{1-\phi_1-\phi_2}\right) + (\chi_{12}-\chi_{23}-\chi_{13})\,\phi_1 + \chi_{23}(1-2\phi_2) - \kappa_2 \nabla^2 \phi_2\right]. \quad (11)$$

where the expression $\phi_3 = 1 - \phi_1 - \phi_2$ has been used.

In addition to mass conservation, the Navier–Stokes equation at Zero Reynolds number (Stokes equation) is solved for momentum conservation. The Stokes equation at constant viscosity is given by

$$0 = -\nabla p + \eta \nabla^2 \mathbf{v} - \phi \nabla \mu, \quad (12)$$

where p is the pressure and η is the solution viscosity. However, composition-dependent viscosity is a key issue in studying the dynamics of phase separation of a polymer–non-solvent–solvent ternary system. Thus we need to modify the Stokes equation for variable viscosity,

$$0 = -\nabla p + \nabla \cdot \left[\eta\left(\phi\right)\left(\nabla \mathbf{v} + \nabla \mathbf{v}^{T}\right)\right]\mathbf{v} - \phi\nabla\mu. \tag{13}$$

The numerical calculation was done using a pseudo-spectral method [3] with periodic boundary conditions. This scheme is efficiently implemented using fast Fourier transforms.

2.3 Results

Figure 5 shows an example of the results of 2D calculation where, during the exchange between the non-solvent and the solvent, precipitation of a dense polymer phase and

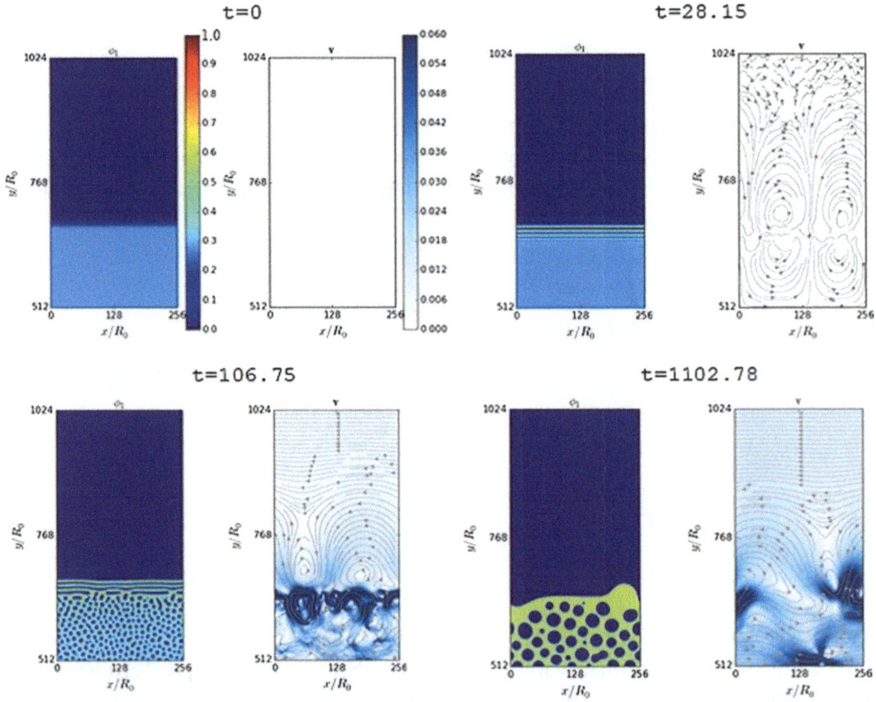

Fig. 5 Volume fraction of polymer and velocity field of the NIPS process as a function of time. The lower region is the polymer solution and the upper region is the non-solvent at initial time t = 0. The time and position values are shown in reduced units

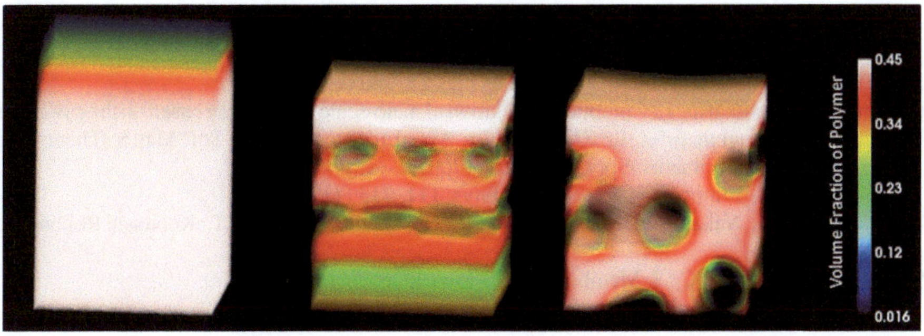

Fig. 6 Snapshot profiles of volume fraction of polymer during 3D NIPS calculation

a porous structure is observed. We should note that the actual calculation was done using periodic boundary conditions with 254 × 1024 mesh, and the polymer solution phase at initial state is located in the central part with respect to the y-axis. The figure shows only the upper part of the total mesh. We confirmed three stages of phase separation from the results, i.e., (1) formation of a surface skin layer at the interface between polymer solution and non-solvent, which is shown in upper right figure of Fig. 5 (reduced time $t = 28.15$), (2) spinodal decomposition during the diffusion of non-solvent, which is shown in lower left figure of Fig. 5 ($t = 106.75$), and (3) coarsening of morphology by the effect of hydrodynamic flow, which is shown in lower right figure of Fig. 5 ($t = 1102.78$).

Figure 6 is an example of 3D calculation, in which only the fraction of polymer is shown. The mesh size is 32 × 32 × 128 and a half region is shown on the z-axis. In this 3D calculation, the same stages of phase separation were observed as in the 2D calculation. Each figure in Fig. 6 shows the stage of phase separation, i.e., formation of skin layer (left), spinodal decomposition (middle), and coarsening of morphology (right).

The viscosity depends upon the fraction of polymer. However, it is not trivial to conduct numerical calculation when the viscosity differences in the multiphase system become large. Currently we are developing a stable numerical scheme to solve the Stokes equation with large viscosity changes. Furthermore, the final porous structure is fixed when the polymer-dense phase solidifies by glass formation or crystallization. The implementation of a realistic scheme to represent such a solidification will be the object of further study.

References

1. Aoyagi, T., Honda, T., Doi, M.: J. Chem. Phys. **117**, 8153 (2002)
2. Cahn, J.W., Hilliard, J.E.: J. Chem. Phys. **28**, 258 (1958)
3. Ceniceros, H.D., Fredrickson, G.H.: Multiscale Model. Simul. **2**, 452 (2004)

4. Fredrickson, G.H. : The Equilibrium Theory of Inhomogeneous Polymers. Oxford University Press (2006)
5. Helfand, E., Wasserman, Z.R.: Macromolecules **11**, 960 (1978)
6. Honda, T., Kawakatsu, T.: Computer simulation of nano-scale phenomena based on the dynamics density functional theories. In: Zvelindovsky, A. (ed.) Nanostructured Soft Matter. Theory and Perspectives, Experiments. Springer, Berlin (2007)
7. Matsen, M.W., Tompson, R.B.: J. Chem. Phys. **111**, 7139 (1999)
8. Takayanagi, A., Nakama, Y., Maeda, F., Oda, R., Hashimoto, S., Honda, T.: Kobunshi Ronbunshu **72**, 104 (2015)
9. Zhou, B., Powell, A.C.: J. Memb. Sci. **268**, 150 (2006)

Highly Parallel Computation of Generalized Eigenvalue Problem in Vibration for Automatic Transmission of Vehicles Using the Sakurai–Sugiura Method and Supercomputers

Takanori Ide, Yuto Inoue, Yasunori Futamura and Tetsuya Sakurai

Abstract In this paper, we discuss highly parallel computational approach for solving eigenvalue problems arising from vibration problem in automatic transmission of vehicles. Vibration performance is an important quality measure of vehicles. Typically, vibration performance of automatic transmission strongly ties up to comfortable driving. Therefore, reduction of vibration is one of the key consideration of new automatic transmission design. However, the computational time for solving eigenvalue problems dominates that for the design optimization routine, and it becomes unacceptably long when we use a precise model which has a large number of degrees of freedom. Therefore, efficient approach to solve the large-scale eigenvalue problem is required. Owing to this situation, in this study, we present a performance of a hierarchical parallel eigensolver using state-of-the-art supercomputers such as the K computer and COMA.

T. Ide (✉)
Engineering Division, AISIN AW CO., LTD., Fujii-cho, Takane 10, Anjo, Aichi 444-1192, Japan
e-mail: i24824_ide@aisin-aw.co.jp

Y. Inoue · Y. Futamura · T. Sakurai
Department of Computer Science, University of Tsukuba, 1-1-1, Tennodai, Tsukuba, Ibaraki 305-8573, Japan
e-mail: yuto@mma.cs.tsukuba.ac.jp

Y. Futamura
e-mail: futamura@cs.tsukuba.ac.jp

T. Sakurai
e-mail: sakurai@cs.tsukuba.ac.jp

H. Itou et al. (eds.), *Mathematical Analysis of Continuum Mechanics and Industrial Applications*, Mathematics for Industry 26,
DOI 10.1007/978-981-10-2633-1_16

1 Introduction

The main purposes of this paper are to show:

1. highly parallel computational approach based on the Sakurai–Sugiura method [1] for generalized eigenvalue problem; and
2. the computational performance using the Sakurai–Sugiura method and supercomputers (the K computer and COMA) in vibration problem for automatic transmission of vehicles.

Our previous paper proposed highly parallel computational approach in vibration problem for automatic transmission of vehicles [2]. Our proposal approach employed the Sakurai–Sugiura method and supercomputers such as RIKEN's K computer and University of Tsukuba's COMA. The K computer is the world's fourth fastest super computer of the TOP500 benchmark as of November 2015 [3]. The K computer has 88,128 nodes in total and has 16 GB memory in one node. COMA is the world's 113th fastest super computer of the TOP500 benchmark as of November 2015. COMA has 393 nodes in total and has 64 GB in one node.

In our previous paper, we showed the comparison of computational performance of conventional approach that uses iterative eigenvalue solver with one-node computational environment. Our proposal approach used the Sakurai–Sugiura method with highly parallel computational environment. Then, we used 3456 nodes in the K computer and 288 nodes in COMA. Furthermore, we showed the node scalability of COMA.

The Sakurai–Sugiura method is designed to attain good scalability in large-scale distributed computing. It is derived using a numerical quadrature of matrix inverses with several quadrature points on the complex plane, and the hierarchical structure of the method is used to utilize a large amount of computational resources. The Sakurai–Sugiura method has been applied for various areas of scientific computation ranging from computational chemistry [4], materials science [5, 6], nuclear physics [7], particle physics [8], to linear collider simulation [9].

In this paper, we will show the detailed theoretical explanation, the numerical result using nodes of the K computer up to 6,912 nodes and the scalability of the K computer. Figure 1 shows the actual cut model of 8th speed FWD automatic transmission [10]. Automatic transmission of vehicles has very complex mechanical structure.

Vibration performance is one of the important requirements for comfortable driving. In addition, reduction of product design study time is another important requirement to address today's competitive market. To achieve this, utilizing computer simulation is required. To obtain high accuracy numerical computation of eigenvalues and eigenvectors, detailed finite element model is required. Therefore, we need to compute the large-scale finite element model. As the consequence, computational time of eigenvalue problem becomes unacceptably long.

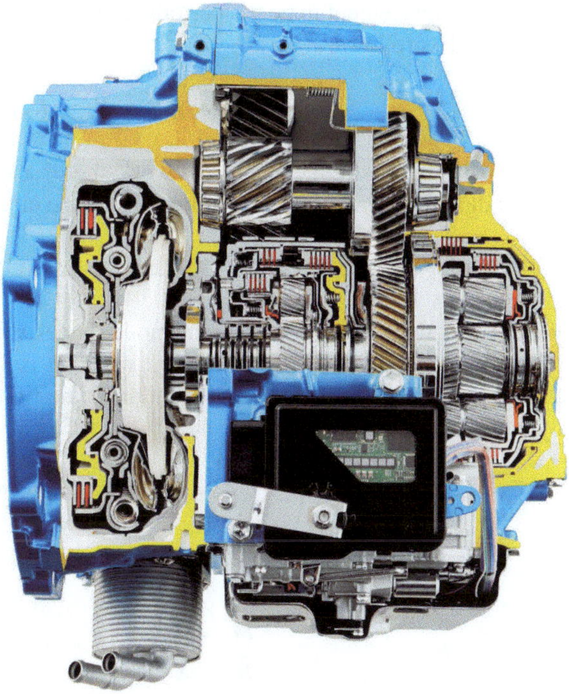

Fig. 1 8th speed FWD automatic transmission

The rest of this paper is organized as follows. In Sect. 2, generalized eigenvalue problem in vibration is discussed. In Sect. 3, the Sakurai–Sugiura method that we employed is discussed. In Sect. 4, the detailed finite element model description of our automatic transmission and computational time by our proposed method are discussed. In Sect. 5, the proposed method and the benefits are summarized in the conclusion.

2 Eigenvalue Problem in Automatic Transmission of Vehicles

We recall generalized eigenvalue problem in vibration problem [11]. The governing equation for the dynamic frequency response is the following:

$$M\ddot{\boldsymbol{u}} + B\dot{\boldsymbol{u}} + K\boldsymbol{u} + \mathrm{i}K_s\boldsymbol{u} = \boldsymbol{p} \tag{1}$$

where M is the mass matrix, B is the viscous damping matrix, K is the stiffness matrix, K_s is the structural damping matrix, $\boldsymbol{p}$ is the load vector, $\boldsymbol{u}$ is the dynamic displacement and i is the imaginary unit. In addition, we assume the load vector and dynamic displacement to be the periodic functions. Then the governing equation (1) is replaced as follows:

$$\left(-\omega^2 M + \mathrm{i}\omega B + K + \mathrm{i}K_s\right) \boldsymbol{u} = \boldsymbol{p}, \tag{2}$$

where ω is angular frequency. In this work, the structural damping K_s is ignored and modal damping is applied. Therefore solving the eigenvalue problem

$$K\boldsymbol{x} = \lambda M\boldsymbol{x} \tag{3}$$

yields the natural frequencies and the mode shapes $\boldsymbol{x}$ that can reduce Eq. (2) to modal space. Here, let $X := [\boldsymbol{x}_1, \boldsymbol{x}_2, \dots, \boldsymbol{x}_m]$, where m is the number of eigenvector used for the reduction. The following equation is the finite element scheme of a modal dynamic frequency problem

$$\left(-\omega^2 \tilde{M} + \mathrm{i}\omega\tilde{B} + \tilde{K}\right) \boldsymbol{z} = C\boldsymbol{z} = \tilde{\boldsymbol{p}} \tag{4}$$

where

$$\begin{aligned}
\tilde{M} &:= X^{\mathrm{T}} M X, \\
\tilde{B} &:= X^{\mathrm{T}} B X, \\
\tilde{K} &:= X^{\mathrm{T}} K X, \\
C &:= -\omega^2 \tilde{M} + \mathrm{i}\omega\tilde{B} + \tilde{K}, \\
\tilde{\boldsymbol{p}} &:= X^{\mathrm{T}} \boldsymbol{p}, \\
\boldsymbol{u} &:= X\boldsymbol{z}.
\end{aligned}$$

The dynamic velocity vector $\boldsymbol{v}$ can be obtained by solving Eq. (4).

3 Sakurai–Sugiura Method

The Sakurai–Sugiura method is an eigenvalue solver which computes eigenvalues in a specified domain Ω on the complex plane and their corresponding eigenvectors. Variants of the method using the Rayleigh–Ritz procedure [12] and block versions [13, 14] of the method have been proposed. In this paper, we refer to these variants as the Sakurai–Sugiura method. We describe an outline of the method in this section.

In the Sakurai–Sugiura method, we consider a contour integration

$$S_k := \frac{1}{2\pi \mathrm{i}} \oint_\Gamma z^k (zM - K)^{-1} M V \mathrm{d}z$$

for solving (3). Here, Γ is a positively oriented Jordan curve whose shape is identical to the boundary of Ω, V is an $n \times L$ random real matrix whose column vectors are linearly independent. Let $S := [S_0, S_1, \ldots, S_{n_m-1}]$. By the Cauchy's integral expression, the column vectors of S become linear combinations of eigenvectors corresponding to eigenvalues inside Γ. Let m_Γ be the number of eigenvalues (counting multiplicity) inside Γ. Range(S) is identical to the subspace spanned by the eigenvectors associated with eigenvalues inside Γ, provided that $n_\mathrm{m} L \leq m_\Gamma$ and L is greater than or equal to the maximum multiplicity of the eigenvalues inside Γ. Then, by the Rayleigh–Ritz procedure for Range(S) or solving a small generalized eigenvalue problem of block Hankel matrices consist of $V^\mathrm{T} S_k$ $(k = 0, 1, \ldots, 2n_m - 1)$, we can compute all eigenvalues in Γ and their corresponding eigenvectors.

For discretizing the contour integral, we use a numerical quadrature

$$\hat{S}_k := \sum_{j=1}^{N} w_j \zeta_j^k (z_j M - K)^{-1} M V,$$

where N is the number of quadrature points, z_j is a quadrature point, w_j is a quadrature weight, and ζ_j is given by scaling z_j such that its absolute value is close to unity for numerical stability. In the Sakurai–Sugiura method, as the most time-consuming part, N linear systems (with multiple right-hand sides)

$$(z_j M - K) Y_j = M V \quad (j = 1, 2, \ldots, N) \tag{5}$$

need to be solved for computing $\hat{S}_k$.

Fortunately, because they can be solved independently, we have a coarse-grained parallelism. In addition, we can solve each linear system by a parallel linear solver. We can also locate multiple (non-overlapped) contour paths for computing the target eigenvalues. Thus, consequently, by using the Sakurai–Sugiura method, we can introduce the following three-level hierarchical parallelism:

Top level: Domains are computed independently
Middle level: Linear systems are solved independently
Bottom level: Each linear system is solved by a parallel linear solver

A schematic illustration for this hierarchical parallelism is shown in Fig. 2.

As in [2], in this study, we use z-Pares [15] as the implementation of the Sakurai–Sugiura method and use MUMPS [16] as the linear solver for (5). The distributed parallelism is implemented using Message Passing Interface (MPI).

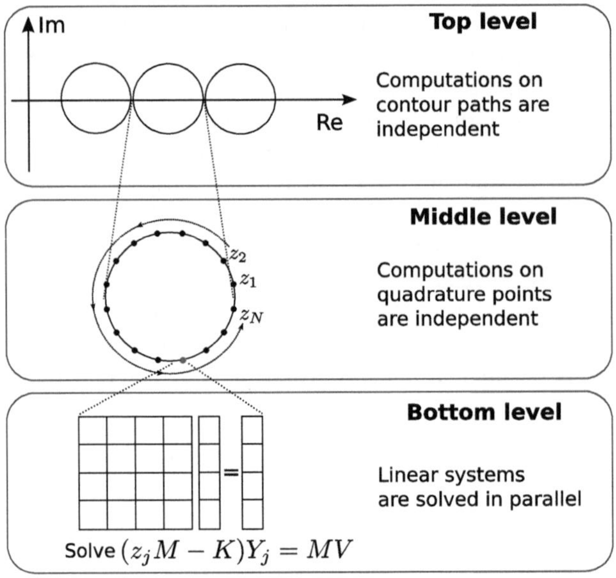

Fig. 2 Hierarchical parallelism of the Sakurai–Sugiura method

4 Numerical Experiments

To demonstrate the effectiveness of the Sakurai–Sugiura method and highly parallel supercomputers, we applied them for 8th speed FWD automatic transmission of vehicles that we considered in [2].

4.1 Finite Element Model

We recall our finite element model [2]. Automatic transmission is one of the complicated structure in today's vehicles and requires precise finite element model. Finite element model is created from CAD data and refined using correlation techniques to match the experimental data. The finite element model needs to be detailed enough to represent high frequency modes. To precisely represent the complicated geometry of the automatic transmission including the variations of wall thickness and ribs, a finite element model is generated. Our finite element model uses solid and shell elements. Figure 3 shows the finite element model of FWD automatic transmission. Our finite element model consists of tetra, hexahedron, triangular and quadrilateral elements with 5,435,438 grids and 16,747,928 DOF's.

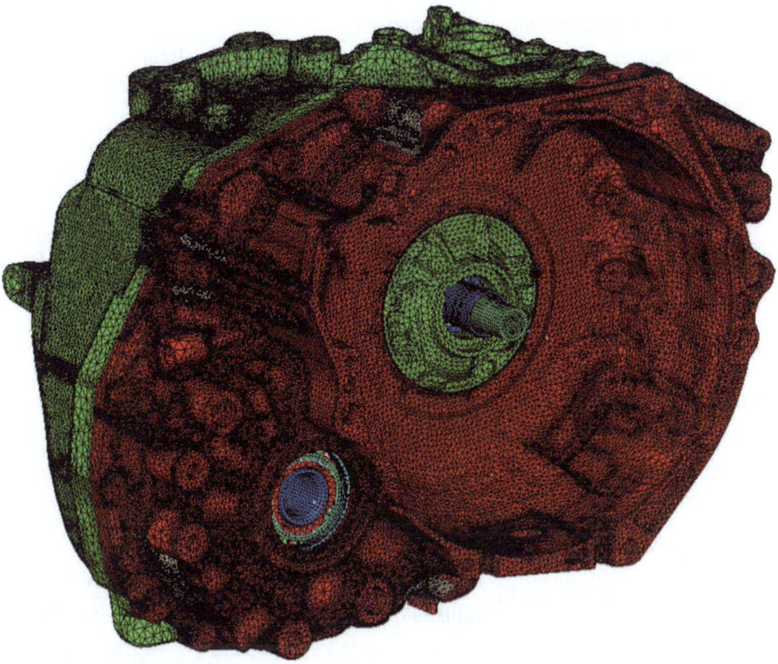

Fig. 3 Finite element model

4.2 Computational Environments

We solved a generalized eigenvalue problem for our finite element model with the targeted frequency from 0 to 6,000 Hz. To validate the efficiency of the Sakurai–Sugiura method and highly parallel supercomputer, we evaluate the performance of an iterative eigensolver in commercial software using a workstation and the performance of the Sakurai–Sugiura method using highly parallel supercomputers (the K computer and COMA).

Table 1 shows the comparison of the computational environments used for performance evaluations in this paper. Note that, though the supercomputer COMA has Intel Xeon Phi 7110P in its nodes, we describe only the CPU part in the table. In Table 2, we show the settings of experiments on the K computer and COMA such as the compilers used and their options.

4.3 Parallel Performance of MUMPS Linear Solver

At the bottom level of the parallelism of the Sakurai–Sugiura method, we run a distributed parallel sparse direct linear solver implemented in MUMPS. Here, as a

Table 1 Comparison of computational environment

	Machine	CPU	Memory (G/Node)	Total number of nodes	Total number of cores	Software
AISIN AW	AW's computer	Intel Xeon E5-2690 (2.90 GHz) 2CPU	512	1	16	Commercial
RIKEN	The K computer	SPARC64 VIIIfx 8core (2.0 GHz) 1CPU	16	88,128	705,024	z-Pares
Univ. Tsukuba	COMA (CPU part)	Intel Xeon E5-2670v2 (2.5 GHz) 2CPU	64	393	7,860	z-Pares

Table 2 Settings on the K computer and COMA

	The K computer	COMA
Fortran compiler	Fujitsu Fortran 1.2.0	Intel ifort 15.0.1
Compiler options	`-Kfast,parallel,openmp`	`-O3 -openmp`
BLAS,LAPACK,SCALAPACK	SSL II	Intel MKL 11.2.0

Table 3 Performance of MUMPS

Number of nodes for MUMPS	72	144	288
Elapsed time for MUMPS (s)	396.1	341.1	335.7
Percentage in total elapsed time (%)	87.7	90.1	94.4

preliminary result, we show parallel performance of MUMPS on the K computer. The result is shown in Table 3.

To measure the parallel performance, we started from 72 nodes due to the memory limitation of the nodes. We increased the number of nodes up to 288. One MPI process was assigned to a node. Eight threads ran on each MPI process. As seen in the table, the parallel performance of MUMPS stagnated at 288 nodes. We also show the percentage of the elapsed time for MUMPS in the total elapsed time of the Sakurai–Sugiura method. We can see that the MUMPS linear solver is the most time-consuming computational kernel of the Sakura–Sugiura method.

Table 4 Number of processes used for each level of parallelism

Total number of processes (nodes)	96	192	384	768	1,536	2,304	3,456	6,912
Number of super-groups of process for top level	1	2	4	4	4	4	4	4
Number of groups of process for middle level	1	1	1	2	4	6	12	12
Number of processes for bottom level	96	96	96	96	96	96	72	144

4.4 Scalability of the Sakurai–Sugiura Method on the K Computer

Here we describe the parallel scalability of the Sakurai–Sugiura method on the K computer. For the parameters of the Sakurai–Sugiura method, we set $L = 100$, $N = 12$ and $n_m = 8$.

In Table 4, we show the number of (groups of) processes used for each level of parallelism on the K computer. As seen in Table 3, the parallel performance of MUMPS stagnates when 288 processes are used for the bottom level. Thus the number of processes for the bottom-level parallelism was increased up to 144. We set the number of quadrature points $N = 12$, thus the number of groups of process for the middle-level parallelism was increased up to 12. We set four sub-intervals, thus the number of super-groups of process for the top-level parallelism was increased up to four. We assign one MPI process per node. We ran eight threads on each MPI process. We decreased the number of nodes used for the bottom level from 96 nodes to 72 nodes when the total number of processes was increased from 2,304 to 3,456. This is because the memory space for eigenvectors are distributed by middle-level parallelism then it enabled MUMPS to use more memory space.

The parallel scalability is shown in Fig. 4. On the K computer, using 96 nodes, the computation time was 11,617 s, and using 6,912 nodes, the computation time was 378 s. Thus the speedup ratio is 30.7.

4.5 Performance Comparison

Figure 5 shows the comparison of computational time for one-node computer with commercial software that AISIN AW currently uses and highly parallel supercomputers (the K computer and COMA) with the Sakurai–Sugiura method. As the baseline, we set AISIN AW currently use computer with commercial software. The result on COMA is recalled from the one described in [2].

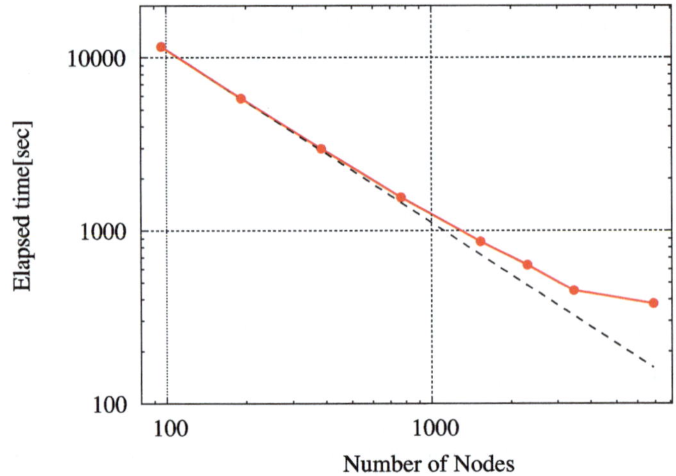

Fig. 4 Parallel scalability of the Sakurai–Sugiura method on the K computer

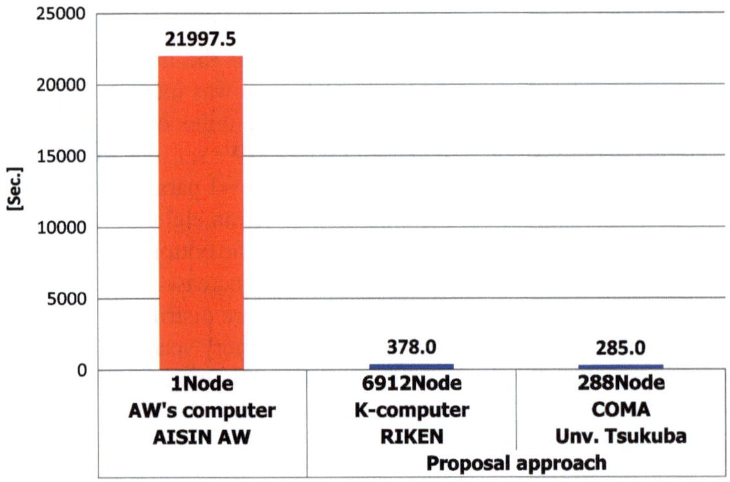

Fig. 5 Comparison of computational time

Comparing the computational time between our currently use computer with commercial software and the K computer with the Sakurai–Sugiura method, the reduction ratio is 98.3 %. Furthermore, comparing the computational time between our currently used computer with commercial software and COMA with the Sakurai–Sugiura method, the reduction ratio is 98.7 %.

5 Conclusions

In this paper, we presented the performance of the Sakurai–Sugiura eigensolver on state-of-the-art supercomputers such as the K computer and COMA for solving a generalized eigenvalue problem arising in a vibration analysis of an automatic transmission of vehicle. The algorithm and the three-level hierarchical parallelism of the Sakurai–Sugiura method were described. In the numerical experiment, we discussed the parallel performance of MUMPS linear solver used at the bottom level of the parallelism of the Sakurai–Sugiura method. Then by using the performance result of MUMPS, we described how to manage the three-level hierarchical parallelism of the Sakurai–Sugiura method on the K computer and showed the total parallel performance of the eigensolver using up to 6,912 nodes of the K computer. We compare the Sakurai–Sugiura method and an eigensolver in currently used commercial software in AISIN AW. The Sakurai–Sugiura method was successfully applied to reduce the computational time for solving an eigenvalue problem arising in the vibration analysis for an automatic transmission. As the consequence, we can expect to drastically reduce the design study time for improving vibration performance for automatic transmissions of vehicles, if we use highly parallel supercomputer such as the K computer and COMA and a suitable eigensolver for such systems.

References

1. Sakurai, T., Sugiura, H.: A projection method for generalized eigenvalue problems using numerical integration. J. Comput. Appl. Math. **159**, 119–128 (2003)
2. Ide, T., Toda, K., Futamura, Y., Sakurai, T.: Highly parallel computation of eigenvalue analysis in vibration for automatic transmission using Sakurai-Sugiura method and K-computer, SAE Technical Paper, 2016–01–1378 (2016)
3. http://www.top500.org/list/2015/11/?page=2
4. Sakurai, T., Tadano, H., Ikegami, T., Nagashima, U.: A parallel eigensolver using contour integration for generalized eigenvalue problems in molecular simulation. Taiwan. J. Math. **14**, 855–867 (2010)
5. Futamura, Y., Sakurai, T., Furuya, S., Iwata, J.-I.: Efficient algorithm for linear systems arising in solutions of eigenproblems and its application to electronic-structure calculations. In: High Performance Computing for Computational Science—VECPAR 2012: 10th International Conference, Kope, Japan, July 17–20, 2012, Revised Selected Papers, pp. 226–235 (2013)
6. Nagai, Y., Shinohara, Y., Futamura, Y., Ota, Y., Sakurai, T.: Numerical construction of a low-energy effective Hamiltonian in a self-consistent Bogoliubov-de Gennes approach of superconductivity. J. Phys. Soc. Jpn. **82**, 094701 (2013)
7. Shimizu, N., Utsuno, Y., Futamura, Y., Sakurai, T., Mizusaki, T., Otsuka, T.: Stochastic estimation of nuclear level density in the nuclear shell model: an application to parity-dependent level density in 58Ni. Phys. Lett. B **753**, 13–17 (2016)
8. Ohno, H., Kuramashi, Y., Tadano, H., Sakurai, T.: A quadrature-based eigensolver with a Krylov subspace method for shifted linear systems for Hermitian eigenproblems in lattice QCD. JSIAM Lett. **2**, 115–118 (2010)
9. Yamazaki, I., Ikegami, T., Tadano, H., Sakurai, T.: Performance comparison of parallel eigensolvers based on a contour integral method and a Lanczos method. Parallel Comput. **39**, 280–290 (2013)

10. http://www.aisin-aw.co.jp/en/products/drivetrain/lineup/index.html
11. Kosaka, I., Leiva, J.P., Watson, B.C., Ide, T.: Improvement of NV Performance for Automatic Transmission Using Large Scale Size Optimization, SAE Technical Paper, 2008–01-0869 (2008)
12. Sakurai, T., Tadano, H.: CIRR: a Rayleigh-Ritz type method with contour integral for generalized eigenvalue problems. Hokkaido Math. J. **36**, 745–757 (2007)
13. Ikegami, T., Sakurai, T., Nagashima, U.: A filter diagonalization for generalized eigenvalue problems based on the Sakurai-Sugiura projection method. J. Comput. Appl. Math. **233**(8), 1927–1936 (2010)
14. Ikegami, T., Sakurai, T.: Contour integral eigensolver for non-Hermitian systems: a Rayleigh-Ritz-type approach. Taiwan. J. Math. **14**, 825–837 (2010)
15. z-Pares, http://zpares.cs.tsukuba.ac.jp/
16. MUMPS, http://mumps.enseeiht.fr/

Mathematical Analysis of Synchronization from the Perspective of Network Science

Hirotada Honda and Atusi Tani

Abstract In this chapter, we discuss the mathematical analysis of synchronization with focusing on that of the Kuramoto–Sakaguchi equation. We also introduce related topics from the perspective of network science. The solvability and existence of vanishing diffusion coefficient are investigated.

1 Introduction

Since the memorable work by Kuramoto [7, 8], theoretical investigations of weakly coupled limit cycle oscillators have been intensively developed over several research fields. For example, in statistical physics, various models are being developed, whereas in network science, the synchronization on random and complex networks [5] are currently attractive topics.

It is noteworthy that the mathematical analysis of this field has been promoted recently [1, 3].

From the viewpoint of the application, the Kuramoto model has been applied in various engineering and industrial areas, for instance, the modeling of neural and brain networks, smart grid [2], and the behavior of TCP throughput in information networks [4].

H. Honda (✉)
NTT Network Technology Laboratories,
3-9-11 Midori-Cho, Tokyo, Musashino-Shi 180-8585, Japan
e-mail: honda.hirotada@lab.ntt.co.jp

A. Tani
Department of Mathematics, Keio University,
3-14-1 Hiyoshi, Yokohama 223-8522, Japan
e-mail: tani@math.keio.ac.jp

H. Itou et al. (eds.), *Mathematical Analysis of Continuum Mechanics and Industrial Applications*, Mathematics for Industry 26,
DOI 10.1007/978-981-10-2633-1_17

2 Functional Equations Based on Kuramoto Models

The Kuramoto–Sakaguchi equation is a model equation of the physical theory of coupled oscillators, which describes the temporal evolution of the probability distribution of each oscillator's phase.

The temporal evolution of the order parameter $r(t)$ and the phase of the mean field $\psi(t)$ at time t is described as

$$r(t)\exp(i\psi(t)) = \int_0^{2\pi}\int_{\mathbf{R}} \exp(i\theta)\rho(\theta,\omega,t)g(\omega)\,\mathrm{d}\theta\mathrm{d}\omega,$$

where $\rho(\theta,\omega,t)$ is the probability density function of phase θ and natural frequency ω at t, and $g(\omega)$ is the probability distribution function of ω. In addition, it is well known that the time evolution of ρ is subject to the following evolution equation:

$$\frac{\partial\rho}{\partial t} + \frac{\partial}{\partial\theta}\Big\{\big[\omega + Kr\sin(\psi-\theta)\big]\rho\Big\} = 0.$$

Combining these yields the following integro-differential equation:

$$\begin{cases} \dfrac{\partial\rho}{\partial t} + \omega\dfrac{\partial\rho}{\partial\theta} + K\dfrac{\partial}{\partial\theta}\left[\rho(\theta,\omega,t)\displaystyle\int_{\mathbf{R}} g(\omega')\mathrm{d}\omega' \int_0^{2\pi} \sin(\phi-\theta)\rho(\phi,\omega',t)\mathrm{d}\phi\right] = 0, \\ \qquad (\theta,\omega,t)\in(0,2\pi)\times\mathbf{R}\times(0,\infty), \\ \dfrac{\partial^i\rho}{\partial\theta^i}\Big|_{\theta=0} = \dfrac{\partial^i\rho}{\partial\theta^i}\Big|_{\theta=2\pi} \quad (i=0,1), \quad (\omega,t)\in\mathbf{R}\times(0,\infty), \\ \rho\big|_{t=0} = \rho_0(\theta,\omega), \quad (\theta,\omega)\in(0,2\pi)\times\mathbf{R}. \end{cases} \tag{1}$$

The parabolic regularization of (1), which is called the *Kuramoto–Sakaguchi equation*, reads as

$$\begin{cases} \dfrac{\partial\rho}{\partial t} - D\dfrac{\partial^2\rho}{\partial\theta^2} + \omega\dfrac{\partial\rho}{\partial\theta} + K\dfrac{\partial}{\partial\theta}\left[\rho(\theta,\omega,t)\displaystyle\int_{\mathbf{R}} g(\omega')\mathrm{d}\omega' \int_0^{2\pi} \sin(\phi-\theta)\rho(\phi,\omega',t)\mathrm{d}\phi\right] = 0, \\ \qquad (\theta,\omega,t)\in(0,2\pi)\times\mathbf{R}\times(0,\infty), \\ \dfrac{\partial^i\rho}{\partial\theta^i}\Big|_{\theta=0} = \dfrac{\partial^i\rho}{\partial\theta^i}\Big|_{\theta=2\pi} \quad (i=0,1), \quad (\omega,t)\in\mathbf{R}\times(0,\infty), \\ \rho\big|_{t=0} = \rho_0(\theta,\omega), \quad (\theta,\omega)\in(0,2\pi)\times\mathbf{R}. \end{cases} \tag{2}$$

Although the original Kuramoto model defined above describes the behavior of oscillators in the complete graph, Kuramoto also provided the model of the non-local coupling oscillators [6, 11]:

$$\begin{cases} \dfrac{\partial \rho}{\partial t} + \dfrac{\partial}{\partial \theta}(\omega \rho) - D\dfrac{\partial^2 \rho}{\partial \theta^2} + \dfrac{\partial}{\partial \theta}\Big(F[\rho, \rho]\Big) = 0, \quad \theta \in (0, 2\pi),\ t > 0,\ (x, \omega) \in \mathbf{R}^2, \\ \dfrac{\partial^i \rho}{\partial \theta^i}\Big|_{\theta=0} = \dfrac{\partial^i \rho}{\partial \theta^i}\Big|_{\theta=2\pi} \quad (i = 0, 1), \quad t > 0,\ (x, \omega) \in \mathbf{R}^2, \\ \rho|_{t=0} = \rho_0, \quad (\theta, x, \omega) \in (0, 2\pi) \times \mathbf{R}^2, \end{cases}$$

where

$$F[\rho_1, \rho_2] \equiv \rho_1(\theta, x, \omega, t) \int_{\mathbf{R}} G(x - y)\mathrm{d}y \int_0^{2\pi} \Gamma(\theta - \phi)\rho_2(\phi, y, \omega, t)\, \mathrm{d}\phi$$

with $G(\cdot)$, the probability distribution function of the coupling strength, and $\Gamma(\cdot)$, the coupling strength. Below, we discuss the original Kuramoto–Sakaguchi equation (2) and its vanishing diffusion version, (1).

3 Existing Arguments and Remaining Issues

As for the mathematical arguments of (2), Lavrentiev and coauthors [9, 10] first proved the global-in-time solvability. Recently, Ha and Xiao [3] discussed the asymptotic stability of the trivial stationary solution to (2) and the convergence of solution to the one for (1) as D tends to zero.

However, the arguments in [9] seem to be unclear since ρ does not decay in the ω-direction for each θ in general. Bearing this in mind, we employ the Hölder spaces with respect to θ with a certain weight in ω, and reconstruct the global-in-time solution to problem (2).

For the convergence as $D \to 0$, Ha and Xiao [3] used the explicit representation of the solution in D, which left their argument only within the space $L_\infty(\Omega)$. In this paper, we apply the compactness argument and show the convergence in the same functions spaces that the solution to (2) belongs to.

In the following, we first prepare the function spaces in the next section, and then state the results of solvability in Sect. 5. In the final section, we observe the convergence of the solution to (2) as D vanishes.

4 Function Spaces

We now introduce the functions spaces and some related notation used throughout this paper. Let $\Omega = (0, 2\pi)$, $\Omega_T = \Omega \times (0, T)$ and $\hat{f}(x, \omega, t) \equiv f(x - \omega t, \omega, t)$. By $C^{r+\alpha}(\Omega)$ with a non-negative integer r and $\alpha \in (0, 1)$, we mean the Banach space of functions from $C^r(\overline{\Omega})$, whose rth derivatives satisfy the Hölder condition with exponent α, i.e., the space of functions with the finite norm

$$|u|_{\Omega}^{(r+\alpha)} = \sum_{k=0}^{r} |D^k u|_{\Omega} + [D^r u]_{\Omega}^{(\alpha)},$$

where $D = \partial/\partial x$, and

$$|u|_{\Omega} = \sup_{x\in\Omega} |u(x)|, \quad [u]_{\Omega}^{(\alpha)} = \sup_{x,y\in\Omega} \frac{|u(x) - u(y)|}{|x - y|^{\alpha}}.$$

By $C^{r+\alpha, \frac{r+\alpha}{2}}(\Omega_T)$ with $r = 0, 1, 2$, we mean the spaces of functions defined in Ω_T and having the finite norms

$$|u|_{\Omega_T}^{(\alpha,\frac{\alpha}{2})} = |u|_{\Omega_T} + [u]_{\Omega_T}^{(\alpha,\frac{\alpha}{2})} \quad (r = 0),$$

where

$$|u|_{\Omega_T} = \sup_{(x,t)\in\Omega_T} |u(x,t)|, \quad [u]_{\Omega_T}^{(\alpha,\frac{\alpha}{2})} = [u]_{x,\Omega_T}^{(\alpha)} + [u]_{t,\Omega_T}^{(\frac{\alpha}{2})},$$

$$[u]_{x,\Omega_T}^{(\alpha)} = \sup_{x,y,t} \frac{|u(x,t) - u(y,t)|}{|x - y|^{\alpha}}, \quad [u]_{t,\Omega_T}^{(\frac{\alpha}{2})} = \sup_{x,y,t} \frac{|u(x,t) - u(x,\tau)|}{|t - \tau|^{\frac{\alpha}{2}}},$$

and for $r = 1, 2$,

$$|u|_{\Omega_T}^{(1+\alpha,\frac{1+\alpha}{2})} = |u|_{\Omega_T} + \left|\frac{\partial u}{\partial x}\right|_{\Omega_T}^{(\alpha,\frac{\alpha}{2})} + [u]_{t,\Omega_T}^{\left(\frac{1+\alpha}{2}\right)},$$

$$|u|_{\Omega_T}^{(2+\alpha,\frac{2+\alpha}{2})} = |u|_{\Omega_T} + \left|\frac{\partial u}{\partial x}\right|_{\Omega_T} + \left|\frac{\partial^2 u}{\partial x^2}\right|_{\Omega_T}^{(\alpha,\frac{\alpha}{2})} + \left|\frac{\partial u}{\partial t}\right|_{\Omega_T}^{(\alpha,\frac{\alpha}{2})} + \left|\frac{\partial u}{\partial x}\right|_{t,\Omega_T}^{(\frac{1+\alpha}{2})},$$

respectively.

Let ε be a fixed number satisfying $0 < \varepsilon < 1/2$ and define the smooth monotone function $h(\omega)$ as

$$h(\omega) = \begin{cases} 2 & (|\omega| < 1); \\ 1 + |\omega|^{2+\varepsilon} & (|\omega| \geq 1). \end{cases}$$

Then, for $\alpha \in (1/2, 1)$, we define

$$\mathscr{V}_T^{2+\alpha} \equiv \left\{ f(\theta,\omega,t) \,\middle|\, h(\omega)\hat{f}(\theta,\omega,t) \in C^{2+\alpha,\frac{2+\alpha}{2}}(\Omega_T),\ \sup_{\omega} \left|h(\omega)\hat{f}(\omega)\right|_{\Omega_T}^{(2+\alpha,\frac{2+\alpha}{2})} < \infty \right\}.$$

For functions independent on t, we define

$$\mathscr{V}^{2+\alpha} \equiv \left\{ f(\theta,\omega) \,\middle|\, h(\omega)f(\theta,\omega) \in C^{2+\alpha}(\Omega),\ \sup_{\omega} |h(\omega)f(\omega)|_{\Omega}^{(2+\alpha)} < \infty \right\}.$$

The set of square-integrable functions with respect to θ equipped with the norm

$$\|f\| \equiv \|f\|_{L_2(\Omega)} \equiv \left(\int_\Omega |f(\theta)|^2 \,\mathrm{d}\theta\right)^{\frac{1}{2}}$$

is denoted by $L_2(\Omega)$, and for those depending also on ω, we introduce

$$|||f||| \equiv \sup_\omega \|f(\omega)\|.$$

Following the definition by Temam [12], we say that; a 2π-periodic function $u(\theta)$ on Ω belongs to the Sobolev space $\mathscr{H}^m$ ($m \in \mathbf{R}$) if the Fourier coefficients $\{a_n\}_{n=-\infty}^{\infty}$ of

$$u(\theta) = \sum_{n=-\infty}^{\infty} a_n e^{in\theta},$$

satisfy

$$\|u\|_m^2 \equiv \sum_{n=-\infty}^{\infty} (1+|n|^2)^m |a_n|^2 < \infty.$$

For periodic functions depending also on ω, we define

$$\overline{\mathscr{H}}^m \equiv \left\{ u(\theta,\omega) = \sum_{n=-\infty}^{\infty} a_n(\omega) e^{in\theta} \,\middle|\, \sup_{\omega\in\mathbf{R}} \|u(\cdot,\omega)\|_m^2 < \infty \right\}.$$

For the Banach space E with a norm $\|\cdot\|_E$, we denote the space of E-valued measurable functions $u(t)$ on the interval (a,b) by $L_p(a,b;E)$, whose norm is defined by

$$\|u\|_{L_p(a,b;E)} \equiv \begin{cases} \left(\int_a^b \|u(t)\|_E^p \,\mathrm{d}t\right)^{1/p} & (1 \le p < \infty), \\ \mathrm{esssup}_{a\le t\le b} \|u(t)\|_E & (p = \infty). \end{cases}$$

5 Local and Global-in-Time Solvability of (1)

The following theorem is one of our main results.

Theorem 1 *Let ε, α and $h(\omega)$ be those defined in the previous section, and assume the following issues:*

(i) $\rho_0 \in \mathscr{V}^{2+\alpha}$;

(ii) $\rho_0 \geq 0, \quad \displaystyle\int_0^{2\pi} \rho_0(\theta, \omega)\, \mathrm{d}\theta = 1.$

Then, there exist a certain $T_* > 0$ *and a unique solution* $\rho \in \mathscr{V}_{T_*}^{2+\alpha}$ *to (2).*

Proof We give an outline of the proof. We introduce the transformation of the coordinate system

$$\theta = \theta' - \omega t', \quad t = t',$$
$$\rho(\theta, \omega, t) = \hat{\rho}(\theta', \omega, t').$$

Then, $\hat{\rho}$ satisfies

$$\frac{\partial \hat{\rho}}{\partial t'} - D\frac{\partial^2 \hat{\rho}}{\partial \theta'^2} = K\frac{\partial}{\partial \theta'}\left[\hat{\rho}(\theta', \omega, t') \int_{\mathbf{R}} g(\omega')\mathrm{d}\omega' \int_0^{2\pi} \sin(\theta' - \phi)\hat{\rho}(\phi, \omega', t')\, \mathrm{d}\phi\right].$$

After multiplying $h(\omega)$ on both sides, and introducing $\tilde{\rho}(\theta', \omega, t') = h(\omega)\hat{\rho}(\theta', \omega, t')$, we have

$$\frac{\partial \tilde{\rho}}{\partial t'} - D\frac{\partial^2 \tilde{\rho}}{\partial \theta'^2} = K\frac{\partial}{\partial \theta'}\left[\tilde{\rho}(\theta', \omega, t') \int_{\mathbf{R}} g(\omega')\mathrm{d}\omega' \int_0^{2\pi} \sin(\theta' - \phi)\frac{\tilde{\rho}(\phi, \omega, t')}{h(\omega')}\, \mathrm{d}\phi\right]. \tag{3}$$

From this, we conclude that there exists a certain $T_* > 0$ such that $\tilde{\rho} \in \mathscr{V}_{T_*}^{2+\alpha}$ by using the iteration argument as that in [9]. □

Next, we state the global-in-time existence of the solution to (2). First, we prepare some lemmas.

Lemma 1 *Suppose that there exists a solution* ρ *to (2) belonging to* $\mathscr{V}_T^{2+\alpha}$ *for any* $T > 0$. *Then the following facts hold:*

- $\rho(\theta, \omega, t) \geq 0$ a.e. on $\theta \in \Omega,\ t \in (0, T),\ \omega \in \mathbf{R}$;
- $\displaystyle\int_\Omega \rho(\theta, \omega, t)\, \mathrm{d}\theta = 1,\ t \in (0, T),\ \omega \in \mathbf{R}.$

The assertion of Lemma 1 is easy to prove (cf. [9]). Now we discuss the global-in-time solvability. By virtue of Theorem 1, we have the local-in-time solution to (2) on the time interval $(0, T_*)$ with T_* specified in Theorem 1. Then, we have the following a-priori estimate. For the sake of simplicity, we hereafter use notations $f^{(k,l)} \equiv \frac{\partial^k}{\partial \theta^k}\frac{\partial^l}{\partial t^l} f$.

Lemma 2 *Let* $T > 0$ *be an arbitrary number. In addition to the assumptions of Theorem 1, we impose* $\rho_0 \in \overline{\mathscr{H}}^4$. *The following inequalities hold for the solution* ρ *to (2) on* $(0, T)$*:*

$$\|\rho^{(k,l)}(t)\| \le c_{(k,l)}(T) \quad (k + 2l \le 4) \tag{4}$$

hold with some constants $c_{(k,l)}(T)$. When $l = 0$, they are independent of T.

Proof We show the estimate of lower order terms as an example. Multiply (2) by ρ, and use Lemma 1 and the periodicity of ρ. Then, we obtain

$$\begin{aligned}
&\int_\Omega \rho(\theta,\omega,t)\frac{\partial}{\partial\theta}\Big(\rho(\theta,\omega,t)\int_{\mathbf{R}} g(\omega')\mathrm{d}\omega \int_\Omega \sin(\phi-\theta)\rho(\phi,\omega,t)\Big)\,\mathrm{d}\theta \\
&\quad = -\frac{1}{2}\int_\Omega \big(\rho(\theta,\omega,t)\big)^2\Big(\int_{\mathbf{R}} g(\omega')\mathrm{d}\omega' \int_\Omega \rho(\phi,\omega,t)\cos(\phi-\theta)\,\mathrm{d}\phi\Big)\,\mathrm{d}\theta \\
&\quad \le \frac{1}{2}\|\rho(\cdot,\omega,t)\|^2.
\end{aligned}$$

On the other hand, tracing the same argument as in [9], we have

$$\|\rho(\cdot,\omega,t)\|^2 \le \frac{1}{2\pi} + C_\varepsilon + \varepsilon\left\|\frac{\partial\rho}{\partial\theta}(\cdot,\omega,t)\right\|^2,$$

for any $\varepsilon > 0$.

Therefore, by virtue of the classical Gronwall inequality, we have [12]

$$\begin{aligned}
|||\rho(t)|||^2 &\le |||\rho_0|||^2 \exp\Big(-2(D-\varepsilon)t\Big) + \frac{C}{D-\varepsilon}\exp\Big(1-\exp\Big(-2(D-\varepsilon)t\Big)\Big) \\
&\le C, \quad \forall t \in (0,T).
\end{aligned}$$

For the estimates of terms with higher order derivatives and $l > 0$, we apply the similar arguments and Poincaré's inequality. □

Now, owing to Lemmas 1 and 2, together with Sobolev's embedding theorem, $\rho\big|_{t=T_*}$ satisfies the same assumptions as ρ_0 in Lemma 2. Hence we are able to extend ρ onto the time interval $(T^*, 2T^*)$ as a solution to (2), from which we deduce estimate (4) on $(0, 2T_*)$. Repeat this procedure finitely many times, and then we have the global-in-time solution. We summarize the above as

Theorem 2 *Under the same assumptions as those in Lemma 2, there exists a unique solution $\rho \in \mathscr{V}_T^{2+\alpha}$ to (2) for arbitrary $T > 0$.*

6 Vanishing Diffusion Limit

In this section, we show the vanishing diffusion limit of ρ as D tends to zero. For the sake of simplicity, we denote the solution of (2) with $D > 0$ by $\rho_{(D)}$, whereas we use $\rho_{(0)}$ to stand for the solution to (1).

Theorem 3 *Let $T > 0$ be an arbitrary number. Under the same assumptions as in Lemma 2, there exists $\rho_{(0)}$ as $D \to 0$ in $\mathcal{V}_T^{2+\alpha}$:*

$$\rho_{(0)} = \lim_{D\to 0} \rho_{(D)} \quad \text{in } \mathcal{V}_T^{2+\alpha}.$$

That is, there exists a solution $\rho_{(0)} \in \mathcal{V}_T^{2+\alpha}$ to (1).

We first prepare some lemmas below.

Lemma 3 *Let $T > 0$, and ρ_0 satisfies the same assumptions as in Lemma 2. Then, the sequence $\{\rho_{(D)}^{(k,l)}\}_{D>0}$ $k + 2l \leq 4$ is bounded in $L_\infty(0, T; \overline{\mathcal{H}}^4) \bigcap C^1(0, T; \overline{\mathcal{H}}^2) \bigcap C^2(0, T; \overline{\mathcal{H}}^0)$.*

Proof What we have to verify are

$$\sup_{t\in(0,T)} \|\rho_{(D)}^{(k,l)}(t)\|_{\overline{\mathcal{H}}^0} \leq c_{k,l}(T), \tag{5}$$

but these are verified by the similar arguments as those in Lemma 2, and we omit them. □

Lemma 4 *The sequence $\{\rho_{(D)}\}_{D>0}$ is the Cauchy sequence in $L_\infty(0, T; \overline{\mathcal{H}}^4) \bigcap C^1(0, T; \overline{\mathcal{H}}^2) \bigcap C^2(0, T; \overline{\mathcal{H}}^0)$.*

Proof We consider the equation satisfied by $\rho_{(D)}$. By subtracting (2) with D replaced by D' from the original one, $\tilde{\tilde{\rho}} \equiv \rho_{(D)} - \rho_{(D')}$ satisfies

$$\begin{aligned}
&\frac{\partial \tilde{\tilde{\rho}}}{\partial t} + \omega \frac{\partial \tilde{\tilde{\rho}}}{\partial \theta} - D \frac{\partial^2 \tilde{\tilde{\rho}}}{\partial \theta^2} - (D - D') \frac{\partial^2 \rho_{(D')}}{\partial \theta^2} \\
&\quad + K \frac{\partial}{\partial \theta} \left[\tilde{\tilde{\rho}}(\theta, \omega, t) \int_{\mathbf{R}} g(\omega') \mathrm{d}\omega' \int_{\Omega} \sin(\phi - \theta) \rho_{(D)}(\phi, \omega', t) \mathrm{d}\phi \right] \\
&\quad + K \frac{\partial}{\partial \theta} \left[\rho_{(D')}(\theta, \omega, t) \int_{\mathbf{R}} g(\omega') \mathrm{d}\omega' \int_{\Omega} \sin(\phi - \theta) \tilde{\tilde{\rho}}(\phi, \omega', t) \mathrm{d}\phi \right] = 0
\end{aligned} \tag{6}$$

Multiplying (6) by $\tilde{\tilde{\rho}}$, integrating by parts over Ω and taking the supremum with respect to ω yield

$$\frac{1}{2} \frac{\mathrm{d}}{\mathrm{d}t} \|\!|\tilde{\tilde{\rho}}(t)|\!\|^2 \leq C \Big(\|\!|\tilde{\tilde{\rho}}(t)|\!\|^2 + |D - D'|^2 \Big).$$

Here we used the estimates for example:

$$K\int_{\Omega}\tilde{\tilde{\rho}}(\theta,\omega,t)\frac{\partial}{\partial\theta}\left[\tilde{\tilde{\rho}}(\theta,\omega,t)\int_{\mathbf{R}}g(\omega')\mathrm{d}\omega'\int_{\Omega}\sin(\phi-\theta)\rho_{(D)}(\phi,\omega',t)\,\mathrm{d}\phi\right]\mathrm{d}\theta$$
$$=-\frac{K}{2}\int_{\Omega}\frac{\partial}{\partial\theta}(\tilde{\tilde{\rho}}(\theta,\omega,t))^2\left(\int_{\mathbf{R}}g(\omega')\mathrm{d}\omega'\int_{\Omega}\sin(\phi-\theta)\rho_{(D)}(\phi,\omega',t)\,\mathrm{d}\phi\right)\mathrm{d}\theta$$
$$\le\frac{K}{2}\|\tilde{\tilde{\rho}}(t)\|^2,$$

$$\int_{\Omega}\tilde{\tilde{\rho}}(\theta,\omega,t)\frac{\partial}{\partial\theta}\left[\rho_{(D')}(\theta,\omega,t)\int_{\mathbf{R}}g(\omega')\mathrm{d}\omega'\int_{\Omega}\sin(\phi-\theta)\tilde{\tilde{\rho}}(\phi,\omega',t)\,\mathrm{d}\phi\right]\mathrm{d}\theta$$
$$\le\sup_{\theta}\left|\rho_{(D')}(t)\right|\left\|\frac{\partial\tilde{\tilde{\rho}}}{\partial\theta}\right\|\left\|\left(\int_{\mathbf{R}}g(\omega')\mathrm{d}\omega'\int_{\Omega}\sin(\phi-\theta)\tilde{\tilde{\rho}}(\phi,\omega',t)\right)\right\|.$$

Thus, by virtue of Gronwall inequality, we have

$$\|\tilde{\tilde{\rho}}(t)\|^2\le C|D-D'|^2te^{Ct},$$

which implies that $\{\rho_{(D)}\}_{D>0}$ makes a Cauchy sequence in $L_\infty(0,T;\overline{\mathscr{H}^0})$.

Similar arguments hold for $\rho_{(D)}^{(k,l)}$ for $(k,l)\ne(0,0)$, and by summing up them multiplied by appropriate constants, we arrive at the desired result. □

Now Theorem 3 follows by virtue of Lemmas 3 and 4.

7 Conclusion and Further Studies

We have introduced some models of weakly coupled oscillators, and discussed the solvability and existence of the vanishing diffusion limit solution to the Kuramoto–Sakaguchi equation.

There exist some open problems, such as the stability of the non-trivial solution both with and without the diffusion, and the bifurcation problems. From the perspective of network science, the analysis of the non-local coupling model and models on various networks such as complex network and dynamic random graph are to be promoted.

References

1. Chiba, H.: A proof of the Kuramoto conjecture for a bifurcation structure of the infinite dimensional Kuramoto model. Ergod. Theor. Dyn. Syst. **35**, 762–834 (2015)
2. Dörfler, F., Bullo, F.: Synchronization and transient stability in power networks and nonuniform Kuramoto oscillators. SIAM J. Control Optim. **50**, 1616–1642 (2012)
3. Ha, S.Y., Xiao, Q.: Remarks on the nonlinear stability of the Kuramoto-Sakaguchi equation. J. Diff. Equ. **259**, 2430–2457 (2015)

4. Han, H., Hollot, C.V., Towsley, D., Chait, Y.: Synchronization of TCP Flows in Networks with Small DropTail Buffers, Proceedings of the 44th IEEE Conference on Decision and Control, and the European Control Conference (2005)
5. Ichinomiya, T.: Frequency synchronization in a random oscillator network. Phys. Rev. E **70**, 026116 (2004)
6. Kawamura, Y., Nakao, H., Kuramoto, Y.: Noise-induced turbulence in nonlocally coupled oscillators. Phys. Rev. E **75**, 036209 (2007)
7. Kuramoto, Y.: International symposium on mathematical problems in theoretical physics. In: Araki H. (ed), Lecture Notes in Physics, vol. 39, pp. 420–422. Springer, New York (1975)
8. Kuramoto, Y.: Chemical Oscillations, Waves, and Turbulence (Springer Series in Synergetics, 19). Springer, Berlin (1984)
9. Lavrentiev, M., Spigler, R.S.: Existence and uniqueness of solutions to the Kuramoto-Sakaguchi nonlinear parabolic integrodifferential equation. Diff. Integr. Equ. **13**, 649–667 (2000)
10. Lavrentiev, M., Spigler, R.S., Tani, A.: Existence, uniqueness, and regularity for the Kuramoto-Sakaguchi equation with unboundedly supported frequency distribution. Diff. Integr. Equ. **27**, 879–892 (2014)
11. Shiogai, Y., Kuramoto, Y.: Wave propagation in nonlocally coupled oscillators with noise. Prog. Thoer. Phys. Suppl. **150**, 435–438 (2003)
12. Temam, R.: Infinite-dimensional dynamical systems in mechanics and physics, 2nd edn., Appl. Math. Sci. **68**, (1997) (Springer)

Index

H. Itou et al. (eds.), *Mathematical Analysis of Continuum Mechanics and Industrial Applications*, Mathematics for Industry 26,
DOI 10.1007/978-981-10-2633-1

MIX
Papier aus verantwortungsvollen Quellen
Paper from responsible sources
FSC® C105338

If you have any concerns about our products,
you can contact us on
ProductSafety@springernature.com

In case Publisher is established outside the EU,
the EU authorized representative is:
Springer Nature Customer Service Center GmbH
Europaplatz 3, 69115 Heidelberg, Germany

Printed by Libri Plureos GmbH
in Hamburg, Germany